Lecture Notes in Chemistry

Volume 94

The Lecture Notes in Chemistry

The series Lecture Notes in Chemistry (LNC) reports new developments in chemistry and molecular science-quickly and informally, but with a high quality and the explicit aim to summarize and communicate current knowledge for teaching and training purposes. Books published in this series are conceived as bridging material between advanced graduate textbooks and the forefront of research. They will serve the following purposes:

- provide an accessible introduction to the field to postgraduate students and nonspecialist researchers from related areas,
- provide a source of advanced teaching material for specialized seminars, courses and schools, and
- be readily accessible in print and online.

The series covers all established fields of chemistry such as analytical chemistry, organic chemistry, inorganic chemistry, physical chemistry including electrochemistry, theoretical and computational chemistry, industrial chemistry, and catalysis. It is also a particularly suitable forum for volumes addressing the interfaces of chemistry with other disciplines, such as biology, medicine, physics, engineering, materials science including polymer and nanoscience, or earth and environmental science.

Both authored and edited volumes will be considered for publication. Edited volumes should however consist of a very limited number of contributions only. Proceedings will not be considered for LNC.

The year 2010 marks the relaunch of LNC.

More information about this series at http://www.springer.com/series/632

Jochen Schirmer

Many-Body Methods for Atoms, Molecules and Clusters

Springer

Jochen Schirmer
Institute of Physical Chemistry
Heidelberg University
Heidelberg, Germany

ISSN 0342-4901 ISSN 2192-6603 (electronic)
Lecture Notes in Chemistry
ISBN 978-3-030-06691-8 ISBN 978-3-319-93602-4 (eBook)
https://doi.org/10.1007/978-3-319-93602-4

Printed on acid-free paper

This Springer imprint is published by the registered company Springer Nature Switzerland AG
The registered company address is: Gewerbestrasse 11, 6330 Cham, Switzerland

Preface

The present book is based on a one-year course of lectures given intermittently during the years from 1990 to 2010 at the University of Heidelberg. The lectures were devised as an in-depth introduction into many-body theory for finite electronic systems, that is, molecules, atoms, and clusters, addressing graduate, doctoral, and postdoctoral students, who were generally interested in quantum-chemical methods and computations. The original course is essentially covered by the first 10 chapters, while 7 additional chapters address further elaborations and extensions.

Many-body methods, or more accurately, field-theoretical many-body methods, have originated in quantum field theory where they were developed as a means to treat the physics of elementary particles. As was soon realized, these methods could be transferred to the treatment of quantum many-body systems in solid-state physics and statistics, not conveying novel physics here but supplying a powerful new formalism and a route toward alternative computational methods. Shortly afterward, this formalism was taken up in the treatment of finite particle systems, first in nuclear physics, and finally in quantum chemistry as well. It is now almost half a century since computational schemes based on field-theoretical many-body theory were developed and successfully applied to finite electronic systems.

In the field-theoretical approach, the many-particle problem is formulated in terms of many-body Green's functions or propagators. These entities are defined as ground-state expectation values of time-dependent operator products, which, in energy representation, take on the form of matrix elements of many-body resolvent operators. They allow for a direct access to the energies and transition moments of generalized excitation processes in the considered system, such as ionization (electron removal), affinities (electron attachment), and neutral electronic excitation.

What is the advantage inherent to these methods and what can they actually do better than the conventional procedures in dealing with small, medium size, and large molecules? An apparent advantage is a direct access to physically relevant quantities such as excitation energies and transition moments, which otherwise have to be assembled from independent computations for the initial ground and final excited states. But there is another, deeper justification, related to characteristic shortcomings in the conventional approach.

In the conventional quantum-theoretical treatment of finite many-electron systems, there are two basic tools: firstly, perturbation theory (PT), which, however, as a computational scheme applies only to the N-electron ground state; and, secondly, the standard numerical procedure of solving the time-independent Schrödinger equation, that is, using suitable basis set expansions for the states of interest and transforming the Schrödinger equation into the secular problem of the corresponding matrix representation of the hamiltonian. The general problem of configuration interaction (CI), as the standard procedure is referred to in quantum chemistry, is the exponentially increasing dimension of the secular matrix, $d = \binom{M}{N}$, both with the size of the systems, reflected in the number of electrons, N, and the demand for accuracy, reflected in the number M of one-particle states underlying the many-electron basis states (CI configurations). This means that a full CI treatment is not viable except for very small systems and limited one-particle basis sets, and one has to resort to approximate CI schemes obtained by truncating the configuration manifold in suitable ways.

Here, however, an unsuspected problem arises which disqualifies the CI as a means of treating extended electron systems. In the CI secular equations, there is an interaction (mixing) of configurations that differ exactly by a double excitation, such as in a singly (S) excited configuration (relative to the reference state) and a triply (T) excited configuration comprising the former single excitation. A corresponding S-T secular matrix element is potentially "non-local"; that is, its magnitude does not decrease or vanish when the involved single and double excitations can be assigned to distant parts of the system or even to separate fragments of a composite system. In truncated CI expansions, the presence of these potentially non-local admixtures causes an uncontrollable error which grows with the spatial extension of the system and, accordingly, is referred to as size-consistency error.

The propagator methods, by contrast, do not suffer from this deficiency. As a common feature, approximation schemes deriving from field-theoretical many-body theory combine perturbation expansions (of the ground-state type) and eigenvalue algebra within a generalized secular problem where in particular any potentially non-local coupling contributions are taken care of in the PT part. As a consequence, the propagator methods are inherently size-consistent and, moreover, more economical, requiring distinctly smaller explicit configuration manifolds in the secular problem than in CI expansions of comparable accuracy.

A brief guide to the tour through the five parts of this book may be helpful. The first two chapters of Part I lay the groundwork for the quantum theory of many-electron systems, addressing states, operators, the evaluation of matrix elements, and, finally, the use of second quantization. Thereupon, the prototypical one-particle Green's function or electron propagator is presented and discussed in Chap. 3.

In the four chapters of Part II, the formalism of diagrammatic perturbation theory is developed, based on three central theorems, the Gell-Mann and Low theorem, Wick's theorem, and the linked-cluster theorem. At the end of that part, the reader should be able to draw and evaluate Feynman diagrams.

However, the diagrammatic arts do not yet establish a procedure to compute the electron propagator or the physical information conveyed therein. So with Chap. 8 in Part III, the focus shifts to the issue of developing computational schemes. Here, the prominent starting point is the Dyson equation, relating the electron propagator to the so-called self-energy. The latter quantity is itself subject to a diagrammatic perturbation expansion, where the diagrams are simpler than those for the electron propagator. The subject of Chap. 9 is the algebraic–diagrammatic construction (ADC), a general procedure to generate systematic higher-order approximations (ADC(n) schemes) to the self-energy, being consistent through order n, and, crucially, reproducing the correct analytical structure of the self-energy. The ADC procedure is quite versatile and can directly be applied to the electron propagator, or, more accurately, to its $(N \pm 1)$-electron parts, as is demonstrated in Chap. 10.

Then, in Chaps. 11 and 12, our tour takes a remarkable turn: The direct ADC approximations can be derived via a radically different route, namely a wave-function-based approach referred to as intermediate state representation (ISR). (An impetuous reader, already familiar with the topics of Chaps. 1 and 2, might take a shortcut directly to Chaps. 11 and 12). The ISR concept bridges the gap between propagator and wave-function methods, lifts certain limitations inherent to the diagrammatic propagator approach, and allows for a rigorous foundation (Chap. 12) of the defining many-body features.

In Part IV, we turn toward the physics of N-electron excitations and the polarization propagator relevant here. Chapter 13 discusses how diagrammatic perturbation theory can be adapted to the polarization propagator. The ADC and ISR concepts for N-electron excitations are presented in Chap. 14, while Chap. 15 reviews the prominent random-phase approximation (RPA), being a paradigmatic model in many-body theory. The final part V takes a look at two related approaches, which may be seen as ISR variants: The equation-of-motion (EOM) methods (Chap. 16) and methods based on the coupled-cluster (CC) ansatz (Chap. 17).

Altogether 9 appendices supplement the main text: Appendix A.1 reviews many-body perturbation theory and recollects some useful algebraic techniques; some more lengthy proofs are deferred to Appendices A.2, A.3, A.4, and A.6; extensions to Chaps. 8, 13, and 16 are given in Appendices A.5, A.7, and A.8, respectively; the final Appendix A.9 compiles various explicit ADC expressions.

As may be permissible in a textbook, perhaps even advisable, the bibliography has been kept relatively short and selective. In topics that are well documented in the literature, only a few key papers or books are quoted. More comprehensive reference is made to subjects or issues that are less familiar or amenable. And, of course, I have tried to indicate the sources wherever the text draws upon exemplary previous presentations.

Many persons have contributed in various ways to the genesis of this book. Foremost, it should be gratefully acknowledged that the students attending the original lectures did, to paraphrase J. A. Wheeler, a great job in educating their lecturer. And particular thanks go to Vitali Averbukh, one of the students then, now holding a faculty position at the Imperial College London, who encouraged—well, more accurately—pressured me to make those lectures publicly available in the form of a book. I am greatly indebted to Lukas Wirz for unremitting technical assistance during the entire project, for translating the bulk of handwritten formulae into LaTex, and for devising numerous diagrams and figures. Over time, quite a few people participated in the critical reading of selected chapters, contributing helpful suggestions and comments. I would like to thank all of them, and, in particular and two for all, Tsveta Miteva, who intrepidly and skillfully shouldered the burden of correcting the very first version of each chapter, and Bridgette Cooper for diligent revisions in the earlier stages of the project. Sincere thanks are due to my former collaborators Alexander B. Trofimov and Frank Mertins, who, perhaps in a less direct but all the more profound way, have shaped various themes of the present book. And finally, I would like to express my gratitude to Lorenz S. Cederbaum for the initiation and longtime common commitment to the challenges of many-body theory.

Heidelberg, Germany Jochen Schirmer

Contents

Part I
Many-Electron Systems and the Electron Propagator

Chapter 1
Systems of Identical Particles

In the first section, we take a look at the basic ingredients in the quantum-theoretical formulation of many-electron systems: wave functions (states) and operators. For an in-depth discussion of physical aspects relevant here, the reader is referred to Chap. XIV in the textbook by A. Messiah [1]. In dealing with many-particle systems, the handling of matrix elements involving Slater determinants is required. Here, the essential tool is a set of simple rules, referred to as Slater–Condon rules, which will be considered in the second section of this chapter.

1.1 Many-Electron Wave Functions

In the following, we will consider a system of N identical (more strictly: indistinguishable) particles, specifically, electrons in an atom or molecule. Each particle is associated with a set of three spatial coordinates, $\boldsymbol{x}$, and a spin variable, σ. Accordingly, an N-particle wave function,

$$\Psi = \Psi(\boldsymbol{x}_1\sigma_1, \ldots, \boldsymbol{x}_N\sigma_N) \tag{1.1}$$

is a function of the N sets of variables, $\boldsymbol{x}_k\sigma_k,\ k = 1, \ldots, N$. The wave function is a representation of an underlying abstract state $|\Psi\rangle$,

$$\Psi(\boldsymbol{x}_1\sigma_1, \ldots, \boldsymbol{x}_N\sigma_N) = \langle \boldsymbol{x}_N\sigma_N| \ldots \langle \boldsymbol{x}_1\sigma_1|\Psi\rangle \tag{1.2}$$

Here, $|\boldsymbol{x}\sigma\rangle$ denotes a (formal) one-particle eigenstate of the position and spin operators. For notational brevity, we shall occasionally combine the spatial and spin variables,

$$\boldsymbol{\xi}_i \equiv \boldsymbol{x}_i\sigma_i \tag{1.3}$$

J. Schirmer, *Many-Body Methods for Atoms, Molecules and Clusters*, Lecture Notes in Chemistry 94, https://doi.org/10.1007/978-3-319-93602-4_1

Using this notation, the wave function takes the form

$$\Psi(\boldsymbol{\xi}_1,\ldots,\boldsymbol{\xi}_N) = \langle\boldsymbol{\xi}_N|\ldots\langle\boldsymbol{\xi}_1|\Psi\rangle \tag{1.4}$$

Integration with respect to the $\boldsymbol{\xi}$-variables is defined according to

$$\int\ldots\int \mathrm{d}\xi_1\ldots\mathrm{d}\xi_N = \sum_{\sigma_1}\cdots\sum_{\sigma_N}\int\ldots\int \mathrm{d}\boldsymbol{x}_1\ldots\mathrm{d}\boldsymbol{x}_{\mathrm{N}} \tag{1.5}$$

To discuss the permutation symmetry, let us introduce **permutation operators**, defined according to

$$\hat{P}\Psi(\xi_1,\ldots,\xi_N) = \Psi(\xi_{P(1)},\ldots,\xi_{P(N)}) \tag{1.6}$$

where P denotes a permutation of the figures $1,\ldots,N$:

$$P:\quad i\rightarrow P(i),\quad i=1,2,\ldots,N \tag{1.7}$$

The set of permutations of N elements forms a group, referred to as the symmetric group $\boldsymbol{S}_N$, and so does the corresponding set of permutation operators. (Here, the group multiplication is the successive application, $\hat{P}'' = \hat{P}\hat{P}'$.) The permutation operators are unitary, that is, $\hat{P}^{-1} = \hat{P}^{\dagger}$ (see Exercise 1.1). Each permutation can be obtained as a product (consecutive application) of transpositions (exchanging two figures). While this way of generating permutations is not unique, the number of transpositions involved is either always even or always odd, depending on the respective permutation. In that sense, a permutation P is said to be "even" or "odd," and a corresponding sign of P can be defined according to

$$(-1)^P = \begin{cases} +1, & \text{even number of transpositions} \\ -1, & \text{odd number of transpositions} \end{cases} \tag{1.8}$$

The **symmetrization postulate** of quantum theory states that the wave functions for a system of N uniform particles must be either totally symmetric (*bosons*) or totally antisymmetric (*fermions*) with respect to any permutation of the particle variables:

$$\hat{P}\Psi(\xi_1,\ldots,\xi_N) = \begin{cases} (-1)^P\Psi(\xi_1,\ldots,\xi_N), & \text{fermions} \\ \Psi(\xi_1,\ldots,\xi_N), & \text{bosons} \end{cases} \tag{1.9}$$

In the following, we shall deal exclusively with fermions, specifically electrons. Here, an immediate consequence of the symmetrization postulate is the **Pauli principle**, stating that the wave function vanishes whenever the coordinates and spins of two (or more) fermions coincide:

$$\Psi(\xi_1,\ldots,\xi_N) = 0 \quad \text{for} \quad \xi_i = \xi_j\ (i\neq j) \tag{1.10}$$

It is useful to expand general N-electron states in terms of products of orthonormal one-particle states

$$|q\gamma\rangle = |q\rangle|\gamma\rangle \tag{1.11}$$

Here, q and $\gamma = \pm\frac{1}{2}$ are spatial and spin quantum numbers, respectively. We shall also use the notation $\gamma = \alpha, \beta$, established in quantum chemistry. Note that the one-particle states are themselves products of spatial states, $|q\rangle$, and spin states, $|\gamma\rangle$, fulfilling the orthonormal conditions

$$\langle q|q'\rangle = \delta_{qq'}, \qquad \langle\gamma|\gamma'\rangle = \delta_{\gamma\gamma'} \tag{1.12}$$

The corresponding wave functions are

$$\varphi_q(\boldsymbol{x}) = \langle\boldsymbol{x}|q\rangle$$
$$\chi_\gamma(\sigma) = \langle\sigma|\gamma\rangle$$

The spin functions may also be written in spinor form,

$$\chi_\alpha = \begin{pmatrix}1\\0\end{pmatrix}, \qquad \chi_\beta = \begin{pmatrix}0\\1\end{pmatrix}$$

A common choice of spatial orbitals is the set of molecular orbitals (MOs) generated by a Hartree–Fock (HF) or self-consistent field (SCF) computation of the N-electron ground state.

For a more abstract representation, it is helpful to replace the pair of a spatial and a spin quantum number by single comprehensive spin-orbital quantum number

$$q\gamma \equiv q \tag{1.13}$$

which, for notational economy, may be labeled by the same latin letter. (Whether q labels a spin-orbital or merely a spatial orbital will be clear from the respective context.) Accordingly, spin-orbital wave functions may be written as

$$\psi_q(\boldsymbol{\xi}) = \langle\boldsymbol{\xi}|q\rangle \tag{1.14}$$

A simple (not yet antisymmetric) product state of N electrons, in which the ith electron "occupies" the spin-orbital q_i, may be written as

$$|\Psi\rangle = |q_1\rangle|q_2\rangle\ldots|q_N\rangle \tag{1.15}$$

The corresponding wave function takes on the form (Hartree product)

$$\Psi(\boldsymbol{\xi}_1,\ldots,\boldsymbol{\xi}_N) = \psi_{q_1}(\boldsymbol{\xi}_1)\ldots\psi_{q_N}(\boldsymbol{\xi}_N) = \langle\boldsymbol{\xi}_N|\ldots\langle\boldsymbol{\xi}_1|q_1\rangle\ldots|q_N\rangle \tag{1.16}$$

As in Eq. (1.4), $|\boldsymbol{\xi}_1\rangle \dots |\boldsymbol{\xi}_N\rangle$ denotes the corresponding Hartree product (in ket form) of the one-particle coordinate eigenfunctions $|\boldsymbol{\xi}_i\rangle$.

For a convenient formulation of antisymmetric product states, it is expedient to introduce the antisymmetrization operator

$$\hat{\mathbb{A}} = \frac{1}{N!} \sum_P (-1)^P \hat{P} \tag{1.17}$$

Here, the sum runs over all $N!$ permutations P (of N elements). The following properties are readily established (see Exercise 1.2):

$$\hat{\mathbb{A}} = \hat{\mathbb{A}}^\dagger \qquad \text{hermiticity} \tag{1.18}$$

$$\hat{P}\hat{\mathbb{A}} = (-1)^P \hat{\mathbb{A}} \tag{1.19}$$

$$\hat{\mathbb{A}}^2 = \hat{\mathbb{A}} \qquad \text{projector} \tag{1.20}$$

With the help of the antisymmetrization operator, we may define normalized antisymmetric product states according to

$$\begin{aligned} |q_1 \dots q_N\rangle &= (N!)^{\frac{1}{2}}\, \hat{\mathbb{A}} |q_1\rangle \dots |q_N\rangle \\ &= (N!)^{-\frac{1}{2}} \sum_P (-1)^P |q_{P(1)}\rangle \dots |q_{P(N)}\rangle \end{aligned} \tag{1.21}$$

The corresponding wave function takes on the form

$$\begin{aligned} \Psi_A(\boldsymbol{\xi}_1, \dots, \boldsymbol{\xi}_N) &= \langle \boldsymbol{\xi}_N | \dots \langle \boldsymbol{\xi}_1 | | q_1 \dots q_N \rangle \\ &= \frac{1}{\sqrt{N!}} \sum_P (-1)^P \psi_{q_{P(1)}}(\boldsymbol{\xi}_1) \psi_{q_{P(2)}}(\boldsymbol{\xi}_2) \dots \psi_{q_{P(N)}}(\boldsymbol{\xi}_N) \\ &= \frac{1}{\sqrt{N!}} \sum_P (-1)^P \psi_{q_1}(\boldsymbol{\xi}_{P(1)}) \psi_{q_2}(\boldsymbol{\xi}_{P(2)}) \dots \psi_{q_N}(\boldsymbol{\xi}_{p(N)}) \end{aligned} \tag{1.22}$$

Note that Ψ_A is normalized (supposing orthonormal one-particle states $\langle q_i|$). Alternatively, one could use the antisymmetrized coordinate eigenstate

$$|\boldsymbol{\xi}_1 \dots \boldsymbol{\xi}_N\rangle = (N!)^{-\frac{1}{2}} \sum_P (-1)^P |\boldsymbol{\xi}_{P(1)}\rangle \dots |\boldsymbol{\xi}_{P(N)}\rangle \tag{1.23}$$

rather than the product of the $\langle \boldsymbol{\xi}_i|$ states. However, there is a subtlety as according to

$$\langle \boldsymbol{\xi}_N \dots \boldsymbol{\xi}_1 | q_1 \dots q_N \rangle = \sqrt{N!}\, \Psi_A(\boldsymbol{\xi}_1, \dots, \boldsymbol{\xi}_N) \tag{1.24}$$

the result has to be multiplied with $1/\sqrt{N!}$ to yield a normalized wave function (see Exercise 1.4).

Owing to the formal equivalence to the definition of matrix determinants, an antisymmetric product state wave function can also be written in the form of a determinant of spin-orbitals, referred to as **Slater determinant**:

$$\Psi_A(\xi_1, \ldots, \xi_N) = \frac{1}{\sqrt{N!}} \begin{vmatrix} \psi_{q_1}(\xi_1) & \psi_{q_1}(\xi_2) & \cdots & \psi_{q_1}(\xi_N) \\ \psi_{q_2}(\xi_1) & \psi_{q_2}(\xi_2) & \cdots & \psi_{q_2}(\xi_N) \\ \vdots & \vdots & \ddots & \vdots \\ \psi_{q_N}(\xi_1) & \psi_{q_N}(\xi_2) & \cdots & \psi_{q_N}(\xi_N) \end{vmatrix} \tag{1.25}$$

A shorthand notation for the Slater determinant (including the normalization factor) is as follows:

$$\Psi_A(\xi_1, \ldots, \xi_N) = \left|\psi_{q_1}(\xi_1)\psi_{q_2}(\xi_2)\cdots\psi_{q_N}(\xi_N)\right| \tag{1.26}$$

The antisymmetric product states fulfill the following properties:

1. Symmetry with respect to permutations:

$$\hat{P}|q_1 \ldots q_N\rangle = (-1)^P |q_1 \ldots q_N\rangle \tag{1.27}$$

2. Pauli principle (the wave function vanishes if two electrons occupy the same one-particle state):

$$|q_1 \ldots q_N\rangle \equiv 0 \qquad \text{if} \quad q_i = q_j,\ i \neq j \tag{1.28}$$

3. Linear combination of spin-orbitals:

$$\begin{aligned} &|q_1 \ldots q_{i-1}(aq + bq')q_{i+1} \ldots q_N\rangle \\ &\qquad = a|q_1 \ldots q_{i-1}q q_{i+1} \ldots q_N\rangle + b|q_1 \ldots q_{i-1}q' q_{i+1} \ldots q_N\rangle \end{aligned} \tag{1.29}$$

The (multi-)linearity of the Slater determinants can be generalized to an arbitrary linear transformation

$$|\tilde{q}_i\rangle = \sum_{j=1}^{N} |q_j\rangle U_{ji}, \qquad U_{ji} \in \mathbb{C} \tag{1.30}$$

of the set of spin-orbitals, yielding the expression

$$|\tilde{q}_1 \ldots \tilde{q}_N\rangle = \det(\boldsymbol{U})|q_1 \ldots q_N\rangle \tag{1.31}$$

for the Slater determinant of the transformed spin-orbitals. Here, $\det(\boldsymbol{U})$ is the determinant of the matrix of elements U_{kl}. Equation (1.31) can readily be derived (Exercise 1.3) using the properties (1.27)–(1.29).

For a complete basis set of one-particle states, the manifold of product states $|q_1 \dots q_N\rangle, q_1 < q_2 < \dots < q_N$ forms a basis of the Hilbert space of antisymmetric N-electron states.

Physical observables are represented by (hermitian) operators. For an N-electron system, the operators usually are of the form

$$\hat{W}_1 = \sum_{i=1}^{N} \hat{w}(i) \qquad \text{one-particle operator} \tag{1.32}$$

$$\hat{W}_2 = \sum_{i<j}^{N} \hat{w}(i, j) = \tfrac{1}{2} \sum_{i \neq j}^{N} \hat{w}(i, j) \qquad \text{two-particle operator} \tag{1.33}$$

Here, $\hat{w}(i)$ is a one-particle operator acting on the coordinates of the ith electron. Likewise, $\hat{w}(i, j)$ denotes a two-particle operator, the action of which depends on the coordinates of both the ith and the jth electrons. While in the elementary physics of interacting electrons only one- and two-particle operators arise, one may, at least formally, introduce r-particle operators $\hat{W}_r$ with $r \geq 3$. Consistent with the indistinguishability of the particles, these operators are symmetric, which means they are invariant with respect to a permutation of the numbering of the electrons:

$$\hat{W}_r = \hat{P}\hat{W}_r\hat{P}^{-1} \quad \text{or} \quad \hat{W}_r\hat{P} = \hat{P}\hat{W}_r, \quad r = 1, 2, \dots \tag{1.34}$$

The (nonrelativistic) hamiltonian for an N-electron atom or molecule may serve as an example:

$$\hat{H} = \sum_{i=1}^{N} \left\{ -\frac{\hbar^2}{2m_e} \Delta^{(i)} - \sum_{a=1}^{K} \frac{e^2 Z_a}{|\boldsymbol{x}_i - \boldsymbol{R}_a|} \right\} + \frac{1}{2} \sum_{i \neq j=1}^{N} \frac{e^2}{|\boldsymbol{x}_i - \boldsymbol{x}_j|} + \sum_{a<b} \frac{Z_a Z_b e^2}{|\boldsymbol{R}_a - \boldsymbol{R}_b|} \tag{1.35}$$

Here, Z_a and $\boldsymbol{R}_a$ denote the nuclear charge numbers and positions. The last term is the nuclear repulsion, which for fixed nuclear positions is simply a constant, not affecting the electronic motion. The electronic hamiltonian (without the nuclear repulsion term),

$$\hat{H} = \hat{T} + \hat{V} \tag{1.36}$$

is composed of a one-particle part

$$\hat{T} = \sum_{i=1}^{N} \hat{t}(i) \tag{1.37}$$

associated with the kinetic energy of the electrons and the electron–nuclei interaction, and the electronic Coulomb repulsion,

$$\hat{V} = \frac{1}{2} \sum_{i \neq j} \hat{v}(i, j) \tag{1.38}$$

being a two-particle operator. It should be noted that $\hat{H}$ is spin-independent; that is, its constituents act exclusively on the spatial coordinates of the electrons. Spin-dependent terms come into play when relativistic effects are taken into consideration.

1.2 Matrix Elements for Many-Electron States

In dealing with many-particle systems, the handling of matrix elements involving Slater determinants is required. The basic tool here is a set of simple rules, referred to as Slater–Condon rules, which we consider in the following.

Let us consider a general Slater determinant

$$|\Phi\rangle = |q_1 \dots q_N\rangle \tag{1.39}$$

corresponding to a specific choice of one-particle states of a given orthonormal basis set $\{|q\rangle\}$. We shall specifically address three cases, namely (a) scalar products, (b) matrix elements of one-particle operators, and (c) matrix elements of two-particle operators.

(a) **Scalar products**:
Let us first consider the scalar product $\langle\Phi|\Phi\rangle$:

$$\langle q_1 \dots q_N | q_1 \dots q_N \rangle = N! \, \langle q_1 | \dots \langle q_N | \hat{\mathbb{A}}^\dagger \hat{\mathbb{A}} | q_1 \rangle \dots | q_N \rangle \tag{1.40}$$

$$= N! \, \langle q_1 | \dots \langle q_N | \hat{\mathbb{A}} | q_1 \rangle \dots | q_N \rangle \tag{1.41}$$

In the second line, we have used $\hat{\mathbb{A}}^\dagger \hat{\mathbb{A}} = \hat{\mathbb{A}}^2 = \hat{\mathbb{A}}$, following from the hermiticity and projector properties (1.18) and (1.20), respectively. Note that the states appearing to the left and right sides of $\hat{\mathbb{A}}$ are Hartree products. To proceed, we use the definition (1.17) of the antisymmetrization operator and evaluate the scalar products of the respective Hartree product states:

$$\langle q_1 \dots q_N | q_1 \dots q_N \rangle = N! \frac{1}{N!} \sum_P (-1)^P \langle q_1 | q_{P(1)} \rangle \dots \langle q_N | q_{P(N)} \rangle$$

$$= \langle q_1 | q_1 \rangle \langle q_2 | q_2 \rangle \dots \langle q_N | q_N \rangle = 1 \tag{1.42}$$

Of all permutations, only the identical permutation, $P(i) = i$, gives rise to a non-vanishing contribution, as in all others there is at least one vanishing overlap factor, $\langle q_k | q_l \rangle = 0$, $q_k \neq q_l$. The final result as given in the second line of Eq. (1.42) establishes the first Slater–Condon rule (SC) for scalar products (**a1**).

The preceding treatment can readily be extended to the case of differing Slater determinants. Let

$$
\begin{aligned}
|\Phi\rangle &= |q_1 \dots q_k \dots q_N\rangle \\
|\Phi'\rangle &= |q_1 \dots q' \dots q_N\rangle, \ q' \neq q_1, \dots, q_N
\end{aligned}
\tag{1.43}
$$

denote two Slater determinants differing at exactly one position. Performing the same algebra as in Eqs. (1.41) and (1.42), the second line of Eq. (1.42) becomes

$$
\langle\Phi'|\Phi\rangle = \langle q_1|q_1\rangle \dots \langle q'|q_k\rangle \dots \langle q_N|q_N\rangle = 0
$$

as even for the identical permutation there is one vanishing scalar product, $\langle q'|q_k\rangle = 0$. In the same manner, Slater determinants differing at two or more positions can be treated. The emerging SC rule **a2** can be stated as follows:
The scalar product of two Slater determinants vanishes if they differ at least in one position (upon appropriate ordering of the orbitals).

(b) **One-particle operators**:
In the discussion of the matrix elements $\langle\Phi'|\hat{W}|\Phi\rangle$ of a general one-particle operator,

$$
\hat{W} = \sum_{i=1}^{N} \hat{w}(i)
$$

we distinguish three cases: (*i*) $|\Phi'\rangle = |\Phi\rangle$; (*ii*) $|\Phi\rangle$ and $|\Phi'\rangle$ differ in one position; (*iii*) they differ in two or more positions.

(*i*) Here, the evaluation of the matrix element proceeds as follows:

$$
\begin{aligned}
\langle\Phi|\hat{W}|\Phi\rangle &= N!\langle q_1| \dots \langle q_N|\hat{\mathbb{A}}^\dagger \hat{W} \hat{\mathbb{A}}|q_1\rangle \dots |q_N\rangle \\
&= N!\langle q_1| \dots \langle q_N|\hat{W} \hat{\mathbb{A}}|q_1\rangle \dots |q_N\rangle \\
&= \sum_P (-1)^P \langle q_1| \dots \langle q_N| \sum_{i=1}^{N} \hat{w}(i)\, |q_{P(1)}\rangle \dots |q_{P(N)}\rangle
\end{aligned}
\tag{1.44}
$$

Besides the hermiticity and projector properties (1.18) and (1.20), we here have used $\hat{\mathbb{A}}\hat{W} = \hat{W}\hat{\mathbb{A}}$, being an immediate consequence of the commutation relation (1.34). Each term in the third line of Eq. (1.44) is a product of $N-1$ overlap factors and a single one-particle integral:

$$
\begin{aligned}
\langle\Phi|\hat{W}|\Phi\rangle &= \sum_{i=1}^{N} \sum_P (-1)^P \langle q_i|\hat{w}|q_{P(i)}\rangle \prod_{j\neq i} \langle q_j|q_{P(j)}\rangle \\
&= \sum_{i=1}^{N} \langle q_i|\hat{w}|q_i\rangle
\end{aligned}
\tag{1.45}
$$

Here,

$$\langle p|\hat{w}|q\rangle = \int \psi_p^*(\boldsymbol{\xi})\hat{w}\,\psi_q(\boldsymbol{\xi})\mathrm{d}\boldsymbol{\xi} \tag{1.46}$$

denotes the one-particle matrix element for the spin-orbitals p and q. As in the case of the scalar products, the product of overlap factors vanishes for all permutations except for the identical permutation, giving rise to the simple final result in the second line of Eq. (1.45). This constitutes the first SC rule (**b1**) for matrix elements of a one-particle operator.

(*ii*) In the case of two Slater determinants (1.43), differing at the kth position, the evaluation of the matrix element $\langle\Phi'|\hat{W}|\Phi\rangle$ is largely analogous to case (*i*) in Eq. (1.44). Again, in summing over the permutations only the identical permutation survives, and only the summation over the one-particle indices remains,

$$\langle\Phi'|\hat{W}|\Phi\rangle = \langle q_1|q_1\rangle \ldots \langle q'|\hat{w}|q_k\rangle \ldots \langle q_N|q_N\rangle + \sum_{i\neq k}^{N}\langle q_i|\hat{w}|q_i\rangle\langle q'|q_k\rangle \prod_{j\neq i,k}\langle q_j|q_j\rangle$$

Here, the term corresponding to $i = k$ has been taken out of the sum. Obviously, the first term on the right-hand side is the only non-vanishing one, because all other summands contain the vanishing overlap factor $\langle q'|q_k\rangle$. The final result (constituting the SC rule **b2**) reads

$$\langle\Phi'|\hat{W}|\Phi\rangle = \langle q'|\hat{w}|q_k\rangle$$

(*iii*) If the Slater determinants differ at two positions, say at k and l ($k < l$),

$$\begin{aligned} |\Phi\rangle &= |q_1 \ldots q_k \ldots q_l \ldots q_N\rangle \\ |\Phi''\rangle &= |q_1 \ldots q' \ldots q'' \ldots q_N\rangle \qquad q' \neq q'' \neq q_1, \ldots, q_N \end{aligned} \tag{1.47}$$

the matrix element is readily seen to vanish,

$$\langle\Phi''|\hat{W}|\Phi\rangle = 0$$

The corresponding SC rule (**b3**) is that for a one-particle operator the matrix element of two Slater determinants vanishes if they differ in two or more positions (upon appropriate reordering).

(c) **Two-particle operators**:
In the following, we consider an arbitrary two-particle operator, written in the form

$$\hat{V} = \sum_{i<j} \hat{v}(i, j) \tag{1.48}$$

As above, we shall distinguish several cases, here (*i*)–(*iv*), corresponding to the number of positions in which the two Slater determinants differ.

(i) Expectation values:
Like in the case of the one-particle operator, we may use the operator identity $\hat{\mathbb{A}}^\dagger \hat{V} \hat{\mathbb{A}} = \hat{V} \hat{\mathbb{A}}$, which allows us to write

$$
\begin{aligned}
\langle \Phi | \hat{V} | \Phi \rangle &= N! \langle q_1 | \ldots \langle q_N | \hat{V} \hat{\mathbb{A}} | q_1 \rangle \ldots | q_N \rangle \\
&= \sum_P (-1)^P \langle q_1 | \ldots \langle q_N | \sum_{i<j} \hat{v}(i,j) \, | q_{P(1)} \rangle \ldots | q_{P(N)} \rangle \\
&= \sum_{i<j} \sum_P (-1)^P \langle q_i q_j | \hat{v} | q_{P(i)} q_{P(j)} \rangle \prod_{k \neq i,j} \langle q_k | q_{P(k)} \rangle
\end{aligned}
\tag{1.49}
$$

where

$$
\langle pq | \hat{v} | rs \rangle = \int \int \mathrm{d}\xi_1 \mathrm{d}\xi_2 \psi_p^*(\xi_1) \psi_q^*(\xi_2) \hat{v}(\xi_1, \xi_2) \psi_r(\xi_1) \psi_s(\xi_2) \tag{1.50}
$$

denotes the two-particle matrix element, involving the four spin-orbitals p, q, r, s; we shall also use the familiar shorthand notation

$$
V_{pqrs} = \langle pq | \hat{v} | rs \rangle \tag{1.51}
$$

For a given pair (i, j) in the third line of Eq. (1.49), the overlap product implies that only two permutations lead to a non-vanishing contribution, namely the identical permutation where

$$
P(i) = i, \qquad P(j) = j
$$

and the transposition exchanging i and j,

$$
P(i) = j, \qquad P(j) = i
$$

The situation can be depicted in the following scheme

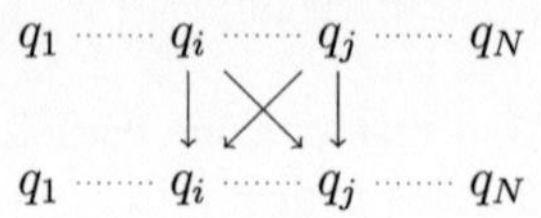

Since a transposition, being an odd permutation, implies the sign $(-1)^P = -1$, one arrives at the result

$$
\langle \Phi | \hat{V} | \Phi \rangle = \sum_{i<j} \left(\langle q_i q_j | \hat{v} | q_i q_j \rangle - \langle q_i q_j | \hat{v} | q_j q_i \rangle \right) = \sum_{i<j} V_{q_i q_j [q_i q_j]} = \tfrac{1}{2} \sum_{i,j} V_{q_i q_j [q_i q_j]} \tag{1.52}
$$

The two integrals in the integrand, referred to as direct and exchange integrals, can be combined in the antisymmetrized two-particle integral

$$V_{pq[rs]} = V_{pqrs} - V_{pqsr} \tag{1.53}$$

Equation (1.52) constitutes the first SC rule (**c1**) for the two-particle matrix elements.
(*ii*) Slater determinants differing at one position:
The matrix element $\langle \Phi' | \hat{V} | \Phi \rangle$ for the two Slater determinants (1.39) and (1.43), differing at the kth position, can be evaluated as above, yielding

$$\langle \Phi' | \hat{V} | \Phi \rangle = \sum_P (-1)^P \langle q_1 | \dots \langle q' | \dots \langle q_N | \sum_{i<j} \hat{v}(i, j) | q_{P(1)} \rangle \dots | q_{P(N)} \rangle$$

where q' is at the kth position of the Slater determinant on the left. If q' does not enter the two-particle integral, that is, if $i, j \neq k$, there will be a vanishing overlap factor, $\langle q' | q_{P(k)} \rangle = 0$, irrespective of the permutation P. This means that the double summation running over the orbital indices $i < j$ becomes a single summation according to

$$\sum_{i<j} \rightarrow \sum_{\substack{i<k \\ j=k}} + \sum_{\substack{j>k \\ i=k}}$$

This gives

$$\langle \Phi' | \hat{V} | \Phi \rangle = \sum_{i<k} \sum_P (-1)^P \langle q_i q' | \hat{v} | q_{P(i)} q_{P(k)} \rangle \prod_{l \neq i,k} \langle q_l | q_{P(l)} \rangle$$
$$+ \sum_{j>k} \sum_P (-1)^P \langle q' q_j | \hat{v} | q_{P(k)} q_{P(j)} \rangle \prod_{l \neq j,k} \langle q_l | q_{P(l)} \rangle \tag{1.54}$$

Let us consider the first term (A) on the right-hand side. The overlap product restricts the permutations to the identical permutation and the transposition

$$P(i) = k, \qquad P(k) = i$$

as indicated in the following scheme:

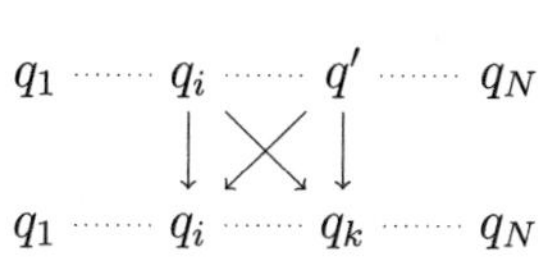

Taking the sign of the transposition into account, the first term on the right-hand side of Eq. (1.54) becomes

$$(A) = \sum_{i<k} (\langle q_i q' | \hat{v} | q_i q_k \rangle - \langle q_i q' | \hat{v} | q_k q_i \rangle)$$

In the same way, the second term (B) in Eq. (1.54) can be treated. Bringing both contributions together, the final result (SC rule **c2**) can be written as

$$\langle \Phi' | \hat{V} | \Phi \rangle = \sum_{i,i \neq k} (\langle q_i q' | \hat{v} | q_i q_k \rangle - \langle q_i q' | \hat{v} | q_k q_i \rangle) = \sum_{i=1}^{N} V_{q_i q' [q_i q_k]} \tag{1.55}$$

Note that the restriction $i \neq k$ in the summation can be dropped because antisymmetric two-particle integral vanishes for $i = k$.

(*iii*) Slater determinants differing at two positions:

The matrix element involving the two determinants (1.47) is evaluated according to

$$\langle \Phi'' | \hat{V} | \Phi \rangle = \sum_P (-1)^P \langle q_1 | \dots \langle q' | \dots \langle q'' | \dots \langle q_N | \sum_{i<j} \hat{v}(i, j) \, |q_{P(1)}\rangle \dots |q_{P(N)}\rangle$$

where q' and q'' are in the positions k and l, respectively. The overlap argument means that only the contribution with $i = k$, $j = l$ does not vanish in the double summation over i, j. Again, the overlap product on the right-hand side restricts the permutations to the identical permutation and the transposition exchanging k and l. The final result (SC rule **c3**) reads

$$\langle \Phi'' | \hat{V} | \Phi \rangle = V_{q' q'' [q_k q_l]} \tag{1.56}$$

(*iv*) In a similar way, one can see that the matrix element of a two-particle operator vanishes if the two determinants differ in three or more positions (upon appropriate ordering of the orbitals). This is SC rule **c4**.

Examples:

As an example, let us consider the ground-state Slater determinant,

$$|\Phi_0\rangle = |1\,2 \dots N\rangle \tag{1.57}$$

in which $|q\rangle$, $q = 1, \dots, N$, are the N energetically lowest Hartree–Fock (HF) spin-orbitals. The expectation value of the hamiltonian (1.35) can readily be evaluated according to the SC rules **b1** and **c1**,

$$\langle \Phi_0 | \hat{H} | \Phi_0 \rangle = \sum_{i=1}^{N} t_{ii} + \tfrac{1}{2} \sum_{i,j=1}^{N} V_{ij[ij]} \tag{1.58}$$

where $t_{ij} = \langle i | \hat{t} | j \rangle$ and V_{ijkl} denote the one- and two-particle integrals, respectively. The Slater determinant

$$|\Phi_{ak}\rangle = |1 \dots (k-1) a (k+1) \dots N\rangle, \; a > N \tag{1.59}$$

corresponds to a singly excited state (with respect to $|\Phi_0\rangle$), in which the kth electron is excited to the virtual orbital a. The matrix element

$$\langle \Phi_{ak}|\hat{V}|\Phi_0\rangle = \sum_{i=1}^{N} V_{ia[ik]} \tag{1.60}$$

may serve as an example for the SC rule **c2**.

To evaluate matrix elements for differing Slater determinants, one has to assure that the orbitals are ordered in the form supposed in the derivation of the SC rules. At the end of the next section, we shall demonstrate how this can be achieved quite conveniently using second quantization.

Spin-Free Expressions:
Finally, we take a look at the derivation of spin-free expressions for many-electron matrix elements, which applies to operators acting on the spatial variables only. Let us again expand the spin-orbital quantum numbers used so far into the pair of spatial and spin quantum numbers:

$$q \rightarrow q\gamma$$

and recall that the spin-orbitals are products of spatial and spin-orbitals,

$$\psi_{q\gamma}(\boldsymbol{\xi}) = \varphi_q(\boldsymbol{x})\chi_\gamma(\sigma) \tag{1.61}$$

For a spatial operator, the one-particle integrals (1.46) simplify according to

$$\begin{aligned}\langle p\gamma|\hat{w}|q\gamma'\rangle &= \langle\gamma|\gamma'\rangle \int \varphi_p^*(\boldsymbol{x})\hat{w}\varphi_q(\boldsymbol{x})\mathrm{d}x \\ &= \delta_{\gamma\gamma'} w_{pq}\end{aligned} \tag{1.62}$$

where w_{pq} is a spatial one-particle integral, and the spin-integral becomes a trivial Kronecker delta. In a similar way, the general two-particle integrals can be evaluated to become

$$\begin{aligned}\langle p\gamma q\sigma|\hat{v}|r\rho s\tau\rangle &= \langle\gamma|\rho\rangle\langle\sigma|\tau\rangle \int\int \mathrm{d}\boldsymbol{x}_1\mathrm{d}\boldsymbol{x}_2\varphi_p^*(\boldsymbol{x}_1)\varphi_q^*(\boldsymbol{x}_2)\frac{e^2}{|\boldsymbol{x}_1-\boldsymbol{x}_2|}\varphi_r(\boldsymbol{x}_1)\varphi_s(\boldsymbol{x}_2) \\ &= \delta_{\gamma\rho}\delta_{\sigma\tau}V_{pqrs}\end{aligned} \tag{1.63}$$

The simplification of the spin–orbit integrals can readily be exploited in the many-electron matrix elements. As an example, let us consider the ground-state expectation value (1.58) of the hamiltonian. Supposing that there are $n = \frac{1}{2}N$ spatial orbitals, each occupied by a spin-α and spin-β electron,

$$|\Phi_0\rangle = |\varphi_{1\alpha}\varphi_{1\beta}\ldots\varphi_{n\alpha}\varphi_{n\beta}\rangle$$

the one-particle part can be evaluated as follows:

$$\sum_{i=1}^{N} t_{ii} = \sum_{p=1}^{n} \sum_{\gamma=\alpha,\beta} t_{p\gamma,p\gamma}$$
$$= \sum_{p=1}^{n} \sum_{\gamma=\alpha,\beta} \delta_{\gamma\gamma} t_{pp} = 2\sum_{p=1}^{n} t_{pp} \tag{1.64}$$

Here, the spin-orbitals have been expanded according to $i \equiv p\gamma$.

To evaluate the Coulomb part, the antisymmetric Coulomb integrals have to be written in the original explicit form, because the direct and exchange integrals differ with respect to the spin-integration,

$$\langle\Phi_0|\hat{V}|\Phi_0\rangle = \tfrac{1}{2}\sum_{i,j=1}^{N} \left(V_{ijij} - V_{ijji}\right) = \tfrac{1}{2}\sum_{p,q=1}^{n} \sum_{\gamma,\gamma'=\alpha,\beta} \left(V_{p\gamma q\gamma' p\gamma q\gamma'} - V_{p\gamma q\gamma' q\gamma' p\gamma}\right)$$
$$= \tfrac{1}{2}\sum_{p,q=1}^{n} \sum_{\gamma,\gamma'=\alpha,\beta} \left(\delta_{\gamma\gamma}\delta_{\gamma'\gamma'} V_{pqpq} - \delta_{\gamma\gamma'} V_{pqqp}\right)$$
$$= \tfrac{1}{2}\sum_{p,q=1}^{n} \left(4V_{pqpq} - 2V_{pqqp}\right) \tag{1.65}$$

The spin-orbital indices i, j in the first line have been expanded according to $i \equiv p\gamma$, $j \equiv q\gamma'$.

Exercises

1.1 Show that the permutation operator is a unitary operator.
1.2 Establish the properties (1.18)–(1.20) for the antisymmetrization operator $\hat{\mathbb{A}}$.
1.3 Derive the relation (1.31) for an orbital transformation in the Slater determinant.
1.4 Consider an antisymmetrized normalized state $|\Psi\rangle$ with the (normalized) wave function $\Psi(\xi_1, \ldots, \xi_N)$. Show that

$$\langle\xi_N \ldots \xi_1|\Psi\rangle = \sqrt{N!}\,\Psi(\xi_1, \ldots, \xi_N) \tag{1.66}$$

where $|\xi_1 \ldots \xi_N\rangle$ is given by Eq. (1.23).
1.5 Evaluate for the state (1.59) the excitation energy (through first order) $E_{ak}(1) = \langle\Phi_{ak}|\hat{H}|\Phi_{ak}\rangle - \langle\Phi_0|\hat{H}|\Phi_0\rangle$.
1.6 Specify the spin-orbitals in $|\Phi_{ak}\rangle$ according to $a \to a\gamma$, $k \to k\gamma'$ as products of spatial orbitals and spin functions, and form one singlet and three triplet states

as suitable linear combinations of the "primitive" states $|\Phi_{a\gamma,k\gamma'}\rangle$, $\gamma, \gamma' = \alpha, \beta$. Evaluate the spin-free expressions for the singlet and triplet excitation energies ${}^1E_{ak}(1)$ and ${}^3E_{ak}(1)$, respectively.

Reference

1. Messiah A (1967) Quantum mechanics. North-Holland Publishing Company, Amsterdam

Chapter 2
Second Quantization

The concept of second quantization (SQ), originally developed in quantum field theory, has proven to be an indispensable tool in many-body theory since it allows one to represent many-electron states and operators in an utmost flexible and compact way. In this chapter, we shall review the SQ formalism at some length. The presentation of the SQ operator algebra in the first three subsections is essentially based on a concise formulation in the appendix of a textbook by Baumgärtner and Schuck [1], who acknowledge unpublished lecture notes by W. Brenig. A combination of SQ and the SC rules, described in Sect. 2.4, leads to a practical means for evaluating many-electron matrix elements. The SQ field operators relate to underlying one-particle states. The transformations of the field operators induced by changes of the one-particle representation are considered in the final subsection.

2.1 Definition of Creation and Destruction Operators

The starting point is a complete set (basis) of orthonormal one-particle states $|q\rangle$,

$$\langle q|q'\rangle = \delta_{qq'}$$
$$\sum_q |q\rangle\langle q| = \hat{\mathbb{1}}$$

and the corresponding basis set of normalized antisymmetric N-electron product states,

$$|q_1 q_2 \dots q_N\rangle, \; q_1 < q_2 < \dots < q_N$$

as introduced in Chap. 1. So far, the electron number N was assumed to be arbitrary but fixed. Now, we extend the scope to allow for variable electron numbers. For this purpose, the concept of the Fock space is introduced,

J. Schirmer, *Many-Body Methods for Atoms, Molecules and Clusters*, Lecture Notes in Chemistry 94, https://doi.org/10.1007/978-3-319-93602-4_2

$$\mathcal{F} = \mathcal{H}_0 \oplus \mathcal{H}_1 \oplus \mathcal{H}_2^A \oplus \cdots \tag{2.1}$$

being the direct sum of the N-electron Hilbert spaces, $N = 0, 1, \ldots,$ where for $N \geq 2$ we may confine us to the subspace of antisymmetric states, $\mathcal{H}_N^A$. For $N = 0$, the Hilbert space $\mathcal{H}_0$ is spanned by exactly one state, referred to as the vacuum state, $|\varnothing\rangle$.

Obviously, $\mathcal{F}$ is a linear vector space, with a scalar product defined within each of the linear subspaces. The definition of the scalar product can easily be generalized,

$$\langle q_1' \ldots q_{N'}' | q_1 \ldots q_N \rangle = 0 \quad \text{if} \quad N \neq N' \tag{2.2}$$

to cover products of states with differing electron numbers.

Having established the mathematical background, we are in the position to introduce "fermion operators," more specifically, **creation** and **destruction operators**. Let us first consider the creation operators, which can be defined by specifying their action on the Fock-space basis states:

$$c_q^\dagger |q_1 \ldots q_N\rangle = |q_1 \ldots q_N q\rangle \tag{2.3}$$

Acting on a product state of N electrons, $c_q^\dagger$ generates a product state of $N + 1$ electrons, "creating" an electron in the one-particle state q. Note that $|q_1 \ldots q_N q\rangle = \mathbf{0}$ unless $q \neq q_1, \ldots, q_N$. Beginning with

$$c_q^\dagger |\varnothing\rangle = |q\rangle \tag{2.4}$$

an N-electron product state can be generated according to

$$|q_1 \ldots q_N\rangle = c_{q_N}^\dagger \cdots c_{q_1}^\dagger |\varnothing\rangle \tag{2.5}$$

by letting the creation operators act successively on the vacuum state.

Next, let us consider the effect of the hermitian conjugate operator, c_q, that is, $c_q = (c_q^\dagger)^\dagger$, acting on an N-electron product state. For this purpose, we expand the state of interest in terms of the Fock-space basis states:

$$\begin{aligned} c_q|q_1 \ldots q_N\rangle &= \sum_{\substack{q_1' < q_2' \cdots < q_{N'}' \\ N'=0,1,\ldots}} |q_1' \ldots q_{N'}'\rangle \langle q_1' \ldots q_{N'}' | c_q | q_1 \ldots q_N\rangle \\ &= \sum_{\substack{q_1' < q_2' \cdots < q_{N'}' \\ N'=0,1,\ldots}} |q_1' \ldots q_{N'}'\rangle \langle q_1 \ldots q_N | c_q^\dagger | q_1' \ldots q_{N'}'\rangle^* \end{aligned} \tag{2.6}$$

In the second line, we have used the relation $\langle \Psi | \hat{B} | \Phi \rangle = \langle \Phi | \hat{B}^\dagger | \Psi \rangle^*$ to replace c_q by $c_q^\dagger$. Since

$$c_q^\dagger |q_1' \ldots q_{N'}'\rangle = |q_1' \ldots q_{N'}' q\rangle$$

is a state of $N'+1$ electrons, the scalar products in the second line of Eq. (2.6) vanish unless $N' = N-1$, and the latter equation is simplified accordingly:

$$c_q|q_1 \ldots q_N\rangle = \sum_{q'_1<q'_2\cdots<q'_{N-1}} |q'_1 \ldots q'_{N-1}\rangle\langle q_1 \ldots q_N|c_q^\dagger|q'_1 \ldots q'_{N-1}\rangle^* \tag{2.7}$$

This means that, acting on an N-electron state, c_q generates an $(N-1)$-electron state, $c_q|q_1 \ldots q_N\rangle \in \mathcal{H}^A_{N-1}$. Since the effect of the c_q operators is to remove ("destroy") an electron, they are referred to as destruction operators. Acting on the vacuum state yields the null vector,

$$c_q|\varnothing\rangle \equiv \mathbf{0} \tag{2.8}$$

Using the Slater–Condon rules **a1** and **a2** for scalar products, Eq. (2.7) can be further evaluated according to

$$\begin{aligned} c_q|q_1 \ldots q_N\rangle &= \sum_{q'_1<q'_2\cdots<q'_{N-1}} |q'_1 \ldots q'_{N-1}\rangle\langle q_1 \ldots q_N|q'_1 \ldots q'_{N-1}q\rangle^* \\ &= \delta_{qq_N}|q_1 \ldots q_{N-1}\rangle - \delta_{qq_{N-1}}|q_1 \ldots q_{N-2}\,q_N\rangle + \delta_{qq_{N-2}}|q_1 \ldots q_{N-3}\,q_{N-1}\,q_N\rangle - \ldots \end{aligned} \tag{2.9}$$

The destruction operator c_q removes an electron in the orbital q provided q is present in $|q_1 \ldots q_N\rangle$, that is, $q \in \{q_1 \ldots q_N\}$. The phases arise from aligning the positions of q and q_i in the two product states if $q = q_i, i = 1, \ldots, N$.

2.2 Anticommutation Relations for Creation and Destruction Operators

To establish the anticommutation relations for the operators, we apply the operators $c_p, c_q^\dagger$ successively to an arbitrary basis state,

$$\begin{aligned} c_p c_q^\dagger|q_1 \ldots q_N\rangle &= c_p|q_1 \ldots q_N q\rangle \\ &= \delta_{pq}|q_1 \ldots q_N\rangle - \delta_{p,q_N}|q_1 \ldots q_{N-1}q\rangle + \ldots \end{aligned} \tag{2.10}$$

and in the reversed order

$$c_q^\dagger c_p|q_1 \ldots q_N\rangle = \delta_{p,q_N}|q_1 \ldots q_{N-1}q\rangle - \delta_{p,q_{N-1}}|q_1 \ldots q_{N-2}q_N q\rangle + \ldots \tag{2.11}$$

Here, Eqs. (2.3) and (2.9) have been used. Comparing the latter two equations, one sees that all the terms of Eq. (2.11) do also appear in Eq. (2.10), though with different signs. This means that these terms cancel each other when both equations are added, and only the first (unmatched) term in Eq. (2.10) survives:

$$(c_p c_q^\dagger + c_q^\dagger c_p)|q_1 \dots q_N\rangle = \delta_{pq}|q_1 \dots q_N\rangle$$

Since this holds for arbitrary basis states, we may conclude the operator identity

$$\{c_q^\dagger, c_p\} = c_p c_q^\dagger + c_q^\dagger c_p = \delta_{pq} \tag{2.12}$$

In a similar way, the relations

$$\{c_p^\dagger, c_q^\dagger\} = 0, \qquad \{c_p, c_q\} = 0 \tag{2.13}$$

can be derived. The anticommutator relations (2.12) and (2.13) establish a versatile tool for handling the many-electron product states (2.5) in an algebraic manner.

2.3 Operators in Second Quantization

The creation and destruction operators of second quantization allow us to represent physical operators in a very advantageous way, as will be discussed in the following. Let us first consider operators of the one-particle type (Eq. 1.32):

$$\hat{W} = \sum_{i=1}^{N} \hat{w}(i) \tag{2.14}$$

In second quantization, the operator $\hat{W}$ can be written in a more general form, which is no longer referring to a specific N-electron space:

$$\hat{W}' = \sum_{p,q} w_{pq} c_p^\dagger c_q \tag{2.15}$$

where $w_{pq} = \langle p|\hat{w}|q\rangle$ denote the one-particle integrals (1.46). To prove the equivalence of the two forms (for a specific electron number N), let us inspect the matrix elements of $\hat{W}'$ with respect to the basis set of N-electron product states. Using the anticommutator algebra established above, the result of applying $\hat{W}'$ to a general product state becomes

$$\begin{aligned}\sum_{p,q} w_{pq} c_p^\dagger c_q |q_1 \dots q_N\rangle &= \sum_p \sum_{q_i} w_{pq_i} c_p^\dagger c_{q_i} |q_1 \dots q_N\rangle \\ &= \sum_p \sum_{q_i} w_{pq_i} |q_1 \dots q_{i-1} p q_{i+1} \dots q_N\rangle\end{aligned}$$

Note that here no overall sign change occurs because a pair of operators is moved to the respective position in the product state. Since, according to the SC rules **a1** and **a2**,

$$\langle q_1 \ldots q_N | q_1 \ldots q_{i-1} p q_{i+1} \ldots q_N \rangle = \delta_{pq_i}$$

the expectation value is given by

$$\langle q_1 \ldots q_N | \hat{W}' | q_1 \ldots q_N \rangle = \sum_{i=1}^{N} w_{q_i q_i} \tag{2.16}$$

which means we have verified the SC rule **b1**. The matrix element for two product states differing at one position, say position k, can readily be evaluated to give

$$\langle q_1 \ldots q_k' \ldots q_N | \hat{W}' | q_1 \ldots q_N \rangle = w_{q_k' q_k} \tag{2.17}$$

This reproduces SC rule **b2**, and, in a similar way, also **b3** can be verified. To conclude, within the respective N-electron space the operators $\hat{W}$ and $\hat{W}'$ are equivalent,

$$\hat{W} \equiv \hat{W}' = \sum_{p,q} w_{pq} c_p^\dagger c_q \tag{2.18}$$

In the following, we will skip the apostrophe used to distinguish the general second-quantization form of the operator from the original one.

In an analogous way, the equivalence of a two-particle operator in the traditional (wave-function) form (Eq. 1.33),

$$\hat{V} = \sum_{i<j=1}^{N} \hat{v}(i, j) \tag{2.19}$$

and the second-quantization form

$$\hat{V} = \frac{1}{2} \sum_{p,q,r,s} V_{pqrs} c_p^\dagger c_q^\dagger c_s c_r \tag{2.20}$$

can be shown, where $V_{pqrs} = \langle pq | \hat{v} | rs \rangle$ denote the two-particle integrals (1.50). Note that the order of the operators, c_s and c_r, on the right-hand side of Eq. (2.20) differs from the order of the corresponding indices in the two-electron integral, V_{pqrs}.

As an example, let us just verify the first SC rule (**c1**) for the two-particle operator. If $\hat{V}$ is applied to a general N-electron product state, we find

$$\hat{V} | q_1 \ldots q_N \rangle = \frac{1}{2} \sum_{p,q,r,s} V_{pqrs} c_p^\dagger c_q^\dagger c_s c_r | q_1 \ldots q_N \rangle = \frac{1}{2} \sum_{p,q} \sum_{q_i < q_j} V_{pq[q_i q_j]} c_p^\dagger c_q^\dagger c_{q_j} c_{q_i} | q_1 \ldots q_N \rangle$$

Here, the case $q_i > q_j$ is accounted for by the antisymmetrized two-electron integral, $V_{pq[q_i q_j]} = V_{pq q_i q_j} - V_{pq q_j q_i}$. Since

$$\langle q_1 \ldots q_N | c_p^\dagger c_q^\dagger c_{q_j} c_{q_i} | q_1 \ldots q_N \rangle = \delta_{pq_i} \delta_{qq_j} - \delta_{pq_j} \delta_{qq_i}$$

the expectation value becomes

$$\langle q_1 \dots q_N|\hat{V}|q_1 \dots q_N\rangle = \frac{1}{2}\sum_{q_i<q_j}(V_{q_iq_j[q_iq_j]} - V_{q_jq_i[q_iq_j]}) = \sum_{q_i<q_j} V_{q_iq_j[q_iq_j]} \quad (2.21)$$

which is exactly as given by Eq. (1.52).

As noted above, three-particle operators, or more general, r-particle operators, $r \geq 3$, do not arise in the context of ab initio many-particle physics. Of course, the SQ representation of operators can readily be extended to the case of $r \geq 3$.

A few remarks concerning the use of the SQ operator representations are in order:

1. As we have already noted, the SQ representation of operators is independent of the particle number, as it should be the case for genuine Fock-space operators. On the other hand, the SQ representation is based on a specific choice of one-particle states. Often, one uses the orbitals obtained from a Hartree–Fock (HF) treatment of the N-electron ground state. This means that the SQ representation may depend on the electron number in an implicit manner. This should be kept in mind when SQ operators based on N-electron HF orbitals are used in computations of systems composed of $N \pm 1$, $N \pm 2, \dots$ electrons.
2. Using the SQ representation, one can readily introduce model hamiltonians to study, e.g., electron correlation in a simplified way. An example is the well-known Hubbard hamiltonian,

$$\hat{H} = -t\sum_{i,\gamma}(c^\dagger_{i\gamma}c_{i+1\gamma} + c^\dagger_{i+1\gamma}c_{i\gamma}) + \sum_i U c^\dagger_{i\alpha}c^\dagger_{i\beta}c_{i\beta}c_{i\alpha} \quad (2.22)$$

Here, the index i labels sites in a one-dimensional model crystal, t is the so-called hopping parameter, and U parameterizes the on-site Coulomb repulsion.
A specific Fock-space operator is the particle number operator

$$\hat{N} = \sum_p c^\dagger_p c_p, \qquad \hat{N}|q_1 \dots q_N\rangle = N\,|q_1 \dots q_N\rangle \quad (2.23)$$

3. Based on an N-electron product ground state, $|\Phi_0\rangle = |1 \dots N\rangle$, which usually will be the HF ground state, excited product states can be conveniently introduced according to

$$\begin{aligned}
&|\Phi_{ak}\rangle = c^\dagger_a c_k|\Phi_0\rangle && 1p\text{-}1h \text{ (single) excitations}\\
&|\Phi_{abkl}\rangle = c^\dagger_a c^\dagger_b c_k c_l|\Phi_0\rangle,\ a<b, k<l && 2p\text{-}2h \text{ (double) excitations}\\
&|\Phi_{abcklm}\rangle = \dots\\
&\vdots
\end{aligned} \quad (2.24)$$

States of $N-1$ or $N+1$ electrons can be written as

$$
\begin{aligned}
&|\Phi_k^{N-1}\rangle = c_k|\Phi_0\rangle && \text{1h excitations}\\
&|\Phi_{akl}^{N-1}\rangle = c_a^\dagger c_k c_l|\Phi_0\rangle,\ k < l && \text{2h-1p excitations}\\
&|\Phi_{abklm}^{N-1}\rangle = \dots\\
&\vdots
\end{aligned} \tag{2.25}
$$

or

$$
\begin{aligned}
&|\Phi_a^{N+1}\rangle = c_a^\dagger|\Phi_0\rangle && \text{1p excitations}\\
&|\Phi_{abk}^{N+1}\rangle = c_a^\dagger c_b^\dagger c_k|\Phi_0\rangle,\ a < b && \text{2p-1h excitations}\\
&\vdots
\end{aligned} \tag{2.26}
$$

respectively. Here, the indices $k, l, m, \dots$ and $a, b, c, \dots$ refer to occupied and unoccupied (virtual) orbitals, respectively, with regard to $|\Phi_0\rangle$. In a similar way, one may represent states of $N \pm 2$, $N \pm 3, \dots$ electrons.

4. The SQ representation of states and operators can be used to evaluate matrix elements in an algebraic way. As an example, consider the matrix element of a one-particle operator taken with respect to a single excitation, $|\Phi_{ak}\rangle$, and the ground state, $|\Phi_0\rangle$:

$$
\begin{aligned}
\langle\Phi_{ak}|\hat{W}|\Phi_0\rangle &= \langle\Phi_0|c_k^\dagger c_a \sum_{p,q} w_{pq} c_p^\dagger c_q|\Phi_0\rangle\\
&= \sum_{p,q} w_{pq}\langle\Phi_0|c_k^\dagger c_a c_p^\dagger c_q|\Phi_0\rangle
\end{aligned}
$$

Now the commutator algebra of the c-operators can be used to move c_a and $c_k^\dagger$ to the right-hand side, yielding

$$
\langle\Phi_0|c_k^\dagger c_a c_p^\dagger c_q|\Phi_0\rangle = \delta_{ap}\delta_{kq}
$$

where we have used $c_a|\Phi_0\rangle = \mathbf{0}$ and $c_k^\dagger|\Phi_0\rangle = \mathbf{0}$. The final result is

$$
\langle\Phi_{ak}|\hat{W}|\Phi_0\rangle = w_{ak} \tag{2.27}
$$

The evaluation of ground-state expectation values for products of c-operators, which is the essential step in the computation of matrix elements, can be treated in a systematic manner, as will be discussed in Chap. 5. However, for more demanding matrix elements this procedure becomes rather cumbersome. A more practical approach consists in using the SQ representation to bring the product states into a form adapted to the SC rules, discussed in the preceding chapter. We will demonstrate the latter approach with the help of a few examples below.

2.4 Combining Second Quantization and Slater–Condon Rules

Let us first examine the simple matrix element $\langle\Phi_{ak}|\hat{W}|\Phi_0\rangle$. We have found it helpful to introduce an illustration of the product ground state,

$$|\Phi_0\rangle = \begin{array}{|c|c|c|}\hline \qquad\qquad & \text{k} & \qquad \\ \hline\end{array} \tag{2.28}$$

in terms of a rectangular box divided into N cells, where the kth cell represents the occupied orbital k in $|\Phi_0\rangle$. Using that graphical representation, the two states on the left- and right-hand side of the matrix element may be placed on top of each other according to

$$c_a^\dagger c_k \begin{array}{|c|c|c|}\hline \qquad\qquad & \text{k} & \qquad \\ \hline\end{array} = \begin{array}{|c|c|c|}\hline \qquad\qquad & \text{a} & \qquad \\ \hline\end{array} \tag{2.29}$$

$$|\Phi_0\rangle = \begin{array}{|c|c|c|}\hline \qquad\qquad & \text{k} & \qquad \\ \hline\end{array} \tag{2.30}$$

On the right-hand side of the first line the operator pair $c_a^\dagger c_k$ has been commuted to the position k, to the effect that the orbital k is replaced by the orbital a. Note that the total number of commutations required to reach position k is even, so that the resulting phase is $+1$. Now the two states are in the form supposed in the SC rule **b2**, yielding

$$\langle\Phi_{ak}|\hat{W}|\Phi_0\rangle = w_{ak} \tag{2.31}$$

In a similar way, one may readily obtain the result

$$\langle\Phi_{abkl}|\hat{V}|\Phi_0\rangle = V_{ba[kl]} \tag{2.32}$$

for the matrix element of a two-particle operator according to SC rule **c3**. In the procedure of commuting operator pairs of the initial product $c_a^\dagger c_b^\dagger c_k c_l$ to the respective positions in $|\Phi_0\rangle$, it is recommended to first commute the operator pair $c_b^\dagger c_k$ to position k, and subsequently the remaining pair $c_a^\dagger c_l$ to the position l.

To see how the procedure works in a more demanding case, let us consider the Coulomb matrix element

$$\langle\Phi_j^{N-1}|\hat{V}|\Phi_{akl}^{N-1}\rangle$$

for two $(N-1)$-electron states. First, we suppose $j \neq k, l$. Again, we place the semi-graphical representations of the two states, (I) and (II), on top of each other,

$$(I) = \qquad\qquad c_j \begin{array}{|c|c|c|c|c|c|c|}\hline \quad & \text{k} & \qquad & \text{j} & \qquad & \text{l} & \quad \\ \hline\end{array} \tag{2.33}$$

$$(II) = c_a^\dagger c_k c_l \begin{array}{|c|c|c|c|c|c|c|}\hline \quad & \text{k} & \qquad & \text{j} & \qquad & \text{l} & \quad \\ \hline\end{array} = c_l \begin{array}{|c|c|c|c|c|c|c|}\hline \quad & \text{a} & \qquad & \text{j} & \qquad & \text{l} & \quad \\ \hline\end{array} \tag{2.34}$$

For the state (II), the operator pair $c_a^\dagger c_k$ can be commuted to the position k without involving a sign change, as is indicated on the right-hand side of the second line. To proceed, we commute the two operators $c_j^\dagger$ and $c_l^\dagger$ in $|\Phi_0\rangle$ to the left. This goes with an unspecified number of sign changes, indicated by the phase factor $(-1)^\nu$, depending on the relative positions of j and l:

$$(I) = c_j c_j^\dagger c_l^\dagger \begin{array}{|c|c|c|c|c|c|c|}\hline \quad & k & \quad & - & \quad & - & \quad \\ \hline\end{array} (-1)^\nu$$

$$(II) = c_l c_j^\dagger c_l^\dagger \begin{array}{|c|c|c|c|c|c|c|}\hline \quad & a & \quad & - & \quad & - & \quad \\ \hline\end{array} (-1)^\nu$$

Since the same phase factor arises in both (I) and (II), it will drop out in forming the matrix element. Straightforward operator algebra then yields

$$(I) = c_l^\dagger \begin{array}{|c|c|c|c|c|c|c|}\hline \quad & k & \quad & - & \quad & - & \quad \\ \hline\end{array} (-1)^\nu$$

$$(II) = c_j^\dagger \begin{array}{|c|c|c|c|c|c|c|}\hline \quad & a & \quad & - & \quad & - & \quad \\ \hline\end{array} (-1)^\nu(-1)$$

where the sign (-1) in the second line arises from (anti-) commuting the operators c_l and $c_j^\dagger$. Now we have reached a form where the two product states differ exactly at two positions. Taking the resulting sign into account, SC rule **c3** gives

$$\langle \Phi_j^{N-1} | \hat{V} | \Phi_{akl}^{N-1} \rangle = -V_{lk[ja]} \tag{2.35}$$

In an analogous way, the case $j = k$ (or $j = l$) can be treated. Here, the two states can readily be reshaped such that they differ exactly at one position, and the SC rule **c2** applies:

$$\langle \Phi_k^{N-1} | \hat{V} | \Phi_{akl}^{N-1} \rangle = -\sum_{i \neq k} V_{il[ia]} \tag{2.36}$$

It is interesting to note that the corresponding matrix element of the full hamiltonian, $\hat{H} = \hat{T} + \hat{V}$, is simply given by

$$\langle \Phi_j^{N-1} | \hat{H} | \Phi_{akl}^{N-1} \rangle = -V_{lk[ja]} \tag{2.37}$$

comprising the case $j = k(l)$, provided the one-particle states are HF orbitals (see Exercise 2.2).

2.5 Change of the One-Particle Representation

The SQ creation and destruction operators are defined with respect to a given choice of one-particle states (orbitals). This means that a change of the underlying one-particle basis will result in a corresponding transformation of the fermion operators.

Let us consider two sets of orthonormal one-particle states, denoted by $|q\rangle$ and $|\tilde{s}\rangle$, respectively, being related by a unitary transformation according to

$$|\tilde{s}\rangle = \sum_q |q\rangle\langle q|\tilde{s}\rangle \tag{2.38}$$

where $\langle q|\tilde{s}\rangle$ is the unitary overlap matrix of the two sets of orbitals. Let us denote the fermion operators associated with the second set of orbitals by $b_s^\dagger$ (and b_s). Applying $b_s^\dagger$ to an arbitrary N-electron basis state (in the q-representation) gives

$$\begin{aligned} b_s^\dagger|q_1 \ldots q_N\rangle &= |q_1 \ldots q_N \tilde{s}\rangle \\ &= \sum_q |q_1 \ldots q_N q\rangle\langle q|\tilde{s}\rangle \\ &= \sum_q \langle q|\tilde{s}\rangle c_q^\dagger |q_1 \ldots q_N\rangle \end{aligned} \tag{2.39}$$

where the expansion (2.38) of $|\tilde{s}\rangle$ has been used in the second line. From Eq. (2.39), we can infer the operator relation

$$b_s^\dagger = \sum_q \langle q|\tilde{s}\rangle c_q^\dagger \tag{2.40}$$

The transformation of the destruction operators is obtained by taking the hermitian conjugate of Eq. (2.40)

$$b_s = \sum_q \langle q|\tilde{s}\rangle^* c_q = \sum_q \langle \tilde{s}|q\rangle c_q \tag{2.41}$$

The inverse transformation are given by

$$c_p^\dagger = \sum_s \langle \tilde{s}|q\rangle b_s^\dagger, \qquad c_p = \sum_s \langle q|\tilde{s}\rangle b_s \tag{2.42}$$

A distinguished representation is based on the (continuous) one-particle eigenstates of the position and spin operators, $|\boldsymbol{\xi}\rangle = |\boldsymbol{x}\sigma\rangle$. Here, the transformations relating to normalized one-particle states, considered so far, are given by

$$|p\rangle = \int d\boldsymbol{\xi}\, |\boldsymbol{\xi}\rangle\, \psi_p(\boldsymbol{\xi}) \tag{2.43}$$

$$|\boldsymbol{\xi}\rangle = \sum_q |q\rangle\, \psi_q^*(\boldsymbol{\xi}) \tag{2.44}$$

where $\psi_p(\boldsymbol{\xi}) = \langle \boldsymbol{\xi}|p\rangle$. Accordingly, one may define operators

$$\hat{\psi}^\dagger(\boldsymbol{\xi}) = \sum_q c_q^\dagger\, \psi_q^*(\boldsymbol{\xi})$$
$$\hat{\psi}(\boldsymbol{\xi}) = \sum_q c_q\, \psi_q(\boldsymbol{\xi}) \tag{2.45}$$

creating or destructing, respectively, a particle with spin σ at the position $\boldsymbol{x}$. These operators, also referred to as *field operators*, obey the commutation relations

$$\{\hat{\psi}^\dagger(\boldsymbol{\xi}), \hat{\psi}(\boldsymbol{\xi}')\} = \delta(\boldsymbol{x} - \boldsymbol{x}')\delta_{\sigma\sigma'}$$
$$\{\hat{\psi}(\boldsymbol{\xi}), \hat{\psi}(\boldsymbol{\xi}')\} = 0, \quad \{\hat{\psi}^\dagger(\boldsymbol{\xi}), \hat{\psi}^\dagger(\boldsymbol{\xi}')\} = 0 \tag{2.46}$$

In terms of field operators, the one- and two-particle operators take on the forms

$$\hat{W} = \int d\boldsymbol{\xi}\; \hat{w}(\boldsymbol{\xi})\hat{\psi}^\dagger(\boldsymbol{\xi})\hat{\psi}(\boldsymbol{\xi})$$
$$\hat{V} = \frac{1}{2}\int\!\!\int d\boldsymbol{\xi}\, d\boldsymbol{\xi}'\; \hat{v}(\boldsymbol{\xi}, \boldsymbol{\xi}')\hat{\psi}^\dagger(\boldsymbol{\xi})\hat{\psi}^\dagger(\boldsymbol{\xi}')\hat{\psi}(\boldsymbol{\xi}')\hat{\psi}(\boldsymbol{\xi}) \tag{2.47}$$

and the N-electron coordinate eigenstate (1.21) can be written as

$$|\boldsymbol{\xi}_1 \ldots \boldsymbol{\xi}_N\rangle = \hat{\psi}^\dagger(\boldsymbol{\xi}_N)\ldots\hat{\psi}^\dagger(\boldsymbol{\xi}_1)|\varnothing\rangle \tag{2.48}$$

A useful mixed representation, associated with the products $|\boldsymbol{x}\rangle\chi_\gamma(\sigma)$ of position operator eigenstates and spin-functions, is given according to

$$\hat{\psi}_\gamma^\dagger(\boldsymbol{x}) = \sum_q c_{q\gamma}^\dagger \varphi_q^*(\boldsymbol{x}), \quad \hat{\psi}_\gamma(\boldsymbol{x}) = \sum_q c_{q\gamma}\varphi_q(\boldsymbol{x}) \tag{2.49}$$

Here, the original spin-orbital quantum number q has been expanded into the pair of a spatial and a spin quantum number, $q \to q\gamma$, and $\varphi_q(\boldsymbol{x})$ is the spatial orbital in the spin-orbital $\psi_{q\gamma}(\boldsymbol{\xi})$ as in Eq. (1.61).

Exercises

2.1 Verify the SC rules **c1–c4** for the two-particle operator (2.20) using the approach discussed in Sect. 2.4.

2.2 (a) Evaluate the matrix element $\langle\Phi_j^{N-1}|\hat{H}|\Phi_{akl}^{N-1}\rangle$ for the cases $k = j$ and $l = j$. Show that for HF orbitals the result is as given by Eq. (2.37).
(b) Evaluate the matrix elements $\langle\Phi_j^{N-1}|\hat{V}|\Phi_{abklm}^{N-1}\rangle$ and $\langle\Phi_{aj}|\hat{V}|\Phi_{bcdklm}\rangle$.

2.3 (a) Evaluate the commutator $[c_j, \hat{V}]$ where $\hat{V}$ is a two-particle operator in the form (2.20).
(b) Evaluate the commutator $[\hat{A}, \hat{B}]$ for the one-particle operators $\hat{A} = \sum_{r,s} a_{rs}c_r^\dagger c_s$, $\hat{B} = \sum_{u,v} b_{uv}c_u^\dagger c_v$.

2.4 Going beyond one-particle physics, the simplest model is the two-electron–two-orbital (2E-2O) system, as, for example, the hydrogen molecule in the minimal-basis approximation. There are two spatial (Hartree–Fock) orbitals, ϕ_g, ϕ_u, assumed to be real and of different symmetry (e.g., with respect to inversion). Let t_{gg}, t_{uu} denote the corresponding matrix elements of the one-particle part of the hamiltonian, and V_{gggg}, V_{uuuu}, V_{gguu}, V_{gugu} the non-vanishing spatial Coulomb integrals.
(a) Determine the Hartree–Fock orbital energies, ϵ_g, ϵ_u, according to Eq. (4.5).
(b) Write the two-electron ground state as a linear combination of the two basis states (CI configurations) of g-symmetry, $|\Phi_0\rangle = |g\alpha g\beta|$, $|\Phi_1\rangle = |u\alpha u\beta|$ and determine the elements of the hamiltonian (secular) matrix,

$$\boldsymbol{h} = \begin{pmatrix} h_{00} & h_{01} \\ h_{10} & h_{11} \end{pmatrix} \tag{2.50}$$

(c) Solve the 2×2 secular problem and determine the ground-state energy, e_0, and eigenvector, $\boldsymbol{x}_0$; use here abbreviations, e.g., $h_{11} = h_{00} + \Delta$, $h_{01} = V$.

2.5 Write the hamiltonian of the 2E-2O model in second quantization.

Reference

1. Baumgärtner G, Schuck P (1968) Kernmodelle. Bibliographisches Institut, Mannheim

Chapter 3
One-Particle Green's Function

The one-particle Green's function or electron propagator, which we shall introduce in this chapter, is the first and simplest member in the hierarchy of many-body Green's functions [1–3]. While the formal definition looks rather abstract and even forbidding, the benefits afforded by an approach based on the electron propagator should become clear after the theory has been more fully described. Before working through the various derivations, the reader might take a first look at Eqs. (3.24), (3.25) in which the essence of the electron propagator is apparent: Its elements are matrix elements of the many-body resolvent operator taken with respect to states of $N+1$ or $N-1$ electrons. This indicates that the physics conveyed by the electron propagator relates to excitations of the system following the addition of one electron (*electron attachment*) or the removal of one electron (*ionization*). We shall refer exclusively to the electron propagator in much of the book, that is, when we develop the formalism of diagrammatic perturbation theory in Chaps. 4–7 and establish practical approximation methods in Chaps. 8–12. The polarization propagator and the physics of N-electron excitations will be considered in the Chaps. 13–15 of Part IV.

3.1 Definition and Relation to Physical Quantities

In the following, we suppose a basis set of one-particle states $|p\rangle$ and the associated creation and destruction operators $c_p^\dagger$, c_p as introduced in Chap. 2. We consider an N-electron system with the hamiltonian

$$\hat{H} = \hat{T} + \hat{V} = \sum t_{pq} c_p^\dagger c_q + \tfrac{1}{2} \sum V_{pqrs} c_p^\dagger c_q^\dagger c_s c_r \tag{3.1}$$

and a non-degenerate (normalized) ground state $|\Psi_0\rangle$ of energy E_0. Moreover, we define time-dependent or Heisenberg operators according to

$$c_p^\dagger[t] = e^{i\hat{H}t} c_p^\dagger e^{-i\hat{H}t}, \quad c_p[t] = e^{i\hat{H}t} c_p e^{-i\hat{H}t} \tag{3.2}$$

J. Schirmer, *Many-Body Methods for Atoms, Molecules and Clusters*, Lecture Notes in Chemistry 94, https://doi.org/10.1007/978-3-319-93602-4_3

For notational brevity here and in the following, atomic units will be supposed where $\hbar = 1$. In the chosen representation, the electron propagator or one-particle Green's function (GF) $\boldsymbol{G}(t, t')$ is a matrix of time-dependent functions (components) defined according to

$$G_{pq}(t, t') = -i\theta(t - t')\langle\Psi_0|c_p[t]c_q^\dagger[t']|\Psi_0\rangle + i\theta(t' - t)\langle\Psi_0|c_q^\dagger[t']c_p[t]|\Psi_0\rangle \tag{3.3}$$

Here, $\theta(t)$ denotes the step function

$$\theta(t) = \begin{cases} 1, & t > 0 \\ 0, & t < 0 \end{cases} \tag{3.4}$$

The notation can be shortened by using the time-ordering operator $\hat{\boldsymbol{T}}$, also referred to as Wick's operator. Acting on a product of time-dependent fermion operators, $\hat{\boldsymbol{T}}$ reorders the factors in such a way that operators with larger times are to the left of those with smaller times; $\hat{\boldsymbol{T}}$ also introduces a sign $(-1)^P$, where P is the permutation transforming the original order into the final one. The result is also referred to as **time-ordered product**. In the case of two operators, the time-ordered product is simply given by

$$\hat{\boldsymbol{T}}\left(c_p[t]c_q^\dagger[t']\right) = \begin{cases} c_p[t]c_q^\dagger[t'], & t > t' \\ -c_q^\dagger[t']c_p[t], & t < t' \end{cases} \tag{3.5}$$

This allows us to write the electron propagator components in the more compact form

$$G_{pq}(t, t') = -i\langle\Psi_0|\hat{\boldsymbol{T}}\left(c_p[t]c_q^\dagger[t']\right)|\Psi_0\rangle \tag{3.6}$$

As seen from the definition (3.3), the electron propagator consists of two parts,

$$\boldsymbol{G}(t, t') = \boldsymbol{G}^+(t, t') + \boldsymbol{G}^-(t, t')$$

As will be shown below, the two parts contain spectral information related to electron attachment ($\boldsymbol{G}^+$) and electron removal or ionization ($\boldsymbol{G}^-$). Accordingly, the two parts are also referred to as $(N+1)$- and $(N-1)$-electron parts, respectively.

Let us have a closer look at the physical content of the $(N+1)$-electron part. As a first step, one may insert the explicit definition of the time-dependent operators (3.2) in Eq. (3.3):

$$\begin{aligned} G_{pq}^+(t, t') &= -i\theta(t - t')\langle\Psi_0|e^{i\hat{H}t}c_p e^{-i\hat{H}t}e^{i\hat{H}t'}c_q^\dagger e^{-i\hat{H}t'}|\Psi_0\rangle \\ &= -i\theta(t - t')e^{iE_0(t-t')}\langle\Psi_0|c_p e^{-i\hat{H}(t-t')}c_q^\dagger|\Psi_0\rangle \end{aligned} \tag{3.7}$$

This shows that the GF components depend only on the difference $t - t'$ of the two time arguments. To proceed, we insert the resolution of the identity in terms of a complete set of energy eigenstates $|\Psi_n^{N+1}\rangle$ of the $(N+1)$-electron system in the ground-state expectation value in Eq. (3.7). This yields

$$G_{pq}^{+}(t,t') = -i\theta(t-t')\sum_n e^{-i(E_n^{N+1}-E_0)(t-t')}\langle\Psi_0|c_p|\Psi_n^{N+1}\rangle\langle\Psi_n^{N+1}|c_q^{\dagger}|\Psi_0\rangle \quad (3.8)$$

This expression is essentially a sum of periodic functions of $t - t'$, where the frequencies can be identified with electron attachment energies, $E_0 - E_n^{N+1}$. In an analogous way, the $(N-1)$-electron part can be written as

$$G_{pq}^{-}(t,t') = i\theta(t'-t)\sum_n e^{i(E_n^{N-1}-E_0)(t-t')}\langle\Psi_0|c_q^{\dagger}|\Psi_n^{N-1}\rangle\langle\Psi_n^{N-1}|c_p|\Psi_0\rangle \quad (3.9)$$

At this point, it is useful to switch to the so-called **energy representation**, obtained by Fourier transform according to

$$G_{pq}(\omega) = \int_{-\infty}^{\infty} e^{i\omega(t-t')} G_{pq}(t,t')\mathrm{d}(t-t') \quad (3.10)$$

The inverse transformation is given by

$$G_{pq}(t,t') = \frac{1}{2\pi}\int_{-\infty}^{\infty} e^{-i\omega(t-t')} G_{pq}(\omega)\mathrm{d}\omega \quad (3.11)$$

As can be seen by inspecting the Fourier transform of one of the time-dependent functions in Eq. (3.8),

$$\begin{aligned} f_n^{+}(\omega) &= \int_{-\infty}^{\infty} e^{i\omega\tau}\left[-i\theta(\tau)e^{-i(E_n^{N+1}-E_0)\tau}\right]\mathrm{d}\tau \\ &= -i\int_0^{\infty} e^{i[\omega-E_n^{N+1}+E_0]\tau}\mathrm{d}\tau \end{aligned} \quad (3.12)$$

the time integral is ill-defined at the upper limit, $t = \infty$. This can be cured in an unambiguous way by augmenting the step function with a convergence factor according to

$$\theta(\tau) \rightarrow \theta(\tau)e^{-\eta\tau} \quad (3.13)$$

where η is a positive infinitesimal. Using this convergence factor, the time integral (3.12) simply becomes

$$f_n^+(\omega) = -i \int_0^\infty e^{i\left[\omega - E_n^{N+1} + E_0 + i\eta\right]\tau} \mathrm{d}\tau = \frac{1}{\omega - E_n^{N+1} + E_0 + i\eta} \tag{3.14}$$

As the reader should verify, the inverse transformation

$$\begin{aligned} g_n^+(\tau) &= \frac{1}{2\pi} \int_{-\infty}^{\infty} e^{-i\omega\tau} f_n^+(\omega) \mathrm{d}\omega \\ &= \frac{1}{2\pi} \int_{-\infty}^{\infty} \frac{e^{-i\omega\tau}}{\omega - E_n^{N+1} + E_0 + i\eta} \mathrm{d}\omega = -i\theta(\tau) e^{-\eta\tau} e^{-i(E_n^{N+1} - E_0)\tau} \end{aligned} \tag{3.15}$$

reproduces the original time function. Note that in the contour integrations required here the integration paths involve (infinite) semi-circles in the upper and lower complex ω-plane for the cases $\tau < 0$ and $\tau > 0$, respectively.

An analogous convergence factor has to be applied to the $(N-1)$-electron part, which suggests to introduce these changes already in the definition (3.3) of the electron propagator:

$$\begin{aligned} G_{pq}(t, t') = &-i\theta(t - t') e^{-\eta(t-t')} \langle \Psi_0 | c_p[t] c_q^\dagger[t'] | \Psi_0 \rangle \\ &+ i\theta(t' - t) e^{\eta(t-t')} \langle \Psi_0 | c_q^\dagger[t'] c_p[t] | \Psi_0 \rangle \end{aligned} \tag{3.16}$$

The energy representation of the electron propagator according to the extended definition (3.16) is given by

$$G_{pq}(\omega) = \sum_n \frac{\langle \Psi_0 | c_p | \Psi_n^{N+1} \rangle \langle \Psi_n^{N+1} | c_q^\dagger | \Psi_0 \rangle}{\omega + E_0 - E_n^{N+1} + i\eta} + \sum_n \frac{\langle \Psi_0 | c_q^\dagger | \Psi_n^{N-1} \rangle \langle \Psi_n^{N-1} | c_p | \Psi_0 \rangle}{\omega + E_n^{N-1} - E_0 - i\eta} \tag{3.17}$$

In this form, also referred to as **spectral representation** or Lehmann representation, the physical content of the electron propagator becomes manifest. The two parts G_{pq}^+ and G_{pq}^- are given by sums of simple poles in the lower and upper complex ω-plane, respectively, where the **electron affinities**

$$A_n = E_0 - E_n^{N+1} \tag{3.18}$$

and the **ionization energies**

$$I_n = E_n^{N-1} - E_0 \tag{3.19}$$

are identified as the negative pole positions $-\omega_n$ of G_{pq}^+ and G_{pq}^-, respectively. In a schematical way, the pole structure of the electron propagator is illustrated in Fig. 3.1.

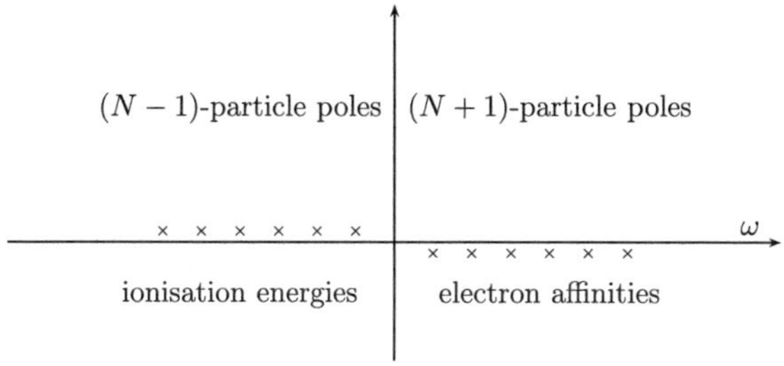

Fig. 3.1 Pole structure of the electron propagator

The corresponding pole strengths are given as products of so-called spectroscopic factors,

$$x_p^{(n)} = \langle \Psi_0 | c_p | \Psi_n^{N+1} \rangle, \qquad n \in \{N+1\} \tag{3.20}$$

$$x_p^{(n)} = \langle \Psi_n^{N-1} | c_p | \Psi_0 \rangle, \qquad n \in \{N-1\} \tag{3.21}$$

As a consequence of the anticommutator relation (2.12), the pole strengths fulfill the following sum rule:

$$\sum_{n \in \{N+1\}} x_p^{(n)} x_q^{(n)*} + \sum_{n \in \{N-1\}} x_p^{(n)} x_q^{(n)*} = \delta_{pq} \tag{3.22}$$

To get an idea of the meaning of the spectroscopic factors, one may inspect the following (simplified) expression for the partial photo-ionization cross section for generating the final ionic state $|\Psi_n^{N-1}\rangle$ and a continuum electron with kinetic energy $\varepsilon = h\nu - I_n$, where $h\nu$ is the energy of the incident light:

$$\sigma_n(\epsilon) \sim \frac{2}{3}\varepsilon \left| \sum_p \langle \varepsilon | \hat{d} | p \rangle x_p^{(n)} \right|^2 \tag{3.23}$$

Here $\langle \varepsilon | \hat{d} | p \rangle$ is the matrix element of the one-particle dipole operator with respect to the orbital $|p\rangle$ and the one-particle scattering state $|\varepsilon\rangle$ of energy ε. The factors $x_p^{(n)}$ weight the "participation" of individual orbitals in the final ionic state. Often there is only one dominant orbital contribution, and the sum on the right-hand side reduces to a single term.

It should be noted that the summation over discrete states supposed in the form (3.17) of the spectral representation can be generalized to comprise the respective continua. In the case of $N-1$ electrons, for example, the propagator component G_{pq}^- would comprise both a discrete summation and an integral of the form

$$\int \mathrm{d}E \frac{\mu_{pq}(E)}{\omega + E - E_0 - i\eta}$$

where $\mu_{pq}(E)$ is a function of the energy E.

By undoing the resolution of the identity in the spectral representation (3.17), one obtains the following compact forms

$$G^{+}_{pq}(\omega) = \langle\Psi_0|c_p\left(\omega - \hat{H} + E_0 + i\eta\right)^{-1} c^{\dagger}_q|\Psi_0\rangle \tag{3.24}$$

$$G^{-}_{pq}(\omega) = \langle\Psi_0|c^{\dagger}_q\left(\omega + \hat{H} - E_0 - i\eta\right)^{-1} c_p|\Psi_0\rangle \tag{3.25}$$

This shows that $G^{\pm}_{pq}(\omega)$ are essentially matrix elements of the many-body resolvent $(\omega - \hat{H})^{-1}$, taken with respect to $(N\pm1)$-electron states $c^{\dagger}_p|\Psi_0\rangle$ and $c_p|\Psi_0\rangle$, respectively.

3.2 Ground-State Expectation Values

Besides spectral information on the $(N\pm1)$-electron systems, the electron propagator allows one to obtain ground-state expectation values of one-particle operators. As basic quantities, let us consider the elements of the one-particle density matrix

$$\rho_{rs} = \langle\Psi_0|c^{\dagger}_s c_r|\Psi_0\rangle \tag{3.26}$$

Comparison with the electron propagator according to Eqs. (3.3) or (3.16) shows that

$$\rho_{rs} = \lim_{t'\to t^{+}} (-i)G_{rs}(t,t') \tag{3.27}$$

where $t' \to t^{+}$ means a limit in which t' approaches t strictly from above, $t' > t$. As a consequence of equating times that way, only the $(N-1)$-electron part survives (since $\theta(t-t') = 0$ for $t' > t$), and we may write

$$\rho_{rs} = -iG_{rs}(t,t^{+}) = -iG^{-}_{rs}(t,t^{+}) \tag{3.28}$$

where t^{+} is used as an abbreviation for the limit $t' \to t, t' > t$. Accordingly, the ground-state expectation value of one-particle operator

$$\hat{A} = \sum_{r,s} a_{rs} c^{\dagger}_r c_s \tag{3.29}$$

can be written as

$$\langle\Psi_0|\hat{A}|\Psi_0\rangle = \sum_{r,s} a_{rs}\rho_{sr} = -i\sum_{r,s} a_{rs} G^{-}_{sr}(t,t^{+}) \tag{3.30}$$

Introducing the matrix

$$\boldsymbol{A} \equiv (a_{rs}) \tag{3.31}$$

of the one-particle matrix elements, the ground-state expectation value can be written according to

$$\langle \Psi_0 | \hat{A} | \Psi_0 \rangle = -i\mathrm{Tr}(\boldsymbol{A}\boldsymbol{G}^-(t, t^+)) \tag{3.32}$$

as the trace of a matrix product.

The constant quantities $G_{rs}(t, t^+)$ obtained by equating the time arguments in the prescribed way can be derived as well from the energy representation of the electron propagator by putting $t' = t + \epsilon$, $\epsilon > 0$, in the Fourier transform (3.10):

$$G_{rs}(t, t^+) = \lim_{\varepsilon \to 0} \frac{1}{2\pi} \int e^{i\omega\varepsilon} G^-_{rs}(\omega) \mathrm{d}\omega, \qquad \varepsilon > 0$$

The factor $e^{i\omega\varepsilon}$ suggests to solve the integral by contour integration, where the contour closes in the upper complex ω-plane, yielding

$$G_{rs}(t, t^+) = \frac{1}{2\pi} \oint G^-_{rs}(\omega) \mathrm{d}\omega \tag{3.33}$$

Since the G^+ part has only poles in the lower complex ω-plane, we may also write

$$\rho_{rs} = \frac{1}{2\pi i} \oint G_{rs}(\omega) \mathrm{d}\omega \tag{3.34}$$

The validity of Eq. (3.33) can easily be verified by performing the contour integration for the spectral representation (3.17) of $\boldsymbol{G}(\omega)$ or $\boldsymbol{G}^-(\omega)$. The equivalent to Eq. (3.32) then reads

$$\langle \Psi_0 | \hat{A} | \Psi_0 \rangle = \frac{1}{2\pi i} \oint \mathrm{Tr}\left(\boldsymbol{A}\boldsymbol{G}^-(\omega)\right) \mathrm{d}\omega \tag{3.35}$$

3.3 Ground-State Energy

The ground-state energy, expressed as the ground-state expectation value of the hamiltonian,

$$E_0 = \langle \Psi_0 | \hat{H} | \Psi_0 \rangle = \sum t_{rs} \langle \Psi_0 | c_r^\dagger c_s | \Psi_0 \rangle + \frac{1}{2} \sum V_{rsuv} \langle \Psi_0 | c_r^\dagger c_s^\dagger c_v c_u | \Psi_0 \rangle \tag{3.36}$$

can also be derived from the electron propagator, even though the second term involves two-particle density matrix elements. This becomes possible because the electron propagator obeys an equation of motion (EOM) to be discussed in the following.

Let us first consider the time derivative of the time-dependent destruction operator:

$$i\frac{\partial}{\partial t}c_p[t] = i\frac{\partial}{\partial t}\left(e^{i\hat{H}t}c_p e^{-i\hat{H}t}\right) = e^{i\hat{H}t}\left[c_p, \hat{H}\right]e^{-i\hat{H}t} \tag{3.37}$$

The commutator appearing in the last expression is readily evaluated (see Exercise 2.3), which gives

$$i\frac{\partial}{\partial t}c_p[t] = \sum_s t_{ps}c_s[t] + \sum_{s,u,v} V_{psuv}c_s^\dagger[t]c_v[t]c_u[t] \tag{3.38}$$

Now the time derivative of the electron propagator with respect to the time argument t can be written as

$$\begin{aligned}
i\frac{\partial}{\partial t}G_{pq}(t,t') &= \delta(t-t')\langle\Psi_0||\left\{c_p[t], c_q^\dagger[t']\right\}\rangle\Psi_0 \\
&\quad - i\langle\Psi_0|\hat{\boldsymbol{T}}\left[\left(i\frac{\partial}{\partial t}c_p[t]\right)c_q^\dagger[t']\right]|\Psi_0\rangle \\
&= \delta_{pq}\delta(t-t') - i\sum_s t_{ps}\langle\Psi_0|\hat{\boldsymbol{T}}\left[c_s[t]c_q^\dagger[t']\right]|\Psi_0\rangle \\
&\quad - i\sum_{s,u,v} V_{psuv}\langle\Psi_0|\hat{\boldsymbol{T}}\left[c_s^\dagger[t]c_v[t]c_u[t]c_q^\dagger[t']\right]|\Psi_0\rangle
\end{aligned} \tag{3.39}$$

The delta function arises from the time derivative of the step function. For equal times, $t = t'$, the anticommutator in the first line becomes $\left\{c_p[t], c_q^\dagger[t']\right\} \to \delta_{pq}$. Note that we have disregarded here the convergence factors $e^{\pm\eta(t-t')}$ of the extended definition (3.16); the additional terms arising from the time derivatives of the convergence factors vanish for $t = t'$, which is of interest below. The last equation can be cast into the form

$$i\frac{\partial}{\partial t}G_{pq}(t,t') - \sum_s t_{ps}G_{sq}(t,t') = \delta_{pq}\delta(t-t') + i\sum_{s,u,v} V_{psuv}G_{vu,qs}(t,t;t',t^+) \tag{3.40}$$

which makes explicit that the EOM of the one-particle GF involves the next higher member in a hierarchy of many-body Green's functions, namely the two-particle GF defined as follows:

$$G_{12,1'2'}(t_1,t_2;t_1',t_2') = (-i)^2\langle\Psi_0|\hat{\boldsymbol{T}}\left[c_1[t_1]c_2[t_2]c_{2'}^\dagger[t_2']c_{1'}^\dagger[t_1']\right]|\Psi_0\rangle \tag{3.41}$$

Here the indices $1, 2, 1', 2'$ are used as an abbreviated notation for general one-particle quantum numbers, e.g., $1 \equiv p$.

To establish a relation to the ground-state energy, we take the sum of the diagonal elements in Eq. (3.39), $\sum_p i\frac{\partial}{\partial t}G_{pp}(t,t')$, and equate the time arguments in the fashion described above. Thus we find

$$\sum_p i\frac{\partial}{\partial t}G_{pp}(t,t^+) - i\langle\Psi_0|\hat{T}|\Psi_0\rangle = 2i\langle\Psi_0|\hat{V}|\Psi_0\rangle \tag{3.42}$$

Note that in the limit $t' \to t, t' > t$ the term $\delta_{pq}\delta(t-t')$ on the right-hand side of Eq. (3.40) must be skipped. Using Eq. (3.32) to relate the ground-state expectation value of $\hat{T}$ to the electron propagator finally yields

$$E_0 = \langle\Psi_0|\hat{T}+\hat{V}|\Psi_0\rangle = \frac{1}{2}\sum_p \frac{\partial}{\partial t}G_{pp}(t,t^+) - \frac{i}{2}\sum_{p,q} t_{pq}G_{qp}(t,t^+) \tag{3.43}$$

The analogous expression in the energy representation reads

$$E_0 = \frac{1}{4\pi i}\oint Tr\,[(\omega\mathbf{1}+\boldsymbol{T})\,\boldsymbol{G}(\omega)]\,\mathrm{d}\omega = \frac{1}{4\pi i}\oint \mathrm{Tr}\left[(\omega\mathbf{1}+\boldsymbol{T})\,\boldsymbol{G}^-(\omega)\right]\mathrm{d}\omega \tag{3.44}$$

As above, this result is obtained from Eq. (3.43) by replacing the time-dependent propagator components with

$$G_{pq}(t,t^+) = \lim_{\varepsilon\to 0}\frac{1}{2\pi}\int e^{i\omega\varepsilon}G^-_{pq}(\omega)\mathrm{d}\omega, \qquad \varepsilon > 0 \tag{3.45}$$

The relation (3.44) can also be derived directly. The contour integrations for the products $\omega G_{pq}(\omega)$ can be readily evaluated to give

$$\begin{aligned}\frac{1}{2\pi i}\oint \omega G^-_{pq}(\omega)\mathrm{d}\omega &= -\sum_n \left(E_n^{N-1}-E_0\right)\langle\Psi_0|c_q^\dagger|\Psi_n^{N-1}\rangle\langle\Psi_n^{N-1}|c_p|\Psi_0\rangle \\ &= \langle\Psi_0|c_q^\dagger[c_p,\hat{H}]|\Psi_0\rangle = -\langle\Psi_0|c_q^\dagger(\hat{H}-E_0)c_p|\Psi_0\rangle\end{aligned} \tag{3.46}$$

This establishes the following *sum rule* for the ionization spectra:

$$\sum_{n\in\{N-1\}}\left(E_n^{N-1}-E_0\right)x_q^{(n)*}x_p^{(n)} = \langle\Psi_0|c_q^\dagger(\hat{H}-E_0)c_p|\Psi_0\rangle \tag{3.47}$$

To arrive at Eq. (3.44), we take the trace on both sides of Eq. (3.46) and evaluate the commutators in the first term of the second line (see Exercise 2.3). This yields

$$\sum_p \frac{1}{2\pi i}\oint \omega G^-_{pp}(\omega)\mathrm{d}\omega = \langle\Psi_0|\hat{T}|\Psi_0\rangle + 2\langle\Psi_0|\hat{V}|\Psi_0\rangle \tag{3.48}$$

Combining this result with the ground-state expectation value of $\hat{T}$ according to Eqs. (3.35), (3.44) is readily retrieved.

3.4 Free One-Particle Green's Function

The electron propagator can be specialized to a system of N non-interacting particles with a hamiltonian of the form

$$\hat{H}_0 = \sum_{i=1}^{N} \hat{h}_0(i) = \sum \varepsilon_r c_r^\dagger c_r \tag{3.49}$$

The orbitals $|r\rangle$ are supposed to be eigenstates of $\hat{h}_0$, and the N-particle ground state $|\Phi_0\rangle$ is given by the Slater determinant of the N energetically lowest orbitals $|r\rangle$. As an example of particular importance, we may consider the HF approximation for an atom or molecule, where $|r\rangle$ are ground-state HF orbitals, and $\hat{h}_0$ is the (one-particle) HF operator.

The time-dependent Heisenberg operators simply become

$$c_p(t) = e^{i\hat{H}_0 t} c_p e^{-i\hat{H}_0 t} = e^{-i\varepsilon_p t} c_p \tag{3.50}$$

since

$$i\frac{\partial}{\partial t} c_p(t) = e^{i\hat{H}_0 t} [c_p, \hat{H}_0] e^{-i\hat{H}_0 t} = \varepsilon_p c_p(t) \tag{3.51}$$

Replacing $\hat{H}$ with $\hat{H}_0$ and $|\Psi_0\rangle$ with $|\Phi_0\rangle$ in the general definition (3.3), one obtains the so-called **free Green's function** (free electron propagator)

$$G^0_{pq}(t,t') = -ie^{-i\varepsilon_p(t-t')}\delta_{pq}\left(\theta(t-t')\bar{n}_p - \theta(t'-t)n_p\right) \tag{3.52}$$

Here $n_p, \bar{n}_q = 1 - n_q$ denote occupation numbers with respect to $|\Phi_0\rangle$,

$$n_p = \begin{cases} 1, & p \le N \\ 0, & p > N \end{cases} \tag{3.53}$$

The corresponding energy representation

$$G^0_{pq}(\omega) = \delta_{pq}\left(\frac{\bar{n}_p}{\omega - \varepsilon_p + i\eta} + \frac{n_p}{\omega - \varepsilon_p - i\eta}\right) \tag{3.54}$$

is obtained via Fourier transformation using convergence factors $e^{\pm\eta(t-t')}$ as discussed above. The free propagator is diagonal, and for each orbital p, there is exactly one pole, located in the lower or upper complex ω-plane depending on whether p is an occupied ($n_p = 1$) or unoccupied ($n_p = 0$) orbital in the ground-state Slater determinant. The ionization energies and electron affinities are simply given by $-\epsilon_p$, that is, the negative orbital energies.

For the diagonal components $G^0_{pp}(t, t')$, the EOM assumes the simple form

$$i\frac{\partial}{\partial t}G^0_{pp}(t, t') - \varepsilon_p G^0_{pp}(t, t') = \delta(t - t') \tag{3.55}$$

As solutions of the latter inhomogeneous differential equation, the functions $G^0_{pp}(t, t')$ are denoted as (mathematical) Green's functions, here for the differential operator $\left(i\frac{\partial}{\partial t} - \epsilon_p\right)$. It is this proximity to mathematical Green's functions which has led to the designation "many-body Green's functions" in the present context. Note that Eq. (3.55) has two distinct solutions, differing with respect to their "causal" behavior. The comparison with the definition (3.52) shows that either the "retarded" solution, $\sim\theta(t - t')$, or the "advanced" solution, $\sim\theta(t' - t)$, is adopted depending on whether $\overline{n}_p = 1$ or $n_p = 1$, respectively.

Exercises

3.1 Spin symmetry of the electron propagator: Write the spin-orbital indices in their expanded form, $p \rightarrow p\gamma$, and show that $G_{p\alpha,q\alpha} = G_{p\beta,q\beta}$ and $G_{p\alpha,q\beta} = 0$, supposing a non-degenerate ground state $|\Psi_0\rangle$.

3.2 Revisit the 2E-2O model of Exercise 2.4 and
   (a) Determine the ionization potentials and electron affinities;
   (b) Evaluate explicitly the spectral representation (3.17) for $G_{g\alpha,g\alpha}(\omega)$.

References

1. Migdal AB (1967) Theory of finite Fermi systems. Wiley-Interscience, New York
2. Abrikosov AA, Gorkov LP, Dzyaloshinski IE (1963) Methods of quantum field theory in statistical physics. Prentice-Hall, Englewood Cliffs
3. Thouless DJ (1961) The quantum mechanics of many-body systems. Academic Press, New York

Part II
Formalism of Diagrammatic Perturbation Theory

How can the electron propagator or other propagators actually be computed? And will the eventual computational schemes offer distinct advantages over the conventional computational methodology based on wave functions and the Schrödinger equation? A prominent route to the computation of propagators is provided by the formalism of diagrammatic perturbation theory, in which the contributions to the PT expansions of the propagator matrix elements are represented in the form of graphical schemes, the famous Feynman diagrams and variations thereof. The following Chaps. 4–6 establish the diagrammatic perturbation theory specifically for the electron propagator. Three basic theorems, namely the Gell-Mann and Low theorem, Wick's theorem, and the linked-cluster theorem constitute the core of the formalism. They are discussed in Chaps. 4 and 5. Here our presentation owes much to Chap. 3 in the exemplary textbook by Fetter and Walecka [1]. Based on those theorems, the rules to draw and evaluate Feynman diagrams are derived and demonstrated in Chap. 6. The final Chap. 7 deals with the time-ordered or Goldstone diagrams, which allow for a direct diagrammatical access to the results of the various time or energy integrations required in the evaluation of the Feynman diagrams.

It should be noted that there are non-diagrammatic approaches as well, such as the algebraic propagator methods reviewed in Chap. 16, the hierarchy of coupled time-dependent equations of motion for many-body Green's functions [2], and the method of functional derivatives [3].

References

1. Fetter AL, Walecka JD (1971) Quantum theory of many-particle systems. Mc Graw-Hill, New York
2. Martin PC, Schwinger J (1959) Phys Rev A 115:1342
3. Kadanoff LP, Baym G (1962) Quantum statistical mechanics. Benjamin, Reading

Chapter 4
Perturbation Theory for the Electron Propagator

As in the familiar Rayleigh–Schrödinger perturbation theory (RSPT) for the N-electron ground state discussed in Appendix A.1, the starting point for perturbation theory is a division of the original hamiltonian (3.1) into two parts,

$$\hat{H} = \hat{T} + \hat{V} = \hat{H}_0 + \hat{H}_I \tag{4.1}$$

where $\hat{H}_0$, defining the zeroth-order part, is a one-particle hamiltonian associated with non-interacting particles, and

$$\hat{H}_I = \hat{H} - \hat{H}_0 = \hat{W} + \hat{V} \tag{4.2}$$

is the interaction part, comprising the bare electron repulsion, $\hat{V}$, and, possibly, a remainder of the one-particle part of the hamiltonian,

$$\hat{W} = \hat{T} - \hat{H}_0 = \sum w_{rs} c_r^\dagger c_s \tag{4.3}$$

In the so-called Møller–Plesset (MP) partitioning,

$$\hat{H}_0 = \sum \epsilon_r c_r^\dagger c_r \tag{4.4}$$

is the Hartree–Fock (HF) hamiltonian, based on the solutions of the HF equations,

$$t_{rs} + \sum_k V_{rk[sk]} n_k = \epsilon_r \delta_{rs} \tag{4.5}$$

where $n_k = 1, 0$ denotes the usual (HF) ground-state occupation numbers. Accordingly, the matrix elements of the one-particle interaction part $\hat{W}$ are given by

$$w_{rs} = t_{rs} - \epsilon_r \delta_{rs} = -\sum_k V_{rk[sk]} n_k \tag{4.6}$$

J. Schirmer, *Many-Body Methods for Atoms, Molecules and Clusters*, Lecture Notes in Chemistry 94, https://doi.org/10.1007/978-3-319-93602-4_4

The non-interacting (HF) ground state, denoted by $|\Phi_0\rangle$, is a solution of the Schrödinger equation

$$\hat{H}_0|\Phi_0\rangle = E_0^{(0)}|\Phi_0\rangle \tag{4.7}$$

where

$$E_0^{(0)} = \sum_{k=1}^{N} \epsilon_k \tag{4.8}$$

is the corresponding ground-state energy.

According to the definition (3.3) of the electron propagator, PT expansions come into play in two ways: firstly, via the ground state, and, secondly, in the time-dependent Heisenberg operators. To accomodate both demands, a procedure based on *time-dependent perturbation theory* (TDPT) has proven advantageous. After a review of TDPT in the ensuing Sect. 4.1, we shall discuss in Sect. 4.2 how this technique can be adapted to the case where the interaction part $\hat{H}_I$ is slowly "switched on" by applying an appropriate time-dependent function. In the so-called *adiabatic limit*, a valid PT formulation of the interacting N-electron ground state is obtained, which is the proposition of the Gell-Mann and Low theorem. The Gell-Mann and Low formulation of the ground state can be extended to ground-state expectation values of time-dependent operators and, moreover, to the elements of the electron propagator, as will be discussed in Sect. 4.3.

4.1 Time-Development Operator in the Interaction Picture

The time development of a quantum state $|\Psi(t)\rangle$ is governed by the time-dependent Schrödinger equation (TDSE)

$$i\frac{\partial}{\partial t}|\Psi(t)\rangle = \hat{H}|\Psi(t)\rangle \tag{4.9}$$

which for a given initial state $|\Psi(t_0)\rangle$ at a time $t = t_0$, uniquely determines $|\Psi(t)\rangle$ for times $t \geq t_0$. For a time-independent hamiltonian, the formal solution of Eq. (4.9) takes the form

$$|\Psi_S(t)\rangle = e^{-i\hat{H}(t-t_0)}|\Psi(t_0)\rangle \tag{4.10}$$

This means that the state at t is obtained from the initial state at $t = t_0$ via a unitary transformation

$$\hat{U}_S(t, t_0) = e^{-i\hat{H}(t-t_0)}. \tag{4.11}$$

The subscript S indicates that we are dealing here with the Schrödinger representation of time-dependent quantum mechanics. As an alternative, one may resort to the so-called **interaction picture**, which is more suitable for the case where the hamiltonian can be divided into a time-independent part and a (possibly) time-dependent perturbation,

$$\hat{H} = \hat{H}_0 + \hat{H}_1(t) \tag{4.12}$$

In the interaction picture, one considers states defined according to

$$|\Psi_I(t)\rangle = e^{i\hat{H}_0 t}|\Psi_S(t)\rangle \tag{4.13}$$

where $|\Psi_S(t)\rangle$ is the time-dependent state of the usual Schrödinger picture, now labeled by the subscript S for clarity. The obvious purpose of this definition is to account for the time development due to $\hat{H}_0$ in a formally explicit fashion. Using the TDSE for $|\Psi_S(t)\rangle$ yields the equation of motion

$$i\frac{\partial}{\partial t}|\Psi_I(t)\rangle = \hat{H}_I(t)|\Psi_I(t)\rangle \tag{4.14}$$

for the states in the interaction picture. Here

$$\hat{H}_I(t) = e^{i\hat{H}_0 t}\hat{H}_1(t)e^{-i\hat{H}_0 t} \tag{4.15}$$

is the perturbation part of the hamiltonian in the interaction picture. In the same way, one may define the interaction picture representation of a general (Schrödinger) operator $\hat{O}_S$:

$$\hat{O}_I(t) = e^{i\hat{H}_0 t}\hat{O}_S e^{-i\hat{H}_0 t} \tag{4.16}$$

As is readily seen, the matrix element of an operator between two states can be written similarly in the Schrödinger and interaction picture,

$$\langle\Psi_I(t)|\hat{O}_I(t)|\Psi_I'(t)\rangle = \langle\Psi_S(t)|\hat{O}_S|\Psi_S'(t)\rangle \tag{4.17}$$

For time-independent operators $\hat{O}_S$, the corresponding interaction-picture operators obey the simple equation of motion

$$i\frac{\partial}{\partial t}\hat{O}_I(t) = \left[\hat{O}_I(t), \hat{H}_0\right] \tag{4.18}$$

When the hamiltonian (4.12) is time independent, the TDSE (4.14) in the interaction picture can be solved in a formal way as follows:

$$\begin{aligned}
|\Psi_I(t)\rangle &= e^{i\hat{H}_0 t}|\Psi_S(t)\rangle \\
&= e^{i\hat{H}_0 t}e^{-i\hat{H}(t-t_0)}|\Psi_S(t_0)\rangle \\
&= e^{i\hat{H}_0 t}e^{-i\hat{H}(t-t_0)}e^{-i\hat{H}_0 t_0}|\Psi_I(t_0)\rangle
\end{aligned}$$

This result may be written in a more compact form

$$|\Psi_I(t)\rangle = \hat{U}(t, t_0)|\Psi_I(t_0)\rangle \tag{4.19}$$

where the unitary operator

$$\hat{U}(t, t_0) = e^{i\hat{H}_0 t} e^{-i\hat{H}(t-t_0)} e^{-i\hat{H}_0 t_0} \tag{4.20}$$

is referred to as the time-evolution operator in the interaction picture.

One may readily verify the following properties of $\hat{U}(t, t')$:

$$\hat{U}(t_0, t_0) = \hat{\mathbb{1}} \tag{4.21}$$

$$\hat{U}(t_1, t_2)\hat{U}(t_2, t_3) = \hat{U}(t_1, t_3) \qquad \text{transitivity} \tag{4.22}$$

$$\hat{U}(t, t_0)^\dagger = \hat{U}(t_0, t) \tag{4.23}$$

$$\hat{U}^\dagger(t, t_0)\hat{U}(t, t_0) = \hat{U}(t, t_0)\hat{U}^\dagger(t, t_0) = \mathbb{1} \qquad \text{unitarity} \tag{4.24}$$

The form (4.19) applies also to the case of time-dependent interaction, $\hat{H}_1 = \hat{H}_1(t)$. Here, the TDSE translates into the following equation of motion for the time-evolution operator $\hat{U}(t, t_0)$ in the interaction picture:

$$i\frac{\partial}{\partial t}\hat{U}(t, t_0) = \hat{H}_I(t)U(t, t_0) \tag{4.25}$$

where $\hat{H}_I(t)$ is given by Eq. (4.15). Note that the properties (4.21)–(4.24) apply to the time-dependent case as well.

Performing time integrations on both sides, Eq. (4.25) can readily be transformed into an integral equation (of Volterra type):

$$\hat{U}(t, t_0) = \hat{\mathbb{1}} - i\int_{t_0}^{t} \mathrm{d}t' \hat{H}_I(t')\hat{U}(t', t_0) \tag{4.26}$$

The advantage of the integral-equation form is that it can be solved in an iterative way:

$$\hat{U}(t, t_0) = \hat{\mathbb{1}} - i\int_{t_0}^{t} \mathrm{d}t_1 \hat{H}_I(t_1) + (-i)^2 \int_{t_0}^{t} \mathrm{d}t_1 \int_{t_0}^{t_1} \mathrm{d}t_2\, \hat{H}_I(t_1)\hat{H}_I(t_2) + \ldots \tag{4.27}$$

This establishes a perturbation expansion of $\hat{U}(t, t_0)$ in terms of powers of $\hat{H}_1(t)$, the nth-order term reading

$$\hat{U}^{(n)}(t,t_0) = (-i)^n \int_{t_0}^{t} \mathrm{d}t_1 \int_{t_0}^{t_1} \mathrm{d}t_2 \ldots \int_{t_0}^{t_{n-1}} \mathrm{d}t_n \hat{H}_I(t_1)\hat{H}_I(t_2)\ldots\hat{H}_I(t_n) \tag{4.28}$$

It should be noted that the interaction operators $\hat{H}_I(t)$ do not commute for different time arguments, that is,

$$[\hat{H}_I(t), \hat{H}_I(t')] \neq 0 \ \text{ for } \ t \neq t' \tag{4.29}$$

Accordingly, the order of the operators in the integrals in Eq. (4.27) is essential. For example, in the integration of the second-order term,

$$\hat{U}^{(2)}(t,t_0) = -\int_{t_0}^{t} \mathrm{d}t_1 \int_{t_0}^{t_1} \mathrm{d}t_2\, \hat{H}_I(t_1)\hat{H}_I(t_2) \tag{4.30}$$

the time-ordering $t_1 \geq t_2$ has to be maintained. We may rewrite this term as

$$\hat{U}^{(2)}(t,t_0) = -\int_{t_0}^{t} \mathrm{d}t_1 \int_{t_0}^{t} \mathrm{d}t_2\, \hat{H}_I(t_1)\hat{H}_I(t_2)\theta(t_1 - t_2) \tag{4.31}$$

where the θ-function guarantees the proper time-ordering and allows one to use a common upper limit for both the t_1 and t_2 integrations. Alternatively, one may write

$$\hat{U}^{(2)}(t,t_0) = -\int_{t_0}^{t} \mathrm{d}t_1 \int_{t_0}^{t} \mathrm{d}t_2\, \hat{H}_I(t_2)\hat{H}_I(t_1)\theta(t_2 - t_1) \tag{4.32}$$

and the latter two forms can be recombined to give

$$\hat{U}^{(2)}(t,t_0) = -\frac{1}{2}\int_{t_0}^{t} \mathrm{d}t_1 \int_{t_0}^{t} \mathrm{d}t_2\, \hat{\mathcal{T}}\left[\hat{H}_I(t_1)\hat{H}_I(t_2)\right] \tag{4.33}$$

Here $\hat{\mathcal{T}}$ is Wick's time-ordering operator (Eq. 3.5), putting operators with larger time arguments to the left of those with smaller time arguments. Note that the re-ordering of operators is not accompanied by any sign changes because the interaction operators are formed by an even number of fermion operators.

The form obtained for the second-order term can readily be generalized to the nth-order term, yielding

$$\hat{U}^{(n)}(t, t_0) = \frac{(-i)^n}{n!} \int_{t_0}^{t} dt_1 \ldots \int_{t_0}^{t} dt_n \, \hat{\mathcal{T}} \left[\hat{H}_I(t_1) \ldots \hat{H}_I(t_n) \right] \tag{4.34}$$

Finally, the perturbation expansion of the time-evolution operator in the interaction picture reads

$$\hat{U}(t, t_0) = \sum_{n=0}^{\infty} \frac{(-i)^n}{n!} \int_{t_0}^{t} dt_1 \ldots \int_{t_0}^{t} dt_n \, \hat{\mathcal{T}} \left[\hat{H}_I(t_1) \ldots \hat{H}_I(t_n) \right] \tag{4.35}$$

In a compact, if somewhat symbolic way, one may also write

$$\hat{U}(t, t_0) = \hat{\mathcal{T}} e^{-i \int_{t_0}^{t} dt' \hat{H}_I(t')}$$

The expansion of the time-evolution operator, according to Eq. (4.27) or Eq. (4.35), provides a basis for time-dependent perturbation theory. In the ensuing Sect. 4.2, we shall use a specific time-dependent approach to re-formulate the usual (time independent) perturbation theory for the ground state and ground-state expectation values of an interacting N-electron system.

4.2 The Gell-Mann and Low Theorem

The starting point for the following derivation is the time-dependent hamiltonian

$$\hat{H}(t) = \hat{H}_0 + e^{-\epsilon|t|} \hat{H}_I \tag{4.36}$$

where the interaction part of the hamiltonian (4.1) is "switched on" (and off) as a function of time. For $t \to \pm\infty$, $\hat{H}(t)$ reduces to $\hat{H}_0$, while at $t = 0$, the original hamiltonian is restored, $\hat{H}(0) = \hat{H}$. The parameter $\epsilon > 0$ controls how fast the interaction is turned on or off (see Fig. 4.1). In the limit $\epsilon \to 0$, referred to as the **adiabatic limit**, one will expect that the ground state $|\Phi_0\rangle$ of the non-interacting system at $t = -\infty$ (assumed to be non-degenerate) evolves into the ground state of the interacting system $|\Psi_0\rangle$ at $t = 0$.

The time-evolution operator associated with the hamiltonian (4.36) can be written as

$$\hat{U}_\epsilon(t, t_0) = \sum_{n=0}^{\infty} \frac{(-i)^n}{n!} \int_{t_0}^{t} dt_1 e^{-\epsilon|t_1|} \ldots \int_{t_0}^{t} dt_n \, e^{-\epsilon|t_n|} \, \hat{\mathcal{T}} \left[\hat{H}_I(t_1) \ldots \hat{H}_I(t_n) \right] \tag{4.37}$$

where

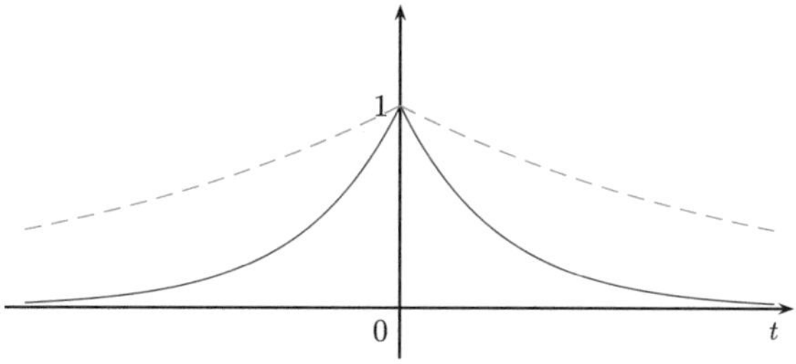

Fig. 4.1 Switching function for two different values of the parameter ϵ

$$\hat{H}_I(t) = e^{i\hat{H}_0 t}\hat{H}_I e^{-i\hat{H}_0 t} \tag{4.38}$$

is the interaction part of the hamiltonian in the interaction picture. The subscript ϵ indicates the dependence of the time-evolution operator on the switching parameter ϵ.

Let us assume that in the infinite past ($t_0 \to -\infty$), the system is in the non-interacting ground state,

$$\hat{H}_0|\Phi_0\rangle = E_0^{(0)}|\Phi_0\rangle \tag{4.39}$$

In this limit, the Schrödinger-picture state becomes $|\Phi_S(t)\rangle = e^{-iE_0^{(0)}t}|\Phi_0\rangle$ and the corresponding interaction-picture state is simply given by $|\Phi_0\rangle$:

$$|\Phi_I(t)\rangle = e^{i\hat{H}_0 t}|\Phi_S(t)\rangle = |\Phi_0\rangle \tag{4.40}$$

The state resulting from $|\Phi_0\rangle$ upon time evolution from $t = -\infty$ to $t = 0$ is given by

$$\begin{aligned}|\Psi_\epsilon(0)\rangle &= \hat{U}_\epsilon(0, -\infty)|\Phi_0\rangle \\ &= \sum_{n=0}^{\infty}\frac{(-i)^n}{n!}\int_{-\infty}^{0} \mathrm{d}t_1 e^{-\epsilon|t_1|}\ldots\int_{-\infty}^{0} \mathrm{d}t_n\, e^{-\epsilon|t_n|}\hat{\mathcal{T}}\left[\hat{H}_I(t_1)\ldots\hat{H}_I(t_n)\right]|\Phi_0\rangle\end{aligned} \tag{4.41}$$

At this point, one might expect that in the adiabatic limit, $\epsilon \to 0$, the state $|\Psi_\epsilon(0)\rangle$ approaches the ground state $|\Psi_0\rangle$ of the interacting system. However, the situation is not that simple. As will be demonstrated in Sect. 4.4, $|\Psi_\epsilon(0)\rangle$ has contributions that diverge as ϵ^{-1}. The divergent contributions can be canceled by multiplying $|\Psi_\epsilon(0)\rangle$ with the inverse of $\langle\Phi_0|\hat{U}_\epsilon(0, -\infty)|\Phi_0\rangle$. This is the essence of the Gell-Mann and Low theorem [4], reading as follows:

Theorem: If the state

$$|\Psi_0'\rangle = \lim_{\epsilon\to 0}\frac{\hat{U}_\epsilon(0, -\infty)|\Phi_0\rangle}{\langle\Phi_0|\hat{U}_\epsilon(0, -\infty)|\Phi_0\rangle} \tag{4.42}$$

exists to all orders of perturbation theory, then it is the ground state of $\hat{H}$ with the eigenvalue

$$E_0 = E_0^{(0)} + \lim_{\epsilon \to 0} \frac{\langle \Phi_0 | \hat{H}_I \hat{U}_\epsilon(0, -\infty) | \Phi_0 \rangle}{\langle \Phi_0 | \hat{U}_\epsilon(0, -\infty) | \Phi_0 \rangle} \tag{4.43}$$

A proof of the theorem is given in Appendix A.2. It is essentially based on an elaborated version [1] of the original proof [4].

A few comments are in order:

1. The Gell-Mann and Low state $|\Psi_0'\rangle$ is formed as the ratio of two perturbation expansions, one for the numerator, the other for the denominator. Both expansions depend on the switching parameter ϵ and comprise terms that diverge in the limit $\epsilon \to 0$. The ratio itself can be expanded, at least formally, in a perturbation expansion. The precondition of the theorem is that in each order n of the latter expansion the limit $\epsilon \to 0$ exists. If that assumption applies, which remains to be shown at a later stage, the adiabatic limit leads to a (formally) well-defined perturbation expansion for $|\Psi_0'\rangle$, and $|\Psi_0'\rangle$ is the interacting ground state. However, the actual convergence properties of that perturbation expansion, depending on the interaction strength, are not subject of the Gell-Mann and Low theorem.
2. Strictly speaking, the proof only guarantees that $|\Psi_0'\rangle$ is an eigenstate, but not necessarily the ground state of the interacting system. In general, however, one may reasonably expect that $|\Psi_0'\rangle$ is the interacting ground state provided the respective non-interacting ground state $|\Phi_0\rangle$ is non-degenerate.
3. The $|\Psi_0'\rangle$ state is not normalized to unity, but satisfies the so-called intermediate normalization,

$$\langle \Phi_0 | \Psi_0' \rangle = 1 \tag{4.44}$$

The proof of the Gell-Mann and Low theorem can readily be transfered to the state

$$|\Psi_0''\rangle = \lim_{\epsilon \to 0} \frac{\hat{U}_\epsilon(0, \infty) | \Phi_0 \rangle}{\langle \Phi_0 | \hat{U}_\epsilon(0, \infty) | \Phi_0 \rangle} \tag{4.45}$$

resulting from a time-reversed adiabatic development $(0 \leftarrow \infty)$. Note that $\hat{U}_\epsilon(0, \infty) = \hat{U}_\epsilon^\dagger(\infty, 0)$, according to Eq. (4.23). If the underlying non-interacting ground state is non-degenerate, the two modes of generating the Gell-Mann and Low state will lead to the same result:

$$|\Psi_0''\rangle = \lim_{\epsilon \to 0} \frac{\hat{U}_\epsilon(0, \infty) | \Phi_0 \rangle}{\langle \Phi_0 | \hat{U}_\epsilon(0, \infty) | \Phi_0 \rangle} = \lim_{\epsilon \to 0} \frac{\hat{U}_\epsilon(0, -\infty) | \Phi_0 \rangle}{\langle \Phi_0 | \hat{U}_\epsilon(0, -\infty) | \Phi_0 \rangle} = |\Psi_0'\rangle \tag{4.46}$$

Both states are subject to intermediate normalization (Eq. 4.44) which precludes the possibility of differing phases.

The time-reversed form (4.45) can be conveniently written as a bra state,

$$\langle\Psi_0''| = \lim_{\epsilon\to 0} \frac{\langle\Phi_0|\hat{U}_\epsilon(\infty,0)}{\langle\Phi_0|\hat{U}_\epsilon(\infty,0)|\Phi_0\rangle} \tag{4.47}$$

to be used in forming ground-state expectation values.

4.3 Expectation Values of Heisenberg Operators

The Gell-Mann and Low formulation of the interacting ground state can be extended to ground-state expectation values of operators. In particular, we are interested in expectation values involving time-dependent Heisenberg operators as encountered in the definition of the electron propagator (3.6).

Let us consider a general Heisenberg operator

$$\hat{O}_H(t) = e^{i\hat{H}t}\hat{O}_S e^{-i\hat{H}t} \tag{4.48}$$

associated with a Schrödinger operator $\hat{O}_S$. Using Eq. (4.16), the $\hat{O}_S$ may be replaced by the corresponding operator $\hat{O}_I(t)$ of the interaction picture:

$$\hat{O}_H(t) = e^{i\hat{H}t}e^{-i\hat{H}_0 t}\hat{O}_I(t)e^{i\hat{H}_0 t}e^{-i\hat{H}t} \tag{4.49}$$

For the time arguments $(t, 0)$, the time-evolution operator in the interaction picture (Eq. 4.20) can be written as

$$\hat{U}(t,0) = e^{i\hat{H}_0 t}e^{-i\hat{H}t} \tag{4.50}$$

Likewise, this result can be derived directly from the equation of motion (4.25). Using Eqs. (4.49) and (4.50), $\hat{O}_H(t)$ can be written in the form

$$\hat{O}_H(t) = \hat{U}(0,t)\hat{O}_I(t)\hat{U}(t,0) = \lim_{\epsilon\to 0}\hat{U}_\epsilon(0,t)\hat{O}_I(t)\hat{U}_\epsilon(t,0) \tag{4.51}$$

Note that for finite time arguments in the time-evolution operator, the limit $\epsilon \to 0$ is unproblematic; that is,

$$\hat{U}(t_1,t_2) = \lim_{\epsilon\to 0}\hat{U}_\epsilon(t_1,t_2) \tag{4.52}$$

which justifies the second part of Eq. (4.51).

In the latter form, the Heisenberg operator is compatible with the Gell-Mann and Low representation (4.42) of the ground state. Noting the normalization (4.44) of $|\Psi_0'\rangle$, the ground-state expectation value of $\hat{O}_H(t)$ can be written as

$$\langle\Psi_0|\hat{O}_H(t)|\Psi_0\rangle = \frac{\langle\Psi_0'|\hat{O}_H(t)|\Psi_0'\rangle}{\langle\Psi_0'|\Psi_0'\rangle} = \lim_{\epsilon\to 0}\frac{\langle\Phi_0|\hat{U}_\epsilon(\infty,t)\hat{O}_I(t)\hat{U}_\epsilon(t,-\infty)|\Phi_0\rangle}{\langle\Phi_0|\hat{U}_\epsilon(\infty,-\infty)|\Phi_0\rangle} \tag{4.53}$$

Here the transitivity relation (4.22) has been used to get

$$\hat{U}_\epsilon(t,0)\hat{U}_\epsilon(0,-\infty) = \hat{U}_\epsilon(t,-\infty) \tag{4.54}$$

and

$$\hat{U}_\epsilon(\infty,0)\hat{U}_\epsilon(0,-\infty) = \hat{U}_\epsilon(\infty,-\infty) \tag{4.55}$$

Finally, the exponential-type perturbation expansions of the time-evolution operators in the numerator on the right-hand side of Eq. (4.53) can be combined within a single perturbation expansion, as is described in more detail in Appendix A.2. The result reads

$$\langle\Psi_0|\hat{O}_H(t)|\Psi_0\rangle = \lim_{\epsilon\to 0}\sum_{n=0}^{\infty}\frac{(-i)^n}{n!}\int_{-\infty}^{\infty} dt_1 e^{-\epsilon|t_1|}\ldots\int_{-\infty}^{\infty} dt_n\, e^{-\epsilon|t_n|} \frac{\langle\Phi_0|\hat{\mathcal{T}}\left[\hat{H}_I(t_1)\ldots\hat{H}_I(t_n)\hat{O}_I(t)\right]|\Phi_0\rangle}{\langle\Phi_0|\hat{U}_\epsilon(\infty,-\infty)|\Phi_0\rangle} \tag{4.56}$$

Again, it should be noted that the limit $\epsilon\to 0$ does not exist independently for the numerator and denominator on the right-hand side.

The formulation given above can readily be extended to the ground-state expectation value of a time-ordered operator product $\hat{\mathcal{T}}\left[\hat{P}_H(t)\hat{Q}_H(t')\right]$, where $\hat{P}_H(t)$ and $\hat{Q}_H(t')$ are Heisenberg operators:

$$\langle\Psi_0|\hat{\mathcal{T}}\left[\hat{P}_H(t)\hat{Q}_H(t')\right]|\Psi_0\rangle = \lim_{\epsilon\to 0}\sum_{n=0}^{\infty}\frac{(-i)^n}{n!}\int_{-\infty}^{\infty} dt_1 e^{-\epsilon|t_1|}\ldots\int_{-\infty}^{\infty} dt_n\, e^{-\epsilon|t_n|} \frac{\langle\Phi_0|\hat{\mathcal{T}}\left[\hat{H}_I(t_1)\ldots\hat{H}_I(t_n)\hat{P}_I(t)\hat{Q}_I(t')\right]|\Phi_0\rangle}{\langle\Phi_0|\hat{U}_\epsilon(\infty,-\infty)|\Phi_0\rangle} \tag{4.57}$$

As an immediate application, we may now write the desired perturbation expansion of the electron propagator as

$$
\begin{aligned}
iG_{pq}(t,t') =& \langle\Psi_0|\hat{\mathcal{T}}\left[c_p[t]c_q^\dagger[t']\right]|\Psi_0\rangle \\
=& \lim_{\epsilon\to 0}\sum_{n=0}^{\infty}\frac{(-i)^n}{n!}\int_{-\infty}^{\infty}\mathrm{d}t_1 e^{-\epsilon|t_1|}\ldots\int_{-\infty}^{\infty}\mathrm{d}t_n e^{-\epsilon|t_n|} \\
& \frac{\langle\Phi_0|\hat{\mathcal{T}}\left[\hat{H}_I(t_1)\ldots\hat{H}_I(t_n)c_p(t)c_q^\dagger(t')\right]|\Phi_0\rangle}{\langle\Phi_0|\hat{U}_\epsilon(\infty,-\infty)|\Phi_0\rangle}
\end{aligned}
\tag{4.58}
$$

where the explicit perturbation expansion of the denominator is given by

$$
\begin{aligned}
\langle\Phi_0|\hat{U}_\epsilon(\infty,-\infty)|\Phi_0\rangle = \sum_{n=0}^{\infty}\frac{(-i)^n}{n!}\int_{-\infty}^{\infty}\mathrm{d}t_1 e^{-\epsilon|t_1|}\ldots\int_{-\infty}^{\infty}\mathrm{d}t_n e^{-\epsilon|t_n|} \\
\langle\Phi_0|\hat{\mathcal{T}}\left[\hat{H}_I(t_1)\ldots\hat{H}_I(t_n)\right]|\Phi_0\rangle
\end{aligned}
\tag{4.59}
$$

Like in the Gell–Mann and Low state (4.42), the resulting expression for $G_{pq}(t,t')$ is seen to be the ratio of two perturbation expansions, both depending on the switching parameter ϵ. Likewise, the adiabatic limit ($\epsilon \to 0$) does not exist independently for the denominator and numerator, but eventually for their ratio. The *linked-cluster theorem*, to be addressed in Sect. 5.3, will show that the denominator cancels a corresponding factor in the numerator, thereby eliminating any diverging contributions in the adiabatic limit.

The essential ingredients in the perturbation expansions for $G_{pq}(t,t')$ are expectation values of time-ordered products of creation and destruction operators in the interaction picture, where the expectation value is to be taken with respect to the non-interacting ground state $|\Phi_0\rangle$. The evaluation of these expectation values is the subject of *Wick's theorem* considered in Chap. 5.

4.4 Comparison with Rayleigh–Schrödinger Perturbation Theory

The Gell–Mann and Low expressions (4.42), (4.43) for the interacting ground state and ground-state energy establish a perturbation theoretical approach, which differs completely from the familiar RSPT procedure. Of course, the resulting perturbation expansions must be identical, and it is instructive to see explicitly how this equivalence comes to pass at lowest orders.

The RSPT expansions for the ground state and ground-state energy can be written in the closed-form expressions presented in Appendix A.1:

$$
|\Psi_0\rangle = |\Phi_0\rangle + \sum_{n=1}^{\infty}\left[\frac{\hat{Q}_0}{E_0^{(0)}-\hat{H}_0}\left(E_0^{(0)}-E_0+\hat{H}_I\right)\right]^n|\Phi_0\rangle \tag{4.60}
$$

$$
\begin{aligned}
E_0 =& \langle \Phi_0 | \hat{H} | \Psi_0 \rangle \\
=& E_0^{(0)} + \langle \Phi_0 | \hat{H}_I | \Phi_0 \rangle + \sum_{n=1}^{\infty} \langle \Phi_0 | \hat{H}_I \left[\frac{\hat{Q}_0}{E_0^{(0)} - \hat{H}_0} \left(E_0^{(0)} - E_0 + \hat{H}_I \right) \right]^n |\Phi_0\rangle
\end{aligned}
\tag{4.61}
$$

where $\hat{Q}_0 = \hat{\mathbb{1}} - |\Phi_0\rangle\langle\Phi_0|$. As further discussed in Appendix A.1, these so far rather formal expansions can be made more explicit by applying the resolution of the identity,

$$
\hat{\mathbb{1}} = \sum_I |\Phi_I\rangle\langle\Phi_I| \tag{4.62}
$$

in terms of excited HF states $|\Phi_I\rangle$, specified in Eq. (2.24).

The first-order wave function, for example, becomes

$$
|\Psi_0^{(1)}\rangle = \sum_{a<b,k<l} \frac{V_{ab[kl]}}{\epsilon_a + \epsilon_b - \epsilon_k - \epsilon_l} |\Phi_{abkl}\rangle \tag{4.63}
$$

Here only the class of double excitations, $|\Phi_{abkl}\rangle$, comes into play, since the matrix elements $\langle\Phi_0|\hat{H}_I|\Phi_I\rangle$ vanish for states of higher excitation classes. Single excitations, on the other hand, do not contribute since

$$
\langle \Phi_0 | \hat{H}_I | \Phi_{ak} \rangle = w_{ak} + \sum_r V_{ar[kr]} n_r = 0 \tag{4.64}
$$

as a result of the HF Eqs. (4.5), (4.6), which is often referred to as Brillouin's theorem. In a similar way, the expansion of the ground-state energy through second order can be written as

$$
E_0 = E_0^{(0)} + \langle \Phi_0 | \hat{H}_I | \Phi_0 \rangle - \sum_{a<b,k<l} \frac{|V_{ab[kl]}|^2}{\epsilon_a + \epsilon_b - \epsilon_k - \epsilon_l} + O(3) \tag{4.65}
$$

Now we come back to the expansions based on the Gell–Mann and Low approach. A more convenient starting point for evaluating low-order contributions in the numerator and denominator on the right-hand sides of Eqs. (4.42) and (4.43) is a closed-form integration [5] in the original expression of the time-evolution operator (see Eq. 4.28),

$$
\begin{aligned}
&\hat{U}_\epsilon(0, -\infty)|\Phi_0\rangle = \\
&|\Phi_0\rangle + \sum_{n=1}^{\infty} (-i)^n \int_{-\infty}^{0} \mathrm{d}t_1 \int_{-\infty}^{t_1} \mathrm{d}t_2 \ldots \int_{-\infty}^{t_{n-1}} \mathrm{d}t_n \, e^{\epsilon(t_1 + \cdots + t_n)} \, \hat{H}_I(t_1) \ldots \hat{H}_I(t_n) |\Phi_0\rangle
\end{aligned}
$$

Since

$$\hat{H}_I(t_j) = e^{i\hat{H}_0 t_j}\hat{H}_I e^{-i\hat{H}_0 t_j} \tag{4.66}$$

the product of two successive time-dependent interaction operators with time arguments $t_j, t_{j'}$, $j' = j + 1$, becomes

$$\hat{H}_I(t_j)\hat{H}_I(t_{j'}) = e^{i\hat{H}_0 t_j}\hat{H}_I e^{-i\hat{H}_0(t_j - t_{j'})}\hat{H}_I e^{-i\hat{H}_0 t_{j'}} \tag{4.67}$$

This suggests to introduce new variables $x_1, \ldots, x_n$ according to

$$\begin{aligned}
x_1 &= t_1 & t_1 &= x_1 \\
x_2 &= t_2 - t_1 & t_2 &= x_1 + x_2 \\
x_3 &= t_3 - t_2 & t_3 &= x_1 + x_2 + x_3 \\
&\vdots & &\vdots \\
x_n &= t_n - t_{n-1} & t_n &= x_1 + x_2 + \cdots + x_n
\end{aligned} \tag{4.68}$$

Here, the second column specifies the inverse transformation. Obviously, one obtains fixed integration limits $(-\infty, 0)$ for each of the x_i integrations. The determinant of the Jacobi matrix is readily evaluated to give

$$\left|\left(\frac{\partial t_i}{\partial x_j}\right)\right| = 1 \tag{4.69}$$

Moreover, there is factor $e^{-iE_0^{(0)}(x_1 + \cdots + x_n)}$ resulting from the last interaction operator acting on the non-interacting ground state, $\hat{H}_I(t_n)|\Phi_0\rangle$. As a result, the n-fold integration in the nth-order term factorizes according to

$$\hat{U}_\epsilon^{(n)}(0, -\infty)|\Phi_0\rangle = (-i)^n \int_{-\infty}^{0} \mathrm{d}x_1 e^{n\epsilon x_1} e^{i(\hat{H}_0 - E_0^{(0)})x_1}\hat{H}_I$$

$$\int_{-\infty}^{0} \mathrm{d}x_2 e^{(n-1)\epsilon x_2} e^{i(\hat{H}_0 - E_0^{(0)})x_2}\hat{H}_I \cdots \int_{-\infty}^{0} \mathrm{d}x_n e^{\epsilon x_n} e^{i(\hat{H}_0 - E_0^{(0)})x_n}\hat{H}_I|\Phi_0\rangle$$

and the individual integrations can readily be performed to give

$$\hat{U}_\epsilon^{(n)}(0, -\infty)|\Phi_0\rangle = \frac{1}{E_0^{(0)} - \hat{H}_0 + ni\epsilon}\hat{H}_I$$

$$\frac{1}{E_0^{(0)} - \hat{H}_0 + (n-1)i\epsilon}\hat{H}_I \cdots \frac{1}{E_0^{(0)} - \hat{H}_0 + i\epsilon}\hat{H}_I|\Phi_0\rangle$$

The resulting perturbation expansion in the numerator of the Gell–Mann and Low state reads

$$\begin{aligned}\hat{U}_\epsilon(0,-\infty)|\Phi_0\rangle = |\Phi_0\rangle &+ \frac{1}{E_0^{(0)}-\hat{H}_0+i\epsilon}\hat{H}_I|\Phi_0\rangle \\ &+ \frac{1}{E_0^{(0)}-\hat{H}_0+2i\epsilon}\hat{H}_I\frac{1}{E_0^{(0)}-\hat{H}_0+i\epsilon}\hat{H}_I|\Phi_0\rangle+\cdots\end{aligned} \tag{4.70}$$

and

$$\langle\Phi_0|\hat{U}_\epsilon(0,-\infty)|\Phi_0\rangle = 1+\langle\Phi_0|\frac{1}{E_0^{(0)}-\hat{H}_0+i\epsilon}\hat{H}_I|\Phi_0\rangle+\cdots \tag{4.71}$$

is the corresponding expansion in the denominator. The expansions (4.70), (4.71), so far being merely formal, can be transformed into explicit perturbation expansions by inserting the resolution of the identity (4.62) in appropriate ways. By contrast to the formal RSPT expansions considered above, the unperturbed ground state $|\Phi_0\rangle$ must not be omitted.

For illustrative purposes, we may consider the simpler case of a one-particle system, to which the Gell–Mann and Low procedure applies as well. Let

$$\hat{h} = \hat{h}_0 + \hat{h}_i \tag{4.72}$$

denote the hamiltonian of a one-particle system. Here, $\hat{h}_0$ is the "unperturbed" part for which the eigenvalue problem

$$\hat{h}_0|\phi_m\rangle = e_m|\phi_m\rangle,\ m = 0, 1, \ldots \tag{4.73}$$

is assumed to be solved; $\hat{h}_i$ is the perturbation, and $v_{mm'} = \langle\phi_m|\hat{h}_i|\phi_{m'}\rangle$ denotes the matrix elements of $\hat{h}_i$. The one-particle analogue to the time-evolution operator (4.37) is obtained by replacing $\hat{H}_I(t)$ with $\hat{h}_I(t) = e^{i\hat{h}_0 t}\hat{h}_i e^{-i\hat{h}_0 t}$. The formal perturbation expansion (4.70) takes the form

$$\hat{U}_\epsilon(0,-\infty)|\phi_0\rangle = |\phi_0\rangle + \frac{1}{e_0-\hat{h}_0+i\epsilon}\hat{h}_i|\phi_0\rangle+\cdots \tag{4.74}$$

which may be further evaluated by inserting $\sum_m|\phi_m\rangle\langle\phi_m|$ on the right-hand side. Note that there is no restriction $m\neq 0$ here. Through first order, the explicit expansion reads

$$\begin{aligned}\hat{U}_\epsilon(0,-\infty)|\phi_0\rangle &= |\phi_0\rangle + \sum_m \frac{v_{m0}}{e_0 - e_m + i\epsilon}|\phi_m\rangle + \dots \\ &= |\phi_0\rangle + \frac{v_{00}}{i\epsilon}|\phi_0\rangle + \sum_{m\neq 0} \frac{v_{m0}}{e_0 - e_m + i\epsilon}|\phi_m\rangle + \dots \end{aligned} \tag{4.75}$$

In the second line, the first term ($m = 0$) has been taken out of the m summation. This term diverges as ϵ^{-1} in the limit $\epsilon \to 0$. Upon multiplication of the numerator with the inverse denominator,

$$\langle\phi_0|\hat{U}_\epsilon(0,-\infty)|\phi_0\rangle^{-1} = 1 - \frac{v_{00}}{i\epsilon} + \dots \tag{4.76}$$

the first-order expansion of the ratio becomes

$$\frac{\hat{U}_\epsilon(0,-\infty)|\phi_0\rangle}{\langle\phi_0|\hat{U}_\epsilon(0,-\infty)|\phi_0\rangle} = |\phi_0\rangle + \sum_{m\neq 0} \frac{v_{m0}}{e_0 - e_m + i\epsilon}|\phi_m\rangle + \dots \tag{4.77}$$

where the singular contributions have canceled. Now the limit $\epsilon \to 0$ can safely be taken, yielding

$$|\psi_0'\rangle = |\phi_0\rangle + \sum_{m\neq 0} \frac{v_{m0}}{e_0 - e_m}|\phi_m\rangle + \dots \tag{4.78}$$

which is seen to reproduce the RSPT first-order result. The explicit perturbation expansion of $|\psi_0'\rangle$ can be extended through second order without undue effort, illustrating here another subtlety in the cancelation of the diverging ϵ terms (see Exercise 4.1).

What we here have seen explicitly at lowest order, is the working of the linked-cluster theorem (see Sect. 5.3), stating that the numerator of the Gell–Mann and Low state factorizes according to

$$\hat{U}_\epsilon(0,-\infty)|\phi_0\rangle = \{\hat{U}_\epsilon(0,-\infty)|\phi_0\rangle\}_L\langle\phi_0|\hat{U}_\epsilon(0,-\infty)|\phi_0\rangle \tag{4.79}$$

The second factor cancels the denominator, whereas the symbolic expression $\{\dots\}_L$, standing for "linked" contributions in the numerator, performs properly in the limit $\epsilon \to 0$.

Exercises

4.1 Extend the perturbation expansions (4.75), (4.76) to second order and verify that the second-order expansion of the ratio (4.77) in the limit $\epsilon \to 0$ reproduces the RSPT expansion for the ground state (using (A.1.8) and the one-particle hamiltonian (4.72)).

4.2 Ground-state PT in the 2E-2O model of Exercise 2.4:
(a) Expand the exact solution for the ground-state energy e_0 in a PT series through fourth order.

(b) Generate the same expansion by means of RSPT for the two-electron ground state (see Appendix A.1).
(c) Perform an analogous analysis through third order for the CI coefficient $x_1 = \langle\Phi_1|\Psi_0\rangle$ in the exact ground state.

References

1. Fetter AL, Walecka JD (1971) Quantum theory of many-particle systems. Mc Graw-Hill, New York
2. Martin PC, Schwinger J (1959) Phys Rev A 115:1342
3. Kadanoff LP, Baym G (1962) Quantum statistical mechanics. Benjamin, Reading
4. Gell-Mann M, Low F (1951) Phys Rev 84:350
5. Goldstone J (1957) Proc R Soc A 239:267

Chapter 5
Introducing Diagrams

The step from the Gell-Mann and Low (GML) formulation of the PT expansion of the electron propagator toward a diagramatic representation is enabled by Wick's theorem. According to this theorem, the expectation values of time-ordered fermion operator products arising in the GML expression can be evaluated in terms of *contractions* of operator pairs (Sect. 5.1). The contractions are related to free electron propagators and can be represented graphically by a directed line between the respective fermion operators. Together with the *wiggly lines* as graphical symbols for the interaction integrals, this allows one to replace the original analytic terms with diagrams (Sect. 5.2).

The contraction concept allows one to distinguish *linked* and *unlinked* contributions (or diagrams) to the propagator PT expansion. The linked-cluster theorem, discussed in Sect. 5.3, states that the denominator in the GML expressions exactly cancels all unlinked parts so that only linked contributions (or diagrams) need to be considered in the PT expansion of the electron propagator.

5.1 Wick's Theorem

The perturbation expansion (4.58) for the electron propagator established in the preceding Chap. 4 involves expectation values of time-ordered fermion operator products, being of the form

$$\langle\Phi_0|\hat{\mathcal{T}}\left[\hat{H}_I(t_1)\dots\hat{H}_I(t_n)c_p(t)c_q^\dagger(t')\right]|\Phi_0\rangle$$

where

$$\hat{H}_I(t) = \sum w_{rs}c_r^\dagger(t)c_s(t) + \frac{1}{2}\sum V_{uvrs}c_u^\dagger(t)c_v^\dagger(t)c_s(t)c_r(t) \tag{5.1}$$

J. Schirmer, *Many-Body Methods for Atoms, Molecules and Clusters*, Lecture Notes in Chemistry 94, https://doi.org/10.1007/978-3-319-93602-4_5

A systematic way to evaluate these quantities is provided by Wick's theorem [1] to be addressed below. The treatment based on Wick's theorem can readily be translated into the concept of Feynman diagrams, as will be discussed in the next section.

The first observation to be made is that the fermion operators $c_p^\dagger$, c_p can be classified according to their respective action on $|\Phi_0\rangle$:

$$c_p^\dagger|\Phi_0\rangle = \begin{cases} |\Phi_p^{N+1}\rangle & \text{for } n_p = 0 \\ 0 & \text{for } n_p = 1 \end{cases} \tag{5.2}$$

$$c_p|\Phi_0\rangle = \begin{cases} |\Phi_p^{N-1}\rangle & \text{for } n_p = 1 \\ 0 & \text{for } n_p = 0 \end{cases} \tag{5.3}$$

An operator is referred to as **physical** if the outcome is an $(N \pm 1)$-state (first case in Eqs. (5.2) and (5.3), respectively) and **unphysical** if the null vector results (second case in Eqs. (5.2) and (5.3), respectively). This allows us to divide the fermion operators into two classes comprising exclusively physical and unphysical operators, respectively:

$$\{\hat{\upsilon}_s\} \equiv \{c_p, c_q^\dagger;\ n_p = 1, n_q = 0\}$$
$$\{\hat{u}_r\} \equiv \{c_p, c_q^\dagger;\ n_p = 0, n_q = 1\}$$

In the following, we shall use the notations $\hat{\upsilon}_i$ and $\hat{u}_i$ for physical and unphysical fermion operators, respectively; general fermion operators will be denoted by $\hat{a}$ or $\hat{b}$. It should be noted that the physical fermion operators anticommute among themselves, and so do the unphysical ones:

$$\{\hat{\upsilon}_i, \hat{\upsilon}_j\} = 0, \quad \{\hat{u}_i, \hat{u}_j\} = 0 \tag{5.4}$$

Consider a product $\hat{a}_i(t_i)\hat{a}_j(t_j)\hat{a}_k(t_k)\ldots$ of time-dependent fermion operators in the interaction picture; for brevity, the time arguments will be skipped in the following. The time-ordering operator, first introduced in Sect. 3.1, generates the **time-ordered product**,

$$\hat{\mathcal{T}}\left[\hat{a}_i\hat{a}_j\hat{a}_k\ldots\right] \equiv (-1)^P \hat{a}_{P(i)}\hat{a}_{P(j)}\ldots$$

where P is a permutation of the factors in the product such that operators with larger time arguments are placed to the left of those with smaller time arguments; $(-1)^P$ is the sign (or parity) of the permutation. In the case of equal (or absent) time arguments, the definition can be generalized to the effect that creation operators $c^\dagger$ are placed to the left of the destruction operators c. Another reordering of the original product is the **normal-ordered product**,

$$\hat{\mathcal{N}}\left[\hat{a}_i\hat{a}_j\hat{a}_k\ldots\right] \equiv (-1)^{P'} \hat{a}_{P'(i)}\hat{a}_{P'(j)}\ldots$$

where P' is a permutation placing physical operators to the left of unphysical ones, $(-1)^{P'}$ being the sign of the permutation. The operator $\hat{\mathcal{N}}$ is referred to as the normal-ordering operator. As an obvious property of the $\hat{\mathcal{N}}$ product, the expectation value taken with respect to the unperturbed ground state vanishes:

$$\langle\Phi_0|\hat{\mathcal{N}}\left[\hat{a}_i\hat{a}_j\hat{a}_k\dots\right]|\Phi_0\rangle = 0 \tag{5.5}$$

Note that this relation also holds if all operators in the product are physical operators.

Now we are in the position to define the **contraction** of two fermion operators as the difference

$$\hat{a}_r^{\bullet}\hat{a}_s^{\bullet} \equiv \hat{\mathcal{T}}\left[\hat{a}_r\hat{a}_s\right] - \hat{\mathcal{N}}\left[\hat{a}_r\hat{a}_s\right] \tag{5.6}$$

between the time-ordered and normal-ordered products. In addition to the dot notation, we shall also use contraction brackets,

$$\overbrace{\hat{a}_r\hat{a}_s} \equiv \hat{a}_r^{\bullet}\hat{a}_s^{\bullet}$$

Obviously, the definition (5.6) is antisymmetric,

$$\hat{a}_r^{\bullet}\hat{a}_s^{\bullet} = -\hat{a}_s^{\bullet}\hat{a}_r^{\bullet} \tag{5.7}$$

To better understand the meaning of a contraction, we will inspect more closely three distinct types:

(i) contractions of two unphysical operators:

$$\begin{aligned}\hat{u}_r(t)^{\bullet}\hat{u}_s(t')^{\bullet} &= \hat{u}_r(t)\hat{u}_s(t')\theta(t-t') - \hat{u}_s(t')\hat{u}_r(t)\theta(t'-t)\\ &\quad - \hat{u}_r(t)\hat{u}_s(t')\left(\theta(t-t') + \theta(t'-t)\right)\\ &= -\{\hat{u}_r(t), \hat{u}_s(t')\}\theta(t'-t) = 0\end{aligned} \tag{5.8}$$

In the second line, we have used $\theta(t-t') + \theta(t'-t) = 1$; the last equation follows from the fact that unphysical operators anticommute (see Eq. 5.4).

(ii) contractions of two physical operators:

$$\hat{v}_r(t)^{\bullet}\hat{v}_s(t')^{\bullet} = 0 \tag{5.9}$$

which follows in a similar way as in (*i*).

(iii) contractions of a physical and an unphysical operator:

$$\begin{aligned}\hat{u}_r(t)^{\bullet}\hat{v}_s(t')^{\bullet} &= \hat{u}_r(t)\hat{v}_s(t')\theta(t-t') - \hat{v}_s(t')\hat{u}_r(t)\theta(t'-t)\\ &\quad + \hat{v}_s(t')\hat{u}_r(t)\left(\theta(t-t') + \theta(t'-t)\right)\\ &= \{\hat{u}_r(t), \hat{v}_s(t')\}\theta(t-t')\end{aligned} \tag{5.10}$$

$$\hat{v}_r(t)^{\bullet}\hat{u}_s(t')^{\bullet} = -\hat{u}_s(t')^{\bullet}\hat{v}_r(t)^{\bullet} \tag{5.11}$$

While contractions of the types (i) and (ii) simply vanish, this is not necessarily the case for the mixed-type contractions (iii). The anticommutator arising in Eq. (5.10) is just a complex number though. This means that contractions are always c-numbers. Using Eq. (5.5), a contraction can be written as the (non-interacting) ground-state expectation value of the corresponding $\hat{\mathcal{T}}$ product,

$$\hat{a}_r^{\bullet}\hat{a}_s^{\bullet} = \langle\Phi_0|\hat{\mathcal{T}}\left[\hat{a}_r\hat{a}_s\right]|\Phi_0\rangle \tag{5.12}$$

As an obvious consequence, contractions of the original fermion operators become

$$c_p(t)^{\bullet}c_q(t')^{\bullet} = c_p^{\dagger}(t)^{\bullet}c_q^{\dagger}(t')^{\bullet} = 0 \tag{5.13}$$

$$c_p(t)^{\bullet}c_q^{\dagger}(t')^{\bullet} = \langle\Phi_0|\hat{\mathcal{T}}\left[c_p(t)c_q^{\dagger}(t')\right]|\Phi_0\rangle = iG_{pq}^0(t,t') \tag{5.14}$$

where $G_{pq}^0(t,t')$ is the free one-particle Green's function (Eq. 3.52). The only non-vanishing contractions are those between a creation and a destruction operator, and such a contraction can be expressed by the free one-particle Green's functions.

To contract two operators in a normal-ordered product, one has to move the operators next to each other, which gives rise to a phase factor $(-1)^{\nu}$ according to the number ν of transpositions needed here. Then, the contraction can be performed and, resulting in a c-number, taken out of the product. For example,

$$\begin{aligned}\hat{\mathcal{N}}\left[\overbracket{\hat{a}_i\hat{a}_j\hat{a}_k}\hat{a}_l\dots\right] &= (-1)\hat{\mathcal{N}}\left[\overbracket{\hat{a}_i\hat{a}_k}\hat{a}_j\hat{a}_l\dots\right]\\ &= (-1)\overbracket{\hat{a}_i\hat{a}_k}\hat{\mathcal{N}}\left[\hat{a}_j\hat{a}_l\dots\right]\end{aligned} \tag{5.15}$$

Wick's theorem [1] establishes a reformulation of a general time-ordered product of fermion operators in terms of normal-ordered products and contractions. It may be stated as follows:

Wick's Theorem
A $\hat{\mathcal{T}}$ product of m fermion operators can be transformed into a sum of $\hat{\mathcal{N}}$ products with all possible contractions of $k = 0, 1, \dots, [m/2]$ operator pairs:

$$\begin{aligned}\hat{\mathcal{T}}\left[\hat{a}_i\hat{a}_j\hat{a}_k\hat{a}_l\dots\hat{a}_r\hat{a}_s\hat{a}_t\right] &= \hat{\mathcal{N}}\left[\hat{a}_i\hat{a}_j\hat{a}_k\hat{a}_l\dots\hat{a}_r\hat{a}_s\hat{a}_t\right]\\ &+\hat{\mathcal{N}}\left[\overbracket{\hat{a}_i\hat{a}_j}\hat{a}_k\dots\right]+\hat{\mathcal{N}}\left[\overbracket{\hat{a}_i\hat{a}_j\hat{a}_k}\dots\right]+\dots\\ &+\hat{\mathcal{N}}\left[\overbracket{\hat{a}_i\hat{a}_j}\overbracket{\hat{a}_k\hat{a}_l}\dots\right]+\dots\\ &\vdots\\ &+\hat{\mathcal{N}}\left[\overbracket{\hat{a}_i\overbracket{\hat{a}_j\overbracket{\hat{a}_k\dots\hat{a}_r}\hat{a}_s}\hat{a}_t}\right]\dots\end{aligned} \tag{5.16}$$

In a somewhat symbolic notation, one may write

$$\hat{\mathcal{T}}\left[\hat{a}_i\hat{a}_j\hat{a}_k\hat{a}_l\ldots\hat{a}_r\hat{a}_s\hat{a}_t\right] = \hat{\mathcal{N}}\left[\hat{a}_i\hat{a}_j\hat{a}_k\hat{a}_l\ldots\hat{a}_r\hat{a}_s\hat{a}_t\right] + \hat{\mathcal{N}}\left[\text{sum over all possible pairs of contractions}\right]$$

Wick's reformulation of a time-ordered product is obtained by moving physical operators (PO) in the time-ordered product successively to the left. This generates additional terms whenever a PO does not anticommute with an operator on its left. A proof of the theorem is given in Appendix A.3.

Wick's theorem establishes an operator identity. While the right-hand side of this identity looks rather complicated, the actual benefit of Wick's theorem becomes apparent when ground-state expectation values of time-ordered products are to be evaluated. According to the property of the $\hat{\mathcal{N}}$ products, only the fully contracted terms contribute to the expectation value:

$$\langle\Phi_0|\hat{\mathcal{T}}\left[\hat{a}_i\hat{a}_j\hat{a}_k\hat{a}_l\ldots\hat{a}_r\hat{a}_s\hat{a}_t\right]|\Phi_0\rangle = \hat{\mathcal{N}}\left[\hat{a}_i\hat{a}_j\hat{a}_k\ldots\hat{a}_r\hat{a}_s\hat{a}_t\right] + \ \ldots \tag{5.17}$$

As indicated on the right-hand side, there is a contribution for each full contraction scheme. In the ensuing section, we will learn how the distinct contraction schemes can be expressed in the form of diagrams.

5.2 Zeroth- and First-Order Feynman Diagrams

The perturbation expansion (4.58) of the electron propagator is of the form

$$iG_{pq}(t,t') = \lim_{\epsilon\to 0}\frac{i\tilde{G}_{pq}(t,t')}{\langle\Phi_0|\hat{U}_\epsilon(\infty,-\infty)|\Phi_0\rangle} \tag{5.18}$$

where the numerator $i\tilde{G}_{pq}(t,t')$ is given by

$$i\tilde{G}_{pq}(t,t') = \sum_{n=0}^{\infty}\frac{(-i)^n}{n!}\int_{-\infty}^{\infty}\mathrm{d}t_1 e^{-\epsilon|t_1|}\ldots\int_{-\infty}^{\infty}\mathrm{d}t_n e^{-\epsilon|t_n|} \langle\Phi_0|\hat{\mathcal{T}}\left[\hat{H}_I(t_1)\ldots\hat{H}_I(t_n)c_p(t)c_q^\dagger(t')\right]|\Phi_0\rangle \tag{5.19}$$

Using Wick's result (5.17) for the ground-state expectation values of time-ordered operator products, the respective nth-order terms in the perturbation expansions, both for the numerator and denominator, can be determined analytically by generating and evaluating all possible (full) contraction schemes. As a by far superior procedure, one can resort to a graphical representation in terms of diagrams, as first introduced by Feynman [2] in the context of quantum electrodynamics (QED).

To introduce the diagrammatic formulation, let us consider the zeroth- and first-order terms in the numerator $i\tilde{G}_{pq}(t,t')$. For simplicity, we will suppose in the following that the interaction hamiltonian (5.1) consists only of the Coulomb part, which in turn means that one has to allow for a non-diagonal free one-particle part $\hat{H}_0$. The case of the full interaction hamiltonian, $\hat{H}_I = \hat{W} + \hat{V}$, comprising also a one-particle part, will be discussed in Sect. 6.2.

Zeroth Order:

$$i\tilde{G}^0_{pq}(t,t') = iG^0_{pq}(t,t') = \langle\Phi_0|\hat{\mathcal{T}}\left[c_p(t)c_q^\dagger(t')\right]|\Phi_0\rangle = c_p(t)^\bullet c_q^\dagger(t')^\bullet \tag{5.20}$$

The zeroth-order term is just the free one-particle Green's function (3.52) discussed in Sect. 3.4, allowing here $\boldsymbol{G}^0$ to be non-diagonal. As the first graphical element, we assign a "free fermion line" to the $c_p(t)^\bullet c_q^\dagger(t')^\bullet$ contraction:

$$\equiv \quad c_p(t)^\bullet c_q^\dagger(t')^\bullet = iG^0_{pq}(t,t') \tag{5.21}$$

The arrow defines the direction of the line: it starts at the lower vertex associated with the creation operator (one-particle index q, time argument t') and ends at the upper vertex associated with the destruction operator (one-particle index p, time argument t).

First Order:

$$i\tilde{G}^{(1)}_{pq}(t,t') =$$

$$(-i)\frac{1}{2}\sum_{u,v,r,s} V_{uvrs}\int_{-\infty}^{\infty} dt_1 e^{-\epsilon|t_1|}\langle\Phi_0|\hat{\mathcal{T}}\left[c_u^\dagger(t_1)c_v^\dagger(t_1)c_s(t_1)c_r(t_1)c_p(t)c_q^\dagger(t')\right]|\Phi_0\rangle$$

The second graphical symbol, associated with the Coulomb interaction (at an internal time argument t_i), is the "wiggly interaction line"

$$\equiv \quad V_{uvrs} \tag{5.22}$$

A wiggly line has two entries and two exits (as indicated by the small arrows), where free fermion lines can begin or end, respectively. With the indices u, v, r, s attached as in (5.22), the interaction line represents the Coulomb integral V_{uvrs}. The order of the indices in the interaction line can be memorized as being *left out, right out, left in, right in*.

The $\hat{\mathcal{T}}$ product of the Coulomb part in $i\tilde{G}_{pq}^{(1)}$ consists of three creation operators, $c_u^{\dagger}, c_v^{\dagger}, c_q^{\dagger}$, and three destruction operators, c_r, c_s, c_p. According to the relations (5.13, 5.14), there are $3! = 6$ distinct non-vanishing contraction schemes, (A), (B),…,(F), which we shall consider successively in the following.

Contraction scheme (A), depicted in the graph below,

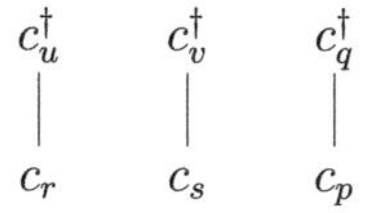

gives rise to the following Feynman diagram:

$A:$

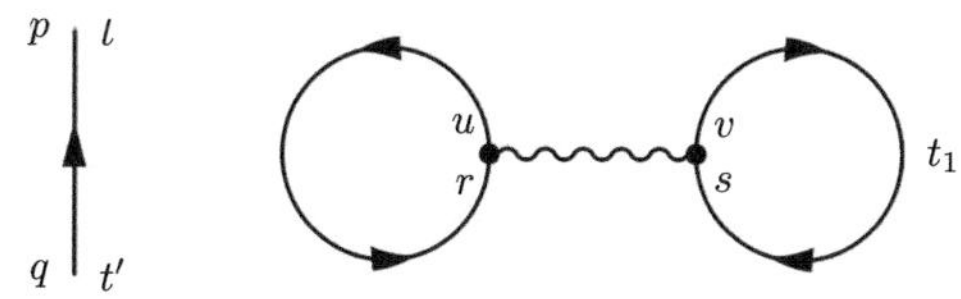

(5.23)

There is a free fermion line as in the zeroth-order term, while the two other contractions connect the operators of the Coulomb interaction, depicted as two free fermion lines beginning and ending at the same Coulomb integral (wiggly line) and at the same internal time t_1. This feature needs further analysis. Let us consider one of the two contractions at equal time,

$$c_s(t_1)^{\bullet} c_v^{\dagger}(t_1)^{\bullet} = \hat{\mathcal{T}}\left[c_s(t_1)c_v^{\dagger}(t_1)\right] - \hat{\mathcal{N}}\left[c_s(t_1)c_v^{\dagger}(t_1)\right] \tag{5.24}$$

For equal times, the time-ordered product places creation operators to the left of destruction operators; that is,

$$\hat{\mathcal{T}}\left[c_s(t_1)c_v^{\dagger}(t_1)\right] = -c_v^{\dagger}(t_1)c_s(t_1) = -c_v^{\dagger}c_s \tag{5.25}$$

Correspondingly, the contraction, being a c-number, becomes

$$c_s(t_1)^{\bullet} c_v^{\dagger}(t_1)^{\bullet} = -\langle\Phi_0|c_v^{\dagger}c_s|\Phi_0\rangle \tag{5.26}$$

The expectation value on the right-hand side can be related to the free one-particle Green's function by equating the time arguments in the following way:

$$-\langle\Phi_0|c_v^{\dagger}c_s|\Phi_0\rangle = \lim_{\tau\to 0} iG_{sv}^0(t_1, t_1+\tau) \equiv iG_{sv}^0(t_1, t_1^+) \tag{5.27}$$

Here $\tau > 0$ and t_1^+ is used as a short-hand notation for the limit $t_1 + \tau \rightarrow t_1$. We summarize this result in the following remark.

Remark 1:
In generalization of the relation (5.14), contractions at equal times are given by

$$c_s(t_1)^{\bullet} c_v^{\dagger}(t_1)^{\bullet} = iG_{sv}^0(t_1, t_1^+) \tag{5.28}$$

Now the full analytical expression for diagram (A) can be written as

$$i\tilde{G}_{pq}^{(1,A)} = (-i)\frac{1}{2}\sum V_{uvrs} \int \mathrm{d}t_1 e^{-\epsilon|t_1|}\, iG_{ru}^0(t_1, t_1^+) iG_{sv}^0(t_1, t_1^+)\, iG_{pq}^0(t, t') \tag{5.29}$$

Relating this expression to diagram (A), we note that it involves summations running over the "internal" one-particle indices, u, v, r, s, and time integration over the "internal" time t_1. The phase factors can be combined to give $-i^4 = -1$. The time integration in (5.29) reduces to an integral of the adiabatic switching function, $\int \mathrm{d}t_1 e^{-\epsilon|t_1|} = \frac{2}{\epsilon}$, being obviously singular in the limit $\epsilon \rightarrow 0$. This outcome is characteristic for contributions such as $\tilde{G}_{pq}^{(1,A)}$, where the diagram (and the corresponding analytical expression) consists of two "unlinked" multiplicative parts.

In a similar way, we may evaluate contraction schemes (B), ..., (F). Contraction scheme (B) represented by the diagram

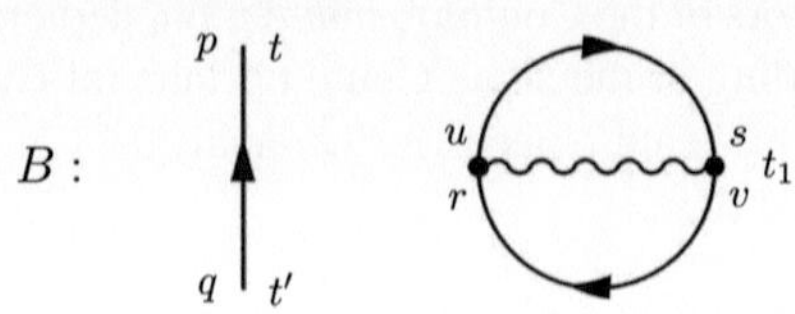

(5.30)

is the second unlinked contribution in first order. The corresponding analytical expression is given by

$$i\tilde{G}_{pq}^{(1,B)} = \frac{1}{2}\sum V_{uvrs} \int \mathrm{d}t_1 e^{-\epsilon|t_1|}\, G_{su}^0(t_1, t_1^+) G_{rv}^0(t_1, t_1^+)\, G_{pq}^0(t, t') \tag{5.31}$$

Note the phase difference with respect to (A), reflecting the different contraction scheme for the four "internal" operators.

Remark 2:
Diagrams (A) and (B) are referred to as **unlinked diagrams** because they consist of two disjoint parts, resulting in products of two factors in their analytical expressions. The unlinked diagrams will be seen to cancel the denominator in the full expression (4.58) as a result of the linked-cluster theorem discussed in the next section.

The next two contributions, represented by the diagrams below,

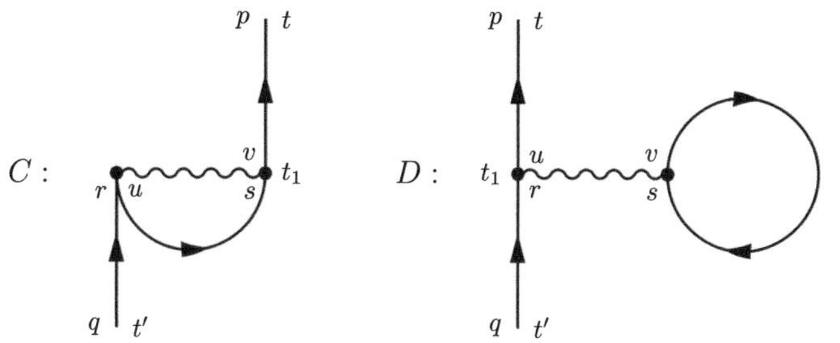

(5.32)

give rise to the expressions

$$i\tilde{G}_{pq}^{(1,C)} = -\frac{1}{2}\sum V_{uvrs}\int \mathrm{d}t_1 e^{-\epsilon|t_1|} G_{pv}^0(t,t_1) G_{rq}^0(t_1,t') G_{su}^0(t_1,t_1^+) \tag{5.33}$$

$$i\tilde{G}_{pq}^{(1,D)} = \frac{1}{2}\sum V_{uvrs}\int \mathrm{d}t_1 e^{-\epsilon|t_1|} G_{pu}^0(t,t_1) G_{rq}^0(t_1,t') G_{sv}^0(t_1,t_1^+) \tag{5.34}$$

In both the diagrams, each of the two external fermion operators is contracted to one of the internal ones. As one may readily verify, here the limit $\epsilon \to 0$ does not affect the outcome of the time integration. We will come back to this issue in Sect. 7.2.

The remaining two diagrams (E) and (F)

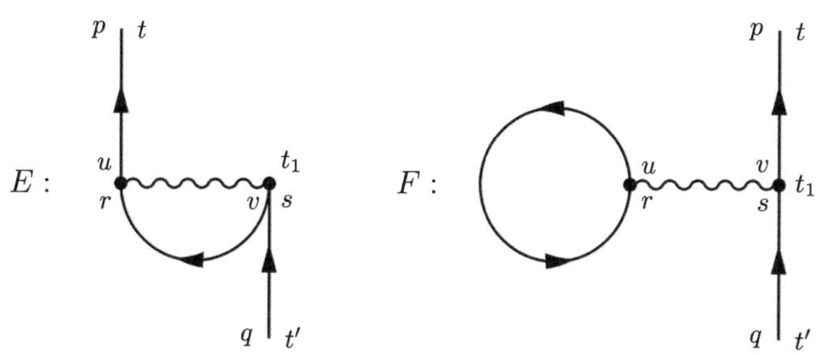

are topologically equivalent to (C) and (D), respectively. The corresponding analytical expressions

$$i\tilde{G}_{pq}^{(1,E)} = -\frac{1}{2}\sum V_{uvrs}\int \mathrm{d}t_1 e^{-\epsilon|t_1|} G_{pu}^0(t,t_1) G_{sq}^0(t_1,t') G_{rv}^0(t_1,t_1^+) \tag{5.35}$$

$$i\tilde{G}_{pq}^{(1,F)} = \frac{1}{2}\sum V_{uvrs}\int \mathrm{d}t_1 e^{-\epsilon|t_1|} G_{pv}^0(t,t_1) G_{sq}^0(t_1,t') G_{ru}^0(t_1,t_1^+) \tag{5.36}$$

are identical to (5.33) and (5.34), respectively, which is seen by replacing the dummy variables $(uvrs)$ with $(vusr)$.

Remark 3:
A linked diagram represents a pair of contraction schemes per interaction line, where the two contraction schemes differ from each other by an out-of-plane rotation of the interaction line. The corresponding analytical expressions are identical, which can be taken into account by applying a factor of 2. Correspondingly, there are 2^n equivalent contraction schemes associated with a (topologically distinct) nth-order diagram. The factor 2^n cancels the factor $(\frac{1}{2})^n$ arising from the Coulomb parts in $\hat{H}_I^n$.

In a similar way, the perturbation expansion of the denominator in Eq. (4.58) can be formulated in a diagrammatic fashion. Through first order, the expansion reads

$$\langle \Phi_0 | \hat{U}_\epsilon(-\infty, \infty) | \Phi_0 \rangle = 1 - \frac{i}{2} \sum V_{uvrs} \int_{-\infty}^{\infty} dt_1 e^{-\epsilon|t_1|} \langle \Phi_0 | \hat{\mathcal{T}} \left[c_u^\dagger(t_1) c_v^\dagger(t_1) c_s(t_1) c_r(t_1) \right] | \Phi_0 \rangle + O(2)$$

There are two non-vanishing contraction schemes in the first-order part, which can be translated into diagrams as follows:

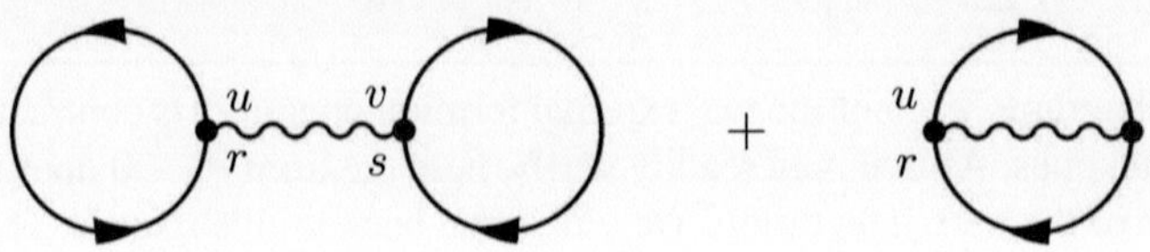

(5.37)

The corresponding analytical expression reads

$$\langle \Phi_0 | \hat{U}_\epsilon(-\infty, \infty) | \Phi_0 \rangle^{(1)} = \frac{i}{2} \sum V_{uvrs} \int dt_1 e^{-\epsilon|t_1|} \left[G^0_{ru}(t_1, t_1^+) G^0_{sv}(t_1, t_1^+) - G^0_{rv}(t_1, t_1^+) G^0_{su}(t_1, t_1^+) \right] \quad (5.38)$$

Obviously, the two first-order diagrams of the denominator are parts of the numerator diagrams (A) and (B), respectively, and the analytical expression is obtained from Eqs. (5.29, 5.31) by omitting the factor $iG^0_{pq}(t, t')$.

5.3 Linked-Cluster Theorem

The first-order perturbation expansions evaluated using Wick's theorem and the corresponding diagrammatic formulation suggests that the numerator on the right-hand side of Eq. (4.58) can be written as a product

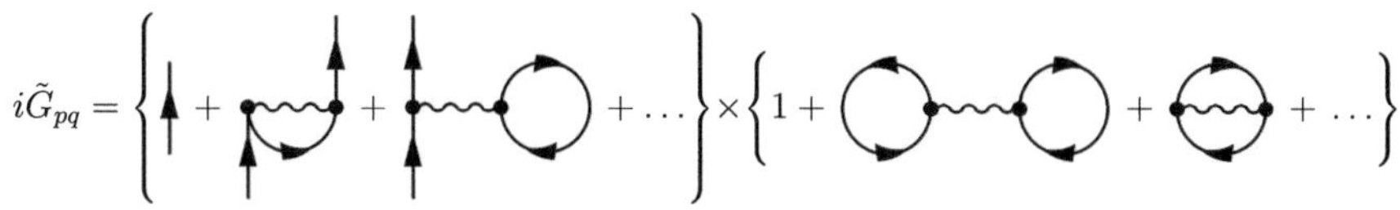

where the first factor collects all linked diagrams in the expansion of $\tilde{G}_{pq}$, that is, diagrams such as (C) and (D), while the second factor constitutes the perturbation expansion of the denominator. This would mean that the second factor cancels the denominator in Eq. (4.58), and the perturbation expansion of G_{pq} comprises linked diagrams only. Through first order, the factorization can readily be verified by inspecting the analytical expressions for $\tilde{G}_{pq}$ and $\langle\Phi_0|\hat{U}_\epsilon(-\infty,\infty)|\Phi_0\rangle$ presented in the preceding section. The general validity of the product form is assured by the linked-cluster theorem (originally derived in diagrammatic PT for the ground state [3, 4]).

Linked-cluster theorem:

$$iG_{pq}(t,t') = \lim_{\epsilon\to 0}\sum_{n=0}^{\infty}\frac{(-i)^n}{n!}\int_{-\infty}^{\infty} dt_1 \ldots \int_{-\infty}^{\infty} dt_n e^{-\epsilon|t_1|\cdots-\epsilon|t_n|}$$
$$\langle\Phi_0|\hat{\mathcal{T}}\left[\hat{H}_I(t_1)\ldots\hat{H}_I(t_n)c_p(t)c_q^\dagger(t')\right]|\Phi_0\rangle_C \tag{5.39}$$

where $\langle\Phi_0|\ldots|\Phi_0\rangle_C$ comprises all linked contributions in the original expectation value. The characterization "linked" is defined by Wick's contraction schemes or the corresponding diagram topology.

As preliminary to the proof, let us consider the nth-order term in Eq. (5.19) and make the following observations:

1. The use of Wick's theorem in the evaluation of the ground-state expectation value (and the diagrammatic formulation based on it) allows one to distinguish linked and unlinked contributions. A linked contribution is associated with a contraction scheme in which all interaction operators are connected directly or indirectly (via other interaction operators) to the external operators $c_p(t)$ and $c_q^\dagger(t')$.
2. An unlinked contribution can be written as a product of two or more factors. For example,

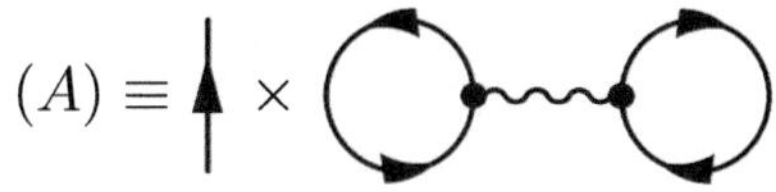

(5.40)

3. The order of the interaction operators $\hat{H}_I(t_j)$ in the $\hat{\mathcal{T}}$ product is arbitrary.
4. The integration variables t_j are dummy variables which can be renamed at will.

Now let us consider a specific nth-order contribution (contraction scheme), where ν $(0 \le \nu \le n)$ interaction parts are linked to the external operators $c_p(t)$ and $c_q^\dagger(t')$, while the remaining $\mu = n - \nu$ interaction parts have no direct or indirect connection to the external operators. Since there are $\binom{n}{\nu}$ ways to choose ν interaction parts out of n ones, there will be altogether $\binom{n}{\nu}$ contraction schemes differing from the original one only by the choice of the interaction parts. Considering the observations 3 and 4, those $\binom{n}{\nu}$ contraction schemes will all result in the same contribution. This allows us to rewrite the nth-order term according to

$$
\begin{aligned}
i\tilde{G}^{(n)}_{pq}(t,t') = \frac{(-i)^n}{n!}\sum_{\nu=0}^{n}\frac{n!}{\nu!\mu!} \\
\int_{-\infty}^{\infty} dt_1 \ldots \int_{-\infty}^{\infty} dt_\nu \, \underbrace{\langle \Phi_0|\hat{\mathcal{T}}\left[\hat{H}_I(t_1)\ldots\hat{H}_I(t_\nu)c_p(t)c_q^\dagger(t')\right]|\Phi_0\rangle_C}_{\text{linked contributions only}} \\
\times \int_{-\infty}^{\infty} dt_{\nu+1} \ldots \int_{-\infty}^{\infty} dt_n \, \underbrace{\langle \Phi_0|\hat{\mathcal{T}}\left[\hat{H}_I(t_{\nu+1})\ldots\hat{H}_I(t_n)\right]|\Phi_0\rangle}_{\mu=n-\nu \text{ interaction operators}}
\end{aligned}
$$

Here the switching functions have been omitted for brevity. Using the usual resummation technique for the exponential series,

$$
\sum_{n=0}^{\infty}\frac{(-i)^n}{n!}\sum_{\nu=0}^{n}\frac{n!}{\nu!(n-\nu)!} \rightarrow \sum_{\nu=0}^{\infty}\frac{(-i)^\nu}{\nu!}\sum_{\mu=0}^{\infty}\frac{(-i)^\mu}{\mu!} \tag{5.41}
$$

the numerator $i\tilde{G}_{pq}(t,t')$ can be written as

$$
\begin{aligned}
i\tilde{G}_{pq}(t,t') = \sum_{\nu=0}^{\infty}\frac{(-i)^\nu}{\nu!} \\
\int_{-\infty}^{\infty} dt_1 \ldots \int_{-\infty}^{\infty} dt_\nu \langle \Phi_0|\hat{\mathcal{T}}\left[\hat{H}_I(t_1)\ldots\hat{H}_I(t_\nu)c_p(t)c_q^\dagger(t')\right]|\Phi_0\rangle_C \\
\times \sum_{\mu=0}^{\infty}\frac{(-i)^\mu}{\mu!}\int_{-\infty}^{\infty} dt_1 \ldots \int_{-\infty}^{\infty} dt_\mu \langle \Phi_0|\hat{\mathcal{T}}\left[\hat{H}_I(t_1)\ldots\hat{H}_I(t_\mu)\right]|\Phi_0\rangle
\end{aligned}
\tag{5.42}
$$

The second factor is seen to be the perturbation expansion of $\langle\Phi_0|\hat{U}_\epsilon(\infty,-\infty)|\Phi_0\rangle$, which concludes the proof.

The linked-cluster theorem guarantees the existence (and triviality) of the adiabatic limit. For any linked contribution to the right-hand side of Eq. (5.39), the limit $\epsilon \rightarrow 0$ can be safely performed within the integrand of the time integration, reducing the switching functions to unity. Accordingly, Eq. (5.39) becomes

$$iG_{pq}(t,t') = \sum_{n=0}^{\infty} \frac{(-i)^n}{n!} \int_{-\infty}^{\infty} \mathrm{d}t_1 \ldots \int_{-\infty}^{\infty} \mathrm{d}t_n$$
$$\langle \Phi_0 | \hat{\mathcal{T}} \left[\hat{H}_I(t_1) \ldots \hat{H}_I(t_n) c_p(t) c_q^\dagger(t') \right] | \Phi_0 \rangle_C \qquad (5.43)$$

A validation of this result will be deferred to Sect. 7.2 and Appendix A.4, addressing the internal time integrations in the Feynman diagrams.

Exercises

5.1 Inspect the contraction patterns in the ground-state expectation value associated with the second-order term, $i\tilde{G}^{(2)}_{pq}(t,t')$, in Eq. (5.19). Select the contraction patterns relating to fully connected contributions (or diagrams).

5.2 Evaluate the ground-state expectation value $\langle \Phi_0 | \hat{V} | \Phi_0 \rangle$ of the Coulomb operator using Wick's theorem.

References

1. Wick GC (1950) Phys Rev 80:268
2. Feynman RP (1949) Phys Rev 76:749
3. Brueckner KA (1955) Phys Rev 100:36
4. Goldstone J (1957) Proc R Soc A 239:267

Chapter 6
Feynman Diagrams

Having established the three pillars of the formalism in the two foregoing chapters, we now can fully implement the diagrammatic approach to the PT expansion of the electron propagator. Specifically, we derive and formulate a list of rules for drawing and evaluating Feynman diagrams (Sect. 6.1). Adopting a HF one-particle representation leads to a considerable reduction of the number of diagrams to be considered, which will be discussed in Sect. 6.2. The systematic construction of higher-order diagrams is greatly facilitated by using the compact Abrikosov notation, in which sets of related Feynman diagrams can be incorporated within individual Abrikosov diagrams (Sect. 6.3). Abrikosov diagrams can be represented by specific matrices, which allows for the algorithmic construction of higher-order diagrams in a systematical way.

6.1 Second-Order Diagrams

To further establish the diagrammatic formulation of the perturbation expansion of the electron propagator, we will now take a look at the second-order contribution,

$$iG^{(2)}_{pq}(t,t') = \frac{(-i)^2}{2!}\int\limits_{-\infty}^{\infty} dt_1 \int\limits_{-\infty}^{\infty} dt_2\, (\tfrac{1}{2})^2 \sum_{\substack{uvrs\\ijkl}} V_{uvrs} V_{ijkl}$$

$$\langle \Phi_0|\hat{\mathcal{T}}[c^\dagger_u(t_1)c^\dagger_v(t_1)c_s(t_1)c_r(t_1)c^\dagger_i(t_2)c^\dagger_j(t_2)c_l(t_2)c_k(t_2)c_p(t)c^\dagger_q(t')]|\Phi_0\rangle_C \qquad (6.1)$$

as deriving from the general expression (5.43). The subscript C on the right-hand side indicates that only linked diagrams are taken into account. The restriction to linked diagrams discards from the outset many of the $5! = 120$ individual contraction schemes arising at second order. According to Remark 3 in Chap. 5,

J. Schirmer, *Many-Body Methods for Atoms, Molecules and Clusters*, Lecture Notes in Chemistry 94, https://doi.org/10.1007/978-3-319-93602-4_6

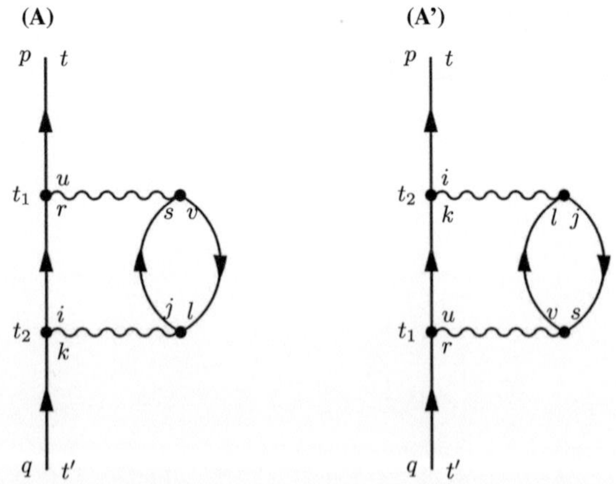

Fig. 6.1 Equivalent contraction schemes comprised within the second-order Feynman diagram (A)

a linked second-order diagram stands for four distinct contraction schemes. Another possibility of representing a class of contraction schemes by a single diagram arises in second order and beyond. As depicted in Fig. 6.1, there are two distinct contraction schemes associated with the second-order diagram (A). The contraction schemes differ in the order of the interaction parts, being (from above) $\hat{H}_I(t_1)\hat{H}_I(t_2)$ in the former and $\hat{H}_I(t_2)\hat{H}_I(t_1)$ in the latter. As is readily seen, the analytical expressions are identical, since the time arguments are dummy variables and can be interchanged ($t_1 \leftrightarrow t_2$), and so can the one-particle indices in the sums ($uvrs \leftrightarrow ijkl$). The two respective contraction schemes can be accounted for by an overall factor of 2, which cancels the factor $\frac{1}{2}$ on the right-hand side of Eq. (6.1). This finding can be readily generalized to nth order:

Remark 4:
A given linked diagram of nth order represents $n!$ equivalent contraction schemes differing only in the order of the interaction parts. This allows one to introduce an overall factor of $n!$, which cancels the corresponding prefactor in the nth order term of the exponential-type perturbation expansion.

The analytical expression associated with diagram (A) reads

$$\begin{aligned} iG^{(A)}_{pq}(t,t') =& (-i)(-1)^{L_A} \int_{-\infty}^{\infty} \mathrm{d}t_1 \int_{-\infty}^{\infty} \mathrm{d}t_2 \sum_{\substack{uvrs\\ ijkl}} V_{uvrs} V_{ijkl} \\ & G^0_{pu}(t,t_1) G^0_{ri}(t_1,t_2) G^0_{sj}(t_1,t_2) G^0_{lv}(t_2,t_1) G^0_{kq}(t_2,t') \end{aligned} \tag{6.2}$$

Here, the phase factor $(-i)$ is obtained according to $(-i)^2 i^5 = -i$; there is an additional phase, $(-1)^{L_A}$, which, at this point, has to be inferred from the underlying contraction scheme.

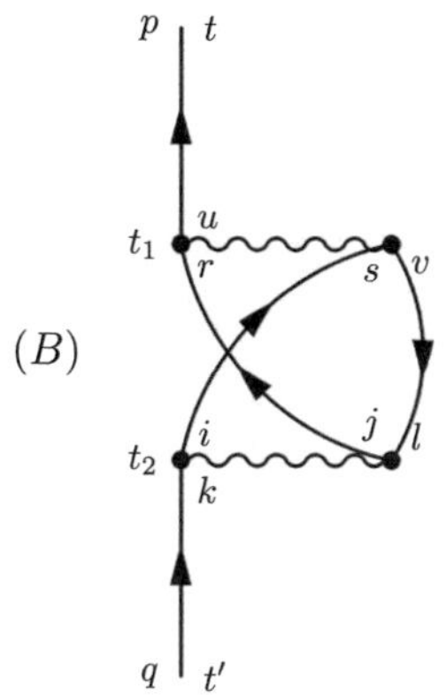

Fig. 6.2 Second-order Feynman diagram (B)

Another second-order diagram is shown in Fig. 6.2. Here, the end points of the two upwards directed fermion lines have been interchanged with respect to diagram (A). The analytical expression

$$iG_{pq}^{(B)}(t,t') = -i(-1)^{L_B} \int_{-\infty}^{\infty} dt_1 \int_{-\infty}^{\infty} dt_2 \sum_{\substack{uvrs\\ijkl}} V_{uvrs}V_{ijkl}$$
$$G_{pu}^{0}(t,t_1)G_{rj}^{0}(t_1,t_2)G_{si}^{0}(t_1,t_2)G_{lv}^{0}(t_2,t_1)G_{kq}^{0}(t_2,t') \quad (6.3)$$

differs from that of (A) by the exchange $i \leftrightarrow j$ of the one-particle indices in the second and third free Green's function. Now let us determine the two phase factors $(-1)^{L_A}$, $(-1)^{L_B}$ by inspecting the respective contraction schemes:

$$A: \quad \langle\Phi_0|c_u^\dagger c_v^\dagger c_s c_r c_i^\dagger c_j^\dagger c_l c_k c_p c_q^\dagger|\Phi_0\rangle \quad \rightarrow (-1)$$

$$B: \quad \langle\Phi_0|c_u^\dagger c_v^\dagger c_s c_r c_i^\dagger c_j^\dagger c_l c_k c_p c_q^\dagger|\Phi_0\rangle \quad \rightarrow (+1)$$

According to the rules for contracting operators in a normal-ordered product (Sect. 5.1), contraction scheme (A) results in an overall phase (-1), being absent in (B). Obviously, the necessity for resorting to the original analytical expression in order to determine the overall phase is a nuisance. Fortunately, the phase can be obtained directly at the diagrammatic level, as will be explained in the following.

Let us consider a succession of fermion lines within a given diagram forming a **closed loop** as schematically depicted in Fig. 6.3: The wiggly interaction line on the right-hand side represents the term $V_{xvyw}c_x^\dagger c_v^\dagger c_w c_y$ or (likewise) $V_{vxwy}c_v^\dagger c_x^\dagger c_y c_w$, where the time arguments have been dropped for brevity. Within the original $\hat{\boldsymbol{T}}$ product, the latter term can be rewritten according to

$$V_{vxwy}c_v^\dagger c_x^\dagger c_y c_w \quad \rightarrow \quad V_{vxwy}c_v^\dagger c_w c_x^\dagger c_y$$

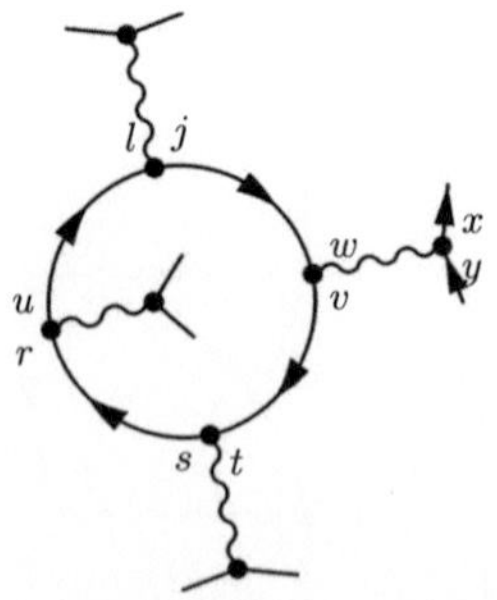

Fig. 6.3 Schematic representation of a closed loop

so that now the two operators $c_v^\dagger$, c_w "attached" to the loop are next to each other. Proceeding in the same way for the other terms involved in the closed loop, we arrive—without sign change—at the following form of the operator product:

$$c_u^\dagger c_r c_s^\dagger c_t c_v^\dagger c_w c_j^\dagger c_l$$

Note that the order of the $\hat{H}_I(t_\nu)$ terms within the $\hat{\mathcal{T}}$ product can arbitrarily be changed. Now the contractions for the three innermost pairs of operators can be taken without any sign change,

$$c_u^\dagger \overbrace{c_r c_s^\dagger} \overbrace{c_t c_v^\dagger} \overbrace{c_w c_j^\dagger} c_l$$

yielding $iG_{rs}^0 iG_{tv}^0 iG_{wj}^0$ according to Eq. (5.14). By contrast, the remaining contraction of the first and last operator introduces a minus sign, $c_u^{\dagger\cdot} c_l^{\cdot} = -c_l^{\cdot} c_u^{\dagger\cdot} = -iG_{lu}^0$. This demonstration can readily be generalized to closed loops of any length, and we may formulate the result as the following rule:

Remark 5:
Each closed loop in a given diagram gives rise to a factor of (-1).

There is one closed loop in diagram (A), and none in diagram (B), that is, $L_A = 1$, $L_B = 0$.

In each diagram, there is one **continuous line**, beginning at the lower vertex (q, t') and ending at the upper vertex (p, t), as depicted in Fig. 6.4. A consideration similar to the case of the closed loops shows that no sign change arises here. For example, the operators attached to the continuous line above can be ordered within the original $\hat{\mathcal{T}}$ product according to

$$c_u^\dagger c_r c_i^\dagger c_l c_v^\dagger c_s c_p c_q^\dagger$$

Obviously, c_p can be moved to the left-hand side without effecting a sign change, and then all contractions can be performed in their standard $cc^\dagger$ form.

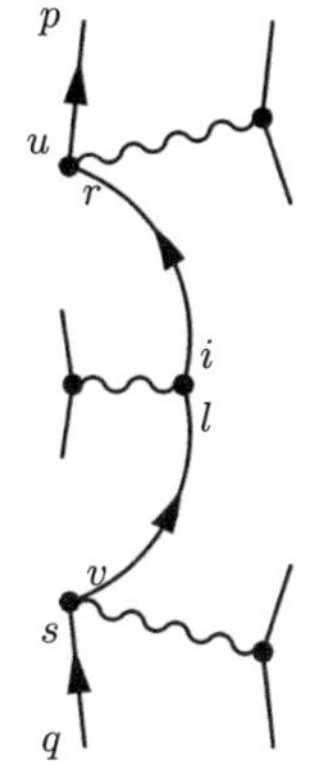

Fig. 6.4 Schematic representation of a continuous fermion line

This completes the diagrammatic approach to the original perturbation expansion of the electron propagator. The diagrammatic formulation can be cast in the following set of diagram rules:

Feynman Diagram Rules for the Electron Propagator

(F1) To generate the nth order contribution to the electron propagator, $G^{(n)}_{pq}(t, t')$, draw all topologically distinct connected diagrams with n wiggly interaction lines and $2n + 1$ [$= (4n - 2)/2 + 2$] directed free fermion or G^0-lines, where the first G^0-line begins at the outer vertex (q, t') and the last one ends at the outer vertex (p, t). At any wiggly interaction line, two G^0-lines begin and two lines end, as depicted in the graph below:

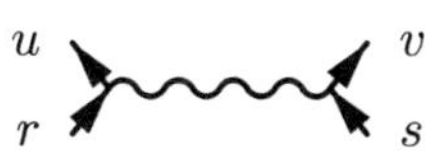

$\equiv V_{uvrs}$

$\equiv iG^0_{pq}(t, t')$

(F2) To evaluate a given diagram, assign one-particle indices and time arguments to the interaction lines (inner vertices), thereby defining the one-particle indices and time arguments of the free fermion lines. The arrows fix the order of the one-particle indices and time arguments in the G^0-functions. Replace the graphical symbols by the corresponding analytical expressions, V_{uvrs} and $G^0_{rs}(t_i, t_j)$, respectively. In the case of a G^0-function with equal time arguments, the limit $G^0(t_i, t_i^+)$ applies according to *Remark 1*.

(F3) Sum over indices and integrate over time arguments of the inner vertices.

(F4) Apply the phase factor i^n, arising according to

$$\left.\begin{array}{ll} (-i) & \text{definition of } G \\ (-i)^n & n\text{-th order} \\ i^{2n+1} & G^0\text{-lines} \end{array}\right\} i^n$$

and multiply by a sign $(-1)^L$, where L is the number of closed loops.

6.2 Diagrams in the Hartree–Fock Representation

So far, we have confined ourselves to the case where the interaction hamiltonian consists only of the Coulomb part. The diagrammatic formulation can readily be extended to the general form

$$\hat{H}_I = \hat{W} + \hat{V}$$

where

$$\hat{W} = \sum w_{rs} c_r^\dagger c_s$$

is a (non-diagonal) one-particle operator (see Eq. 4.3). As a graphical symbol associated with the one-particle interaction, we use the **one-particle interaction cross** with one entry and one exit,

$$\equiv w_{rs} \tag{6.4}$$

According to

$$\frac{1}{n!}\hat{H}_I^n = \frac{1}{n!}(\hat{V} + \hat{W})^n = \frac{1}{n!}\sum_{\nu=0}^{n}\binom{n}{\nu}\hat{V}^\nu \hat{W}^{n-\nu}$$

a general nth order diagram may have ν wiggly interaction lines and $\mu = n - \nu$ crosses, $\nu = 0, \ldots, n$. Remark 4 can be generalized in an obvious way, as the $(\nu!\mu!)$ orderings of the wiggly lines and crosses are equivalent.

The diagram rules can easily be adapted to the general case. The first rule, addressing the generation of diagrams, can be restated explicitly as follows:

(F1') To generate the nth-order contribution $G_{pq}^{(n)}(t, t')$, draw all topologically distinct connected diagrams with ν wiggly interaction lines and $\mu = n - \nu$ interaction crosses, $\nu = 0, \ldots, n$. The number of free Green's function lines here is $2\nu + \mu + 1$; the first G^0-line begins at the outer vertex (q, t') and the last one ends at (p, t). At any wiggly interaction line, two G^0-lines begin and two end; at any interaction cross, one G^0-line begins and one ends.

Rules (F2) and (F3), dealing with the evaluation of diagrams, apply essentially in their original form; the phase factor of rule (F4) has to be adapted according to the number of G^0-lines $(2\nu + \mu + 1)$.

In first order, we now have three diagrams, namely the "oyster" and "tadpole" diagrams (5.32), and a diagram with a one-particle interaction (cross):

$$G^{(1)}_{pq}(t,t') \equiv$$ 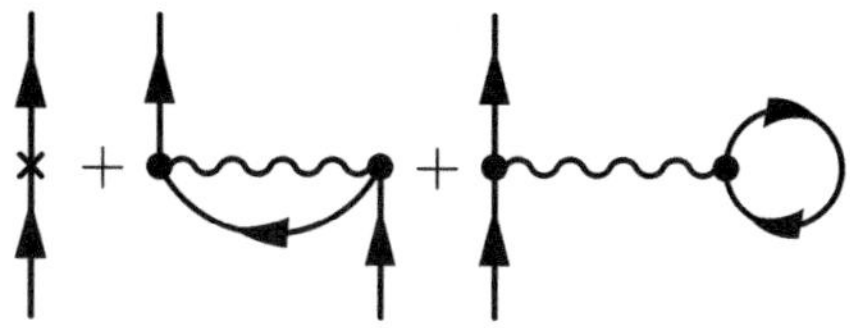 (6.5)

The corresponding analytical expression can be written in the compact form

$$G^{(1)}_{pq}(t,t') = \int dt_1 \sum_{rs} s_{rs}\, G^0_{pr}(t,t_1)G^0_{sq}(t_1,t') \tag{6.6}$$

where the matrix element

$$\begin{aligned} s_{rs} =& w_{rs} + \sum_{uv} iG^0_{vu}(t_1,t_1^+)V_{ruvs} - \sum_{uv} iG^0_{vu}(t_1,t_1^+)V_{rusv} \\ =& w_{rs} + \sum_{uv} V_{ru[sv]}\langle\Phi_0|c^\dagger_u c_v|\Phi_0\rangle \end{aligned} \tag{6.7}$$

is obtained by combining the first-order interaction terms of the cross, oyster, and tadpole diagrams.

The possibility of combining related cross, oyster, and tadpole terms into an effective one-particle interaction applies at higher order as well. This suggests to introduce a corresponding graphical symbol, referred to as "encircled cross,"

 $\equiv s_{rs}$ (6.8)

where the effective one-particle matrix element s_{rs} is given by Eq. (6.7). Now one may replace crosses with encircled crosses and discard any diagrams with an oyster or tadpole part.

Matters simplify considerably in the **HF representation** (Eqs. 4.4, 4.6), which will be supposed in the following. Here, the free Green's function is diagonal,

$$G^0_{pq}(t,t') = \delta_{pq}G^0_p(t,t') \tag{6.9}$$

which means that the free fermion lines can be specified by single one-particle indices. Using

$$-iG^0_{vu}(t_1,t_1^+) = \langle\Phi_0|c^\dagger_u c_v|\Phi_0\rangle = \delta_{uv}n_v$$

Equation (6.7) takes on the form

$$s_{rs} = w_{rs} + \sum_{v} V_{rv[sv]} n_v \tag{6.10}$$

According to Eq. (4.6), the matrix elements of the one-particle interaction part are given by

$$w_{rs} = -\sum_{v} V_{rv[sv]} n_v$$

which means that the effective first-order matrix elements simply vanish,

$$s_{rs} = 0 \tag{6.11}$$

As a consequence, any diagram containing a one-particle cross, or an oyster, or a tadpole part can be discarded. In particular, the first-order contribution to the electron propagator vanishes:

$$G^{(1)}_{pq}(t, t') = 0 \tag{6.12}$$

so that the perturbation expansion through second order becomes

$$G_{pq}(t, t') = G^0_{pq}(t, t') + G^{(2)}_{pq}(t, t') + O(3)$$

Figure 6.5 shows the two second-order Feynman diagrams contributing to $G^{(2)}_{pq}(t, t')$. Using the Feynman diagram rules, diagrams (A) and (B) can readily be translated into an analytical expression:

$$G^{(2)}_{pq}(t, t') = \sum_{r,u,v} \int_{-\infty}^{\infty} \mathrm{d}t_1 \int_{-\infty}^{\infty} \mathrm{d}t_2 \left(V_{pruv} V_{uvqr} - V_{prvu} V_{uvqr}\right)$$
$$G^0_p(t, t_1) G^0_u(t_1, t_2) G^0_v(t_1, t_2) G^0_r(t_2, t_1) G^0_q(t_2, t') \tag{6.13}$$

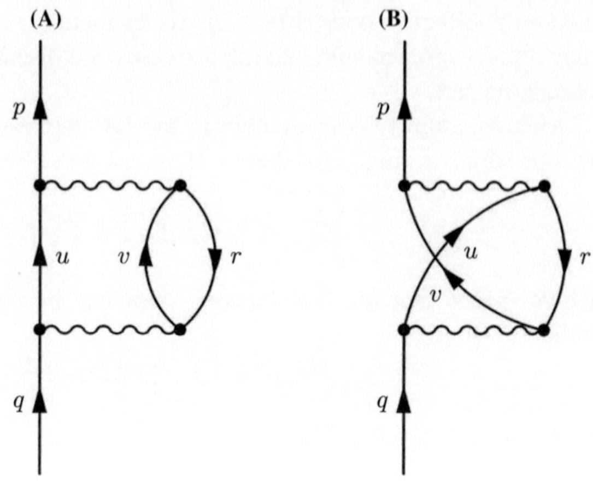

Fig. 6.5 The two second-order Feynman diagrams for the electron propagator assuming the HF representation

Here, the individual contributions (A) and (B) correspond to the two terms obtained by expanding the round brackets.

6.3 Diagrams in Abrikosov Form

The bracket on the right-hand side of Eq. (6.13) could have been written in a more compact form according to

$$(V_{pruv} - V_{prvu})V_{uvqr} = V_{pr[uv]}V_{uvqr}$$

where the antisymmetrized Coulomb integral $V_{pr[uv]}$ combines the integral terms of the (A) and (B) diagrams. This suggests to introduce a corresponding graphical symbol,

$$\equiv \pm V_{uv[rs]} \tag{6.14}$$

referred to as the (two-particle) **interaction dot**. As will be discussed below, it is possible to replace the original Feynman diagrams by diagrams in Abrikosov notation [1] (or simply Abrikosov diagrams) using interaction dots rather than wiggly lines. Obviously, an interaction dot does not determine the sign of the associated antisymmetrized Coulomb integral since the order of the two incoming or two outgoing fermion lines is ambiguous. As a consequence, the overall sign of an Abrikosov diagram can only be defined by resorting to one of the underlying Feynman diagrams.

Replacing the upper wiggly line by interaction dots, the two second-order Feynman diagrams (A) and (B) merge into one diagram shown in Fig. 6.6. Now we may proceed to replace also the lower wiggly line by an interaction dot. The resulting Abrikosov diagram (Fig. 6.7) combines the two second-order Feynman diagrams of Fig. 6.5 within a single diagram. However, replacing $V_{pr[uv]}V_{uvqr}$ by $V_{pr[uv]}V_{uv[qr]}$

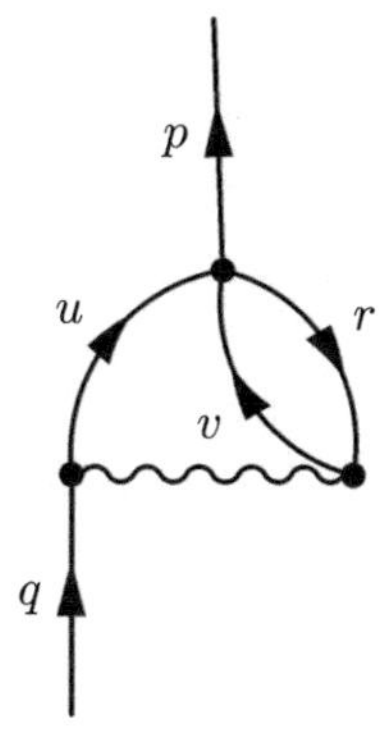

Fig. 6.6 Introducing the "interaction dot"

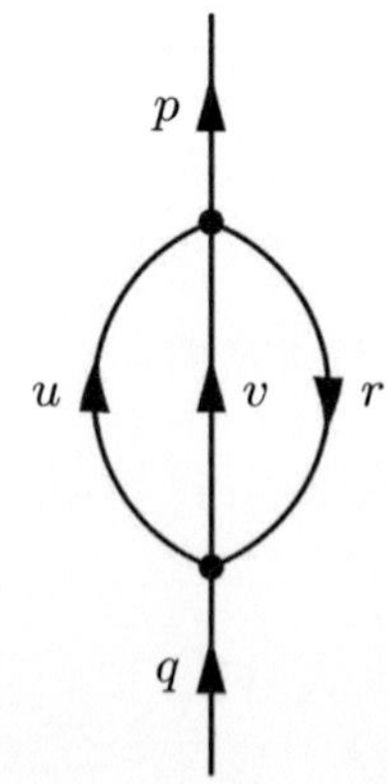

Fig. 6.7 Second-order Abrikosov diagram

in the analytical expression, (6.13), would result in $2 \times G^{(2)}_{pq}(t,t')$, as can readily be checked. To compensate for this "overcounting" of the original Feynman diagrams, a factor $\frac{1}{2}$ has to be introduced. The more general rule here is that whenever two equivalent free fermion lines occur in an Abrikosov diagram a factor $\frac{1}{2}$ has to be applied. The analytical expression associated with the second-order Abrikosov diagram (Fig. 6.7) reads

$$G^{(2)}_{pq}(t,t') = \frac{1}{2} \sum_{r,u,v} \int_{-\infty}^{\infty} dt_1 \int_{-\infty}^{\infty} dt_2 \, V_{pr[uv]} V_{uv[qr]}$$
$$G^0_p(t,t_1) G^0_u(t_1,t_2) G^0_v(t_1,t_2) G^0_r(t_2,t_1) G^0_q(t_2,t') \tag{6.15}$$

As already mentioned, the order within the incoming and outgoing pairs of free fermion lines is not determined for an interaction dot and can, in fact, be chosen at will. Then, for a given choice, e.g., the one adopted on the right-hand side of Eq. (6.15), the overall sign of the analytical expression has to be determined in such a way that one (and thus any) Feynman diagram, here (A) and (B), associated with the given Abrikosov diagram is reproduced correctly.

The Abrikosov notation can readily be established at higher order. Let us consider the following section of a given Feynman diagram (X),

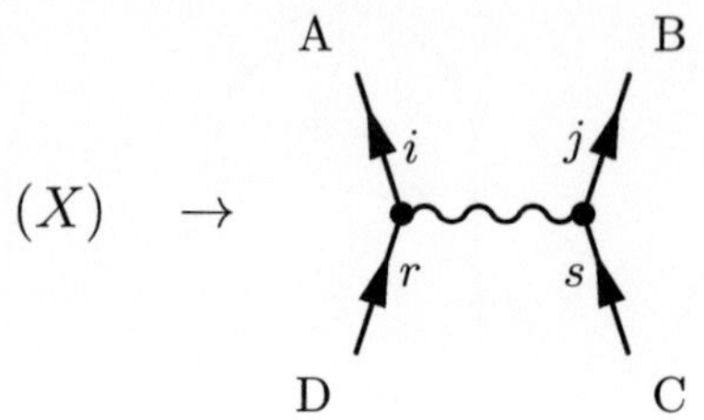

showing a wiggly interaction line ($\equiv V_{ijrs}$) with its four free fermion lines. The letters A, B, C, and D indicate the clipped connections to the rest of the diagram. The fermion line $D \to A$ on the left side of the vertex is part of an extended fermion line, being either a continuous line or a closed loop, and the same applies to $C \to B$ on the right-hand side. Exchanging the two incoming free fermion lines as indicated below,

$(X') \quad \to$

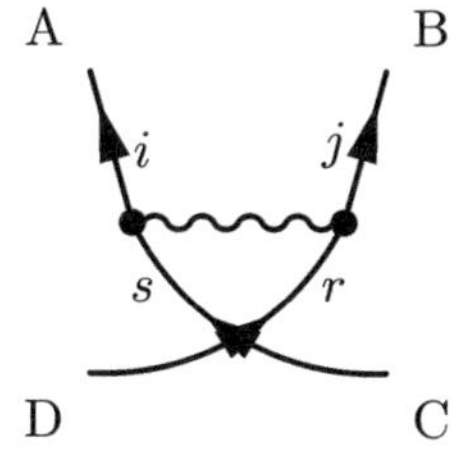

leads to another valid Feynman diagram (X'), where the wiggly interaction line corresponds to the contribution V_{ijsr}. In addition, the exchange introduces a sign change; that is, $(X')= (-1)\ (X)$. This is seen by inspecting what happens to the extended fermion lines in diagram (X) upon switching the entries in the vertex. The two extended fermion lines in diagram (X) can be

(*i*) a continuous line (e.g., on the left) and a closed loop (e.g., on the right);
(*ii*) two separate closed loops;
(*iii*) a common continuous line (containing both $D \to A$ and $C \to B$);
(*iv*) a common closed loop (containing both $D \to A$ and $C \to B$).

In all four cases, the exchange of the two incoming lines leads to a change in the number of closed loops by one unit. For example, case (*ii*) (two closed loops) turns into case (*iv*) (one common closed loop). In any event, (X) and (X') will differ in their overall sign, which means they can always be combined into one diagram by replacing the wiggly interaction line with the Abrikosov interaction dot:

$(X) + (X') \quad \to$

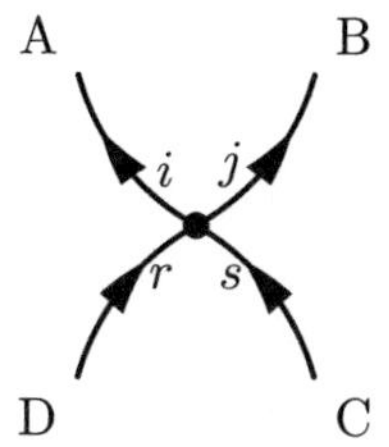

$\leftarrow \quad V_{ij[rs]}$ (6.16)

In such a way, one can replace successively all wiggly interaction lines by interaction dots, thereby combining each two Feynman diagrams into one Abrikosov diagram per interaction point. However, as we have seen in second order, there may result an overcounting of the original Feynman diagrams. This happens when two intermediate diagrams, say (Y) and (Y'), to be combined are identical. For example, let us consider the following section of a Feynman diagram (X),

$(X) \rightarrow$

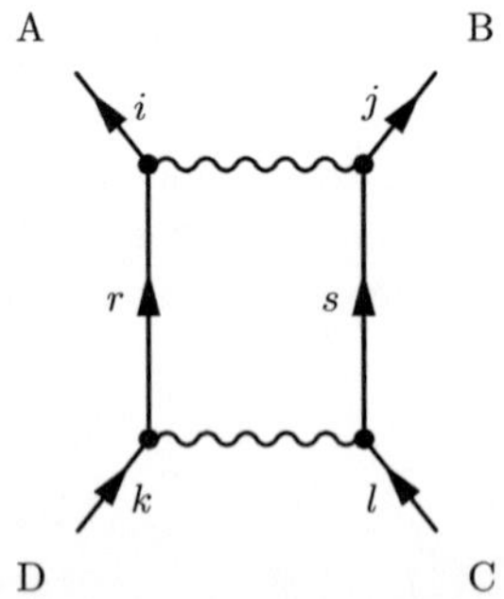

where as above the letters A,...,D indicate the clipped connections to the rest of the diagram. Exchanging the free fermion lines r, s in the upper wiggly interaction leads to diagram (X'),

$(X') \rightarrow$

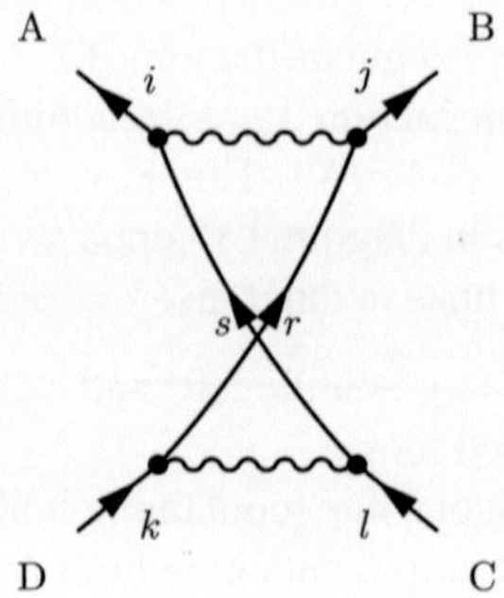

and (X) and (X') can be combined into one Abrikosov diagram according to (6.16). However, a corresponding exchange of the free fermion lines k, l in the lower wiggly interaction merely reproduces the original Feynman diagrams, as here $(X) \rightarrow (X')$ and $(X') \rightarrow (X)$. This means that replacing the second wiggly interaction by an Abrikosov dot leads to double counting, $2 \times ((X) + (X'))$. In the present case, the reason for double counting is that r and s are "equivalent" free fermion lines in the Abrikosov notation. The corresponding rule is to introduce a factor of $\frac{1}{2}$ for each pair of equivalent free fermion lines. It should be noted that pairs of equivalent fermion lines are not the only cause for double counting. At fourth order, for example, one encounters an Abrikosov diagram that is topologically invariant with respect to the permutation of two (inner) interaction points (diagram 8 in Fig. 9.1). Here, a factor of $\frac{1}{2}$ applies to compensate double counting of the associated Feynman diagrams. This means that the Abrikosov notation should not be used without recourse to the underlying Feynman diagrams. In the following, we compile the rules for drawing and evaluating diagrams in the Abrikosov form:

Rules for Abrikosov Diagrams

(A1) Draw all topologically distinct connected diagrams with n interaction dots

$$\equiv V_{uv[rs]}$$

and $2n+1$ directed (solid) free Green's function or G^0-lines starting at the outer vertex (q,t') and ending at the outer vertex (p,t). At each interaction dot, two G^0-lines start and two G^0-lines end; assign a time argument to each interaction dot.

(A2) Attach one-particle indices and time arguments to the G^0-lines; the arrows define the order of the time arguments. Replace the graphical symbols (free fermion lines and interaction dots) by the respective analytical expressions.

(A3) Sum over indices and integrate over time arguments of the inner vertices.

(A4) The overall phase of an Abrikosov diagram can only be fixed by inspecting one of the Feynman diagrams comprised in the Abrikosov diagram. The phase is to be adapted in such a way that this Feynman diagram is reproduced correctly by the Abrikosov expression.

(A5) Apply a factor of $\frac{1}{2}$ for each pair of (topologically) equivalent G^0-lines to compensate for double counting of Feynman diagrams. Double counting may arise for other reasons at fourth and higher order, and this possibility must be checked at the level of Feynman diagrams.

As we have already seen, there is a single second-order Abrikosov diagram (Fig. 6.7). Figure 6.8 shows the three Abrikosov diagrams of third order. Can one be sure that there are not more diagrams in third order? And how would one proceed in fourth and higher order? In the following, we will briefly sketch how Abrikosov diagrams can be constructed in an essentially systematic fashion through a given order of perturbation theory.

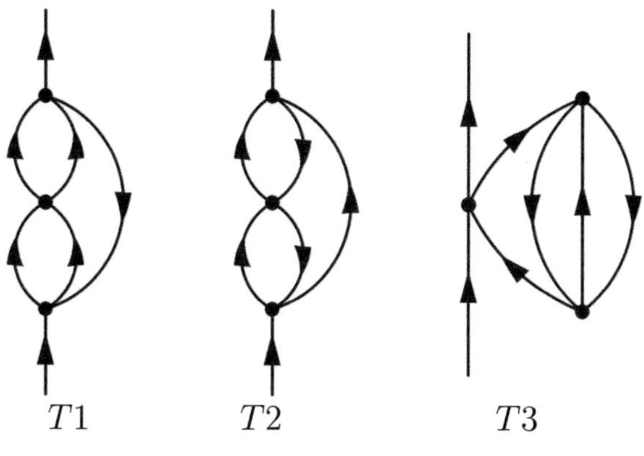

Fig. 6.8 Third-order Abrikosov diagrams

Systematic Construction of Abrikosov Diagrams

Specifically, we will consider "ordinary" diagrams, that is, diagrams beginning with a (1–3)-branching,

and ending with a corresponding (3–1)-junction,

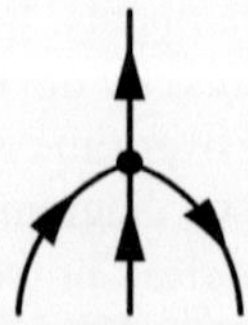

Obviously, the second-order diagram (Fig. 6.7) and the $T1$ and $T2$ third-order diagrams (Fig. 6.8) are ordinary diagrams, whereas the $T3$ diagram is of a different type to be addressed separately.

1. Begin a graph with a (1–3)-branching at the bottom: 

2. Add successively (inner) interaction points, and draw all possible extensions of the previous graphs using three basic operations:

   - junctions of two free fermion lines, 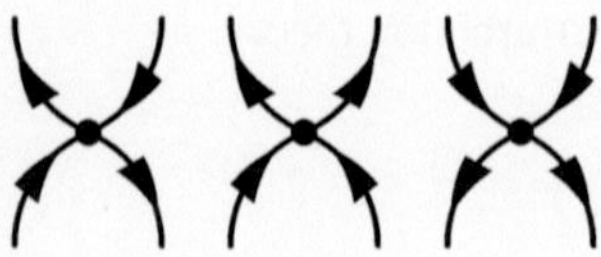

   - (1–3)-branchings, 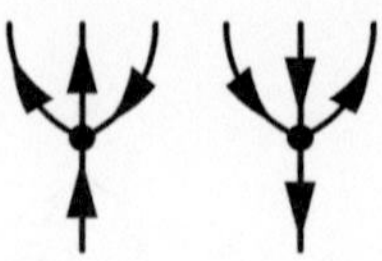

   - (3–1)-mergings, 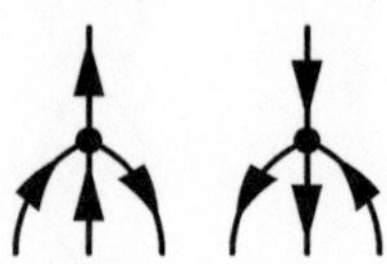

3. Inspect the graphs at order level $n-1$. Those graphs, which can be "closed" with a (3–1)-merging, give rise to ordinary nth order diagrams.

4. Finally, check whether there are topologically equivalent diagrams (e.g., two diagrams differing only in a permutation of interaction points) and discard redundant diagrams.

Let us use these recipes through third order. At the order level $n = 1$, there is just the initial (1–3)-branching

$n = 1:$

The complete set of graphs at second order is as follows:

$n = 2:$

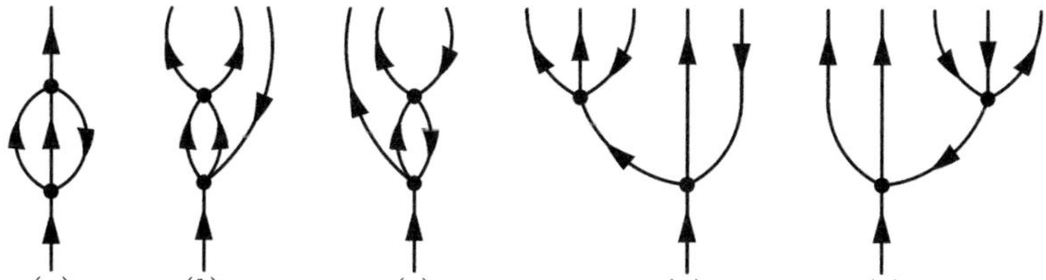

(a) (b) (c) (d) (e)

Here, graph (a) is the second-order diagram obtained by closing the graph of order level $n = 1$ with a (3–1)-junction. The graphs (b) and (c) result by joining two free fermion lines in the second interaction point, the arrow pairs being up-up and up-down, respectively; the two possible (1–3) branchings give rise to graphs (d) and (e).

Given the set of second-order graphs, we now may "harvest" the third-order Abrikosov diagrams. Obviously, graphs (b) and (c) can be closed with a (3–1)-merging, which gives rise to the two ordinary diagrams T1 and T2, respectively.

Yet, there is another possibility of generating a valid third-order diagram. At the top of graph (d), there are five free fermion lines on the loose, three being directed upwards, and two downwards. Obviously, two of the up fermion lines can be merged with the two down ones by an additional interaction point, as depicted below:

The result of such a (4–0)-merging is a viable third-order diagram, ending with the upwards directed fermion line not involved in the merger. Consider the case where the (4–0)-merging spares the last up fermion line (counted from the left). The resulting diagram looks like this:

Obviously, the diagram is topologically equivalent to the non-ordinary $T3$ diagram (as can be seen by shifting the lowest interaction point upwards and placing it at a level between the second and third interaction points). Thus, we have recovered all three third-order diagrams. What about a second possibility of closing graph (d), namely by omitting the first or second up fermion line in the (4–0)-merging? The third-order diagram resulting here is seen to be topologically equivalent to $T2$ (by permuting two inner interaction points). In a similar way, the two distinct (4–0)-mergings of graph (e) result in diagrams topologically equivalent to $T2$ and $T3$, respectively.

Let us note that the rules for the systematic construction of Abrikosov diagrams could be readily extended to comprise also (4–0)-merging and the corresponding (4–0)-branching,

Such an extension would allow one to exhaust the full set of Abrikosov diagrams, though the procedure becomes a lot more cumbersome. On the other hand, the (4–0)-merging/branching operations arise only in non-ordinary diagrams such as $T3$ and in ordinary diagrams with permuted inner vertices. The latter are redundant, and the non-ordinary diagrams can be dealt with in a more specific way, as described in the following.

The $T3$ diagram can be seen as being constructed in the following way: take the second-order diagram (Fig. 6.7), bend the two outer free fermion lines together such that the two outer vertices can be joined and fixed to an interaction point, and connect the interaction point to two free fermion lines. Analytically, this amounts to

$$T3 \equiv (-i) \int_{-\infty}^{\infty} dt_1 G_p^0(t, t_1) G_q^0(t_1, t') \sum_{rs} V_{pr[qs]} G_{rs}^{(2)}(t_1, t_1) \tag{6.17}$$

where joining the two outer vertices corresponds to equating the time arguments in the second-order Green's function, $t = t' = t_1$. Note that there is no need to specify the (infinitesimal) time-ordering of the first and second time argument, since

$G^{(2)}_{rs}(t_1, t_1^+) = G^{(2)}_{rs}(t_1^+, t_1)$, as will be further discussed in the subsequent chapter. Since $G^{(2)}_{rs}(t_1, t_1)$ is not time-dependent, Eq. (6.17) may be written in the form of a first-order diagram,

$T3 \quad \equiv$ 

$$\equiv \int_{-\infty}^{\infty} dt_1 G^0_p(t, t_1) G^0_q(t_1, t') X_{pq} \tag{6.18}$$

where

$$X_{pq} = (-i) \sum_{rs} V_{pr[qs]} G^{(2)}_{rs}(t_1, t_1) \tag{6.19}$$

represents an effective one-particle potential. As will be discussed in Sect. 8.2, X_{pq} is a contribution to the *static self-energy* (SSE) part. In the same way, any diagram of second and higher order can be inserted in a free fermion line in the form of an effective potential, symbolically depicted in the following graph:

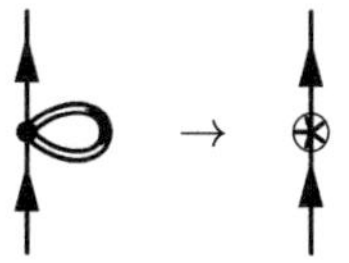

(6.20)

where the double line stands for any second- or higher-order contribution to $G_{rs}(t, t')$. Obviously, such SSE insertions can be made in any free fermion line within a given diagram. Diagrams with SSE insertions occur for the first time at third order, and they exhibit necessarily (4–0)-branchings or mergings. This allows us to define ordinary diagrams more strictly as diagrams which can be brought into a form without any (4–0)-branchings or mergings.

Matrix Representation of Abrikosov Diagrams

The systematic generation of higher-order Abrikosov diagrams can also be based on a simple matrix representation of diagrams, which, in turn, allows for an algorithm-based implementation [2]. The idea is to enumerate the vertices (interaction points and outer vertices), and let the matrix element (i, j) specify the number of fermion lines running from vertex i to vertex j. In the case of the second-order diagram, this mapping between diagrams and matrices reads

$\leftrightarrow$

	1	2	3	4
1	0	1	0	0
2	0	0	2	0
3	0	1	0	1
4	0	0	0	0

More specifically, the matrix representation of diagrams can be based on the following rules:

Matrix representation of Abrikosov Diagrams:

(M1) Label the $n + 2$ vertices by the numbers $1, 2, \ldots, n + 2$, beginning with the outer vertex at the bottom (1) and ending with the outer vertex at the top $(n + 2)$.

(M2) Consider the quadratic $(n + 2)$-dimensional matrices $\mathbf{\Gamma}$ with entries $\gamma_{ij} = 0, 1$, or 2 according to the number of free fermion lines that run from vertex i to vertex j.

Let us make a few rather obvious observations:

1. Any Abrikosov diagram can be mapped uniquely to a matrix according to (M1) and (M2): $D \longrightarrow \mathbf{\Gamma}(D)$
2. Different diagrams are mapped to different matrices; that is, the mapping is 1-1 on the domain $\mathbf{\Gamma}(\{D\})$.
3. The matrices $\mathbf{\Gamma}$ have the following properties:

   (a) The matrix elements in the first column and last row vanish.
   (b) Diagonal matrix elements vanish, $\gamma_{ii} = 0$.
   (c) First row and last column sum up to 1.
   (d) Rows and columns associated with inner vertices ($2 \leq i \leq n + 1$) sum up to 2.

4. A matrix fulfilling the properties (a)–(d) does not necessarily translate into a valid Abrikosov diagram. For example, the matrix

$$\begin{array}{|cccc|} \hline 0 & 0 & 0 & 1 \\ 0 & 0 & 2 & 0 \\ 0 & 2 & 0 & 0 \\ 0 & 0 & 0 & 0 \\ \hline \end{array}$$

   represents an unlinked diagram.
5. A permutation of the inner vertices in a diagram is reflected by a corresponding permutation of columns and rows in the matrix.

To determine the full set of nth-order Abrikosov diagrams, one may proceed as follows: Generate all $(n + 2)$-dimensional matrices according to rule (M2) complying with properties (a)–(d). Draw the corresponding diagrams according to rule (M1); discard invalid diagrams and redundant diagrams, e.g., different permutations of inner vertices.

In view of property (a), one may of course resort to simplified $(n + 1)$-dimensional matrices $\mathbf{\Gamma}'$, in which the first column and last row have been discarded.

Exercises

6.1 Evaluate the analytical expressions for the third-order diagrams $T1$ and $T3$ in Fig. 6.8.

6.2 Generate all third-order Abrikosov diagrams using the matrix representation discussed in Sect. 6.3.

References

1. Abrikosov AA, Gorkov LP, Dzyaloshinski IE (1963) Methods of quantum field theory in statistical physics. Prentice-Hall, Englewood Cliffs, N.J
2. Trofimov AB, Stelter G, Schirmer J (1999) J Chem Phys 111:9982

Chapter 7
Time-Ordered or Goldstone Diagrams

The analytical expressions deriving from the Feynman or Abrikosov diagrams are not yet entirely explicit, since, at nth order, they involve an n-fold time integral with regard to the time arguments of the inner vertices. The evaluation of these integrals, involving products of time-dependent free Green's functions, is rather cumbersome. In the so-called energy (or ω-) representation, the time variables and time integrations are replaced by energy variables and integrations as the result of appropriate Fourier transformations. As will be discussed in Sect. 7.1, the ω-representation is somewhat simpler than the original time representation, but does not solve the integration problem, as one is still left with n ω-integrations. Fortunately, the diagrammatic formulation can be extended as to allow one to obtain the result of the inner time or ω-integrations in an explicit, albeit fragmented form. This is accomplished by inspecting the set of $(n+2)!$ time-ordered or Goldstone diagrams associated with a given nth-order Feynman (or Abrikosov) diagram. The rules for drawing and evaluating the time-ordered diagrams are presented in Sect. 7.2. A derivation of the Goldstone rules is given in Appendix A.4.

7.1 Energy Representation of Diagrams

As discussed in Chap. 3, one can switch back and forth between the time and energy (or ω-) representations of the electron propagator via Fourier transformations,

$$G_{pq}(\omega) = \int\limits_{-\infty}^{\infty} \mathrm{d}(t-t')\, e^{i\omega(t-t')} G_{pq}(t,t')$$

J. Schirmer, *Many-Body Methods for Atoms, Molecules and Clusters*, Lecture Notes in Chemistry 94, https://doi.org/10.1007/978-3-319-93602-4_7

$$G_{pq}(t,t') = \frac{1}{2\pi}\int_{-\infty}^{\infty} \mathrm{d}\omega\, e^{-i\omega(t-t')} G_{pq}(\omega) \tag{7.1}$$

using here the fact that $G_{pq}(t,t')$ depends only on the difference $t - t'$ of the time arguments.

In a similar way, any diagram $D_{pq}(t,t')$ contributing to $G_{pq}(t,t')$ can be transformed to the ω-representation,

$$D_{pq}(\omega) = \int_{-\infty}^{\infty} \mathrm{d}(t-t') e^{i\omega(t-t')} D_{pq}(t,t') \tag{7.2}$$

Let us consider the second-order contribution, represented by the second-order Abrikosov diagram (Fig. 6.7, Eq. 6.15):

$$G^{(2)}_{pq}(\omega) = \frac{1}{2}\sum_{r,u,v} V_{pr[uv]} V_{uv[qr]} \int_{-\infty}^{\infty} \mathrm{d}(t-t') e^{i\omega(t-t')}$$
$$\int_{-\infty}^{\infty} \mathrm{d}t_1 \int_{-\infty}^{\infty} \mathrm{d}t_2\, G^0_p(t,t_1) G^0_u(t_1,t_2) G^0_v(t_1,t_2) G^0_r(t_2,t_1) G^0_q(t_2,t')$$

We may replace the time-dependent free Green's functions by their Fourier transforms,

$$G^0_r(t,t') = \frac{1}{2\pi}\int_{-\infty}^{\infty} \mathrm{d}\omega e^{-i\omega(t-t')} G^0_r(\omega) \tag{7.3}$$

where

$$G^0_r(\omega) = \frac{\bar{n}_r}{\omega - \omega_r + i\eta} + \frac{n_r}{\omega - \omega_r - i\eta} \tag{7.4}$$

This leads to the following expression,

$$G^{(2)}_{pq}(\omega) = \frac{1}{2}\sum_{r,u,v} V_{pr[uv]} V_{uv[qr]} \int_{-\infty}^{\infty} \mathrm{d}(t-t') e^{i\omega(t-t')} \int_{-\infty}^{\infty} \mathrm{d}t_1 \int_{-\infty}^{\infty} \mathrm{d}t_2$$
$$\int_{-\infty}^{\infty} \frac{\mathrm{d}\omega_1}{2\pi} \cdots \int_{-\infty}^{\infty} \frac{\mathrm{d}\omega_5}{2\pi}\, e^{-i\omega_1(t-t_1)} e^{-i\omega_2(t_1-t_2)} e^{-i\omega_3(t_1-t_2)} e^{-i\omega_4(t_2-t_1)} e^{-i\omega_5(t_2-t')}$$
$$G^0_p(\omega_1) G^0_u(\omega_2) G^0_v(\omega_3) G^0_r(\omega_4) G^0_q(\omega_5)$$

comprising altogether three time and five ω-integrations. The three time integrations and three of the ω-integrations can be successively performed as described in the following.

1. Integrate over inner time arguments t_1, t_2:

$$\frac{1}{2\pi}\int_{-\infty}^{\infty} \mathrm{d}t_1 e^{it_1(\omega_1-\omega_2-\omega_3+\omega_4)} = \delta(\omega_1-\omega_2-\omega_3+\omega_4)$$

$$\frac{1}{2\pi}\int_{-\infty}^{\infty} \mathrm{d}t_2 e^{it_2(\omega_2+\omega_3-\omega_4-\omega_5)} = \delta(\omega_2+\omega_3-\omega_4-\omega_5)$$

   The delta functions resulting here impose energy conservation for the ω-variables at each inner vertex:

$$\omega_1+\omega_4 = \omega_2+\omega_3 \tag{7.5}$$

$$\omega_2+\omega_3 = \omega_4+\omega_5 \tag{7.6}$$

   This, in turn, implies

$$\omega_1 = \omega_5 \tag{7.7}$$

2. The two delta functions obtained in the first step allow one to perform two ω-integrations. Choosing ω_4 and ω_5 results in replacing ω_4 by $\omega_2+\omega_3-\omega_1$ and ω_5 by ω_1.
3. Integrate over $t-t'$:

$$\frac{1}{2\pi}\int_{-\infty}^{\infty} \mathrm{d}(t-t')e^{i(\omega-\omega_1)(t-t')} = \delta(\omega-\omega_1)$$

4. Integrate over ω_1, using the delta function arising in the preceding step, which means to replace ω_1 by ω.

   The final result of the foregoing "integral algebra" reads

$$G^{(2)}_{pq}(\omega) = \tfrac{1}{2}\sum_{r,u,v} V_{pr[uv]}V_{uv[qr]}I_{uvr}(\omega)G^0_p(\omega)G^0_q(\omega) \tag{7.8}$$

where

$$I_{uvr}(\omega) = \int\frac{\mathrm{d}\omega_1}{2\pi}\int\frac{\mathrm{d}\omega_2}{2\pi}\, G^0_u(\omega_1)G^0_v(\omega_2)G^0_r(\omega_1+\omega_2-\omega) \tag{7.9}$$

Note that here the original integration variables ω_2, ω_3 have been changed to ω_1, ω_2.

The procedure of replacing the time by ω-integrations can readily be generalized to arbitrary nth-order diagrams. This leads to the following modifications in the original diagram rules.

Feynman/Abrikosov Diagrams in Energy Representation

1. Assign ω-variables to the $2n + 1$ free fermion lines and use the corresponding energy representations, $G_r^0(\omega)$. Chose the (global) energy variable ω for the first (incoming) outer free fermion line ($G_q^0(\omega)$), and inner ω-variables, $\omega_1, \omega_2, \ldots$, for the other free fermion lines.
2. Use energy conservation at each of the n inner vertices: $\omega_i + \omega_{i'} = \omega_o + \omega_{o'}$, where i, i' (o, o') refer to incoming (outgoing) free fermion lines at a given vertex. These n conditions ensure that there are n independent inner ω-variables, $\omega_1, \ldots, \omega_n$, and the second (outgoing) outer free fermion line becomes $G_p^0(\omega)$.
3. Perform the n inner ω-integrations $\int \frac{d\omega_1}{2\pi} \ldots \int \frac{d\omega_n}{2\pi}$.

The choice of ω-variables in the second-order diagram is depicted below:

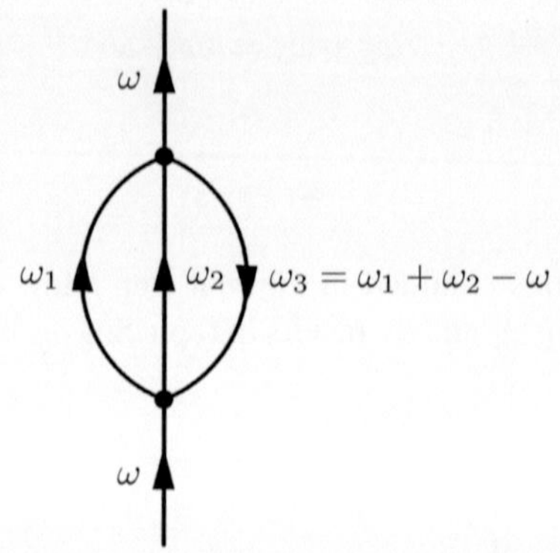

As the second-order diagram shows, the energy representation (7.8) is somewhat simpler than the original time representation since here the two outer free Green's functions $G_p^0(\omega)$ and $G_q^0(\omega)$ can be factored out. Still, one is left with two integrations, now involving energy rather than time variables.

The two ω-integrations in Eq. (7.9) can be evaluated using the calculus of complex integration. Let us first consider the integration over ω_2, involving the product $G_v^0(\omega_2)G_r^0(\omega_1 + \omega_2 - \omega)$ of two free Green's functions. Since this product behaves asymptotically as ω_2^{-2}, we may extend the integration contour by an infinite semicircle either in the upper or lower complex ω_2-plane. According to the general form (7.4) of the free Green's functions, the product $G_v^0(\omega_2)G_r^0(\omega_1 + \omega_2 - \omega)$ can be expanded into the linear combination of four products of each two simple ω_2-poles:

$$
\begin{aligned}
G_v^0(\omega_2)G_r^0(\omega_1+\omega_2-\omega) =& \bar{n}_v\bar{n}_r(\omega_2-\epsilon_v+i\eta)^{-1}(\omega_1+\omega_2-\omega-\epsilon_r+i\eta)^{-1}\\
&+n_v n_r(\cdots-i\eta)^{-1}(\cdots-i\eta)^{-1}\\
&+\bar{n}_v n_r(\cdots+i\eta)^{-1}(\cdots-i\eta)^{-1}\\
&+n_v\bar{n}_r(\cdots-i\eta)^{-1}(\cdots+i\eta)^{-1}
\end{aligned}
$$

In the first and second term, both poles are located either in the lower or upper complex ω_2-plane, which means that they do not contribute to the ω_2-integral (as the contour can be chosen such that both poles are excluded). The third and fourth term each have one pole in the upper and one pole in the lower complex plane, and the theorem of residues can be used to yield the following result:

$$
\begin{aligned}
I(\omega_1-\omega) &= \int\frac{d\omega_2}{2\pi}G_v^0(\omega_2)G_r^0(\omega_1+\omega_2-\omega)\\
&=\frac{(-i)\,\bar{n}_v n_r}{\omega_1-\omega-\epsilon_r+\epsilon_v-i\eta}+\frac{(+i)\,n_v\bar{n}_r}{\omega_1-\omega-\epsilon_r+\epsilon_v+i\eta}
\end{aligned}
$$

In the same manner, the remaining ω_1-integration,

$$
I_{uvr}(\omega)=\int\frac{d\omega_1}{2\pi}G_u^0(\omega_1)I(\omega_1-\omega)
$$

can be evaluated to give

$$
I_{uvr}(\omega)=\frac{n_r\bar{n}_u\bar{n}_v}{\omega+\epsilon_r-\epsilon_u-\epsilon_v+i\eta}+\frac{\bar{n}_r n_u n_v}{\omega+\epsilon_r-\epsilon_u-\epsilon_v-i\eta} \tag{7.10}
$$

Using this result in Eq. (7.8), the energy representation of the second-order electron propagator takes the explicit form

$$
G_{pq}^{(2)}(\omega)=G_p^0(\omega)G_q^0(\omega)\Sigma_{pq}^{(2)}(\omega) \tag{7.11}
$$

where

$$
\Sigma_{pq}^{(2)}(\omega)=\frac{1}{2}\sum_{r,u,v}V_{pr[uv]}V_{uv[qr]}\left(\frac{n_r\bar{n}_u\bar{n}_v}{\omega+\epsilon_r-\epsilon_u-\epsilon_v+i\eta}+\frac{\bar{n}_r n_u n_v}{\omega+\epsilon_r-\epsilon_u-\epsilon_v-i\eta}\right) \tag{7.12}
$$

As will be discussed in the ensuing Chap. 8, the quantity $\Sigma_{pq}^{(2)}(\omega)$ represents the second-order self-energy part. Note that $\Sigma_{pq}^{(2)}(\omega)$ is a sum of simple poles located in the lower (first part) and upper (second part) complex ω-plane.

7.2 Performing the Inner Time Integrations

As we have seen in the preceding section, the energy representation of the Feynman diagrams has simplified the integration problem but not solved it: one still has to deal with an n-fold ω-integration for each nth-order diagram. In the following, we will address a diagrammatic technique, by which the result of the inner time or ω-integrations is derived directly from the so-called time-ordered or Goldstone diagrams (named after Ref. [1]) associated with a given Feynman (or Abrikosov) diagram. Somewhat surprisingly, the concept of time-ordered diagrams can hardly be considered as widely known. While the rules for drawing and evaluating Goldstone diagrams can be found in the literature (see Refs. [2, 3]), actual derivations seem to be missing. For a demonstration of how time-ordered diagrams come into play, we begin with the simple case of the first-order Feynman diagram for a one-particle interaction. Then, the general rules for drawing and evaluating the time-ordered diagrams will be presented and applied to the second-order Abrikosov diagram for the electron propagator. A stringent derivation of the Goldstone diagram rules is given in Appendix A.4.

Fourier and Internal Time Integrations in a Simple Example

Let us consider the simple Feynman diagrams involving one-particle interactions only, shown in Fig. 7.1. Here, as can be seen by an analysis analogous to Sect. 7.1, the energy representation leads directly to a final analytical expression without inner ω-integrations. In time representation, on the other hand, the nth-order analytical expression involves n inner time integrations. In first order, the energy representation simply reads

$$D(\omega) = w_{pq} G_p^0(\omega) G_q^0(\omega) \tag{7.13}$$

where w_{pq} denotes the matrix elements of the one-particle interation. In time representation, on the other hand, the diagram gives rise to the expression

$$D(t, t') = \int_{-\infty}^{\infty} \mathrm{d}t_1 w_{pq} G_p^0(t, t_1) G_q^0(t_1, t') \tag{7.14}$$

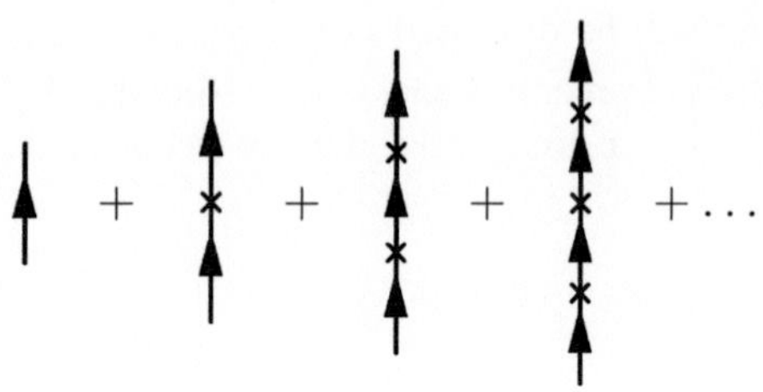

Fig. 7.1 Diagrammatic PT expansion for a one-particle interaction

To make contact with the Goldstone formulation, we will dwell on the latter expression and evaluate $D(\omega)$ from $D(t, t')$. However, rather than applying the usual form of the Fourier transform,

$$D(\omega) = \int_{-\infty}^{\infty} \mathrm{d}(t-t') e^{i\omega(t-t')} D(t, t') \tag{7.15}$$

based on the fact that $D(t, t')$ depends only on the difference $t - t'$ of the outer time arguments, we shall now employ two independent Fourier transforms for each time variable t, t':

$$D(\omega_1, \omega_2) = \int_{-\infty}^{\infty} \mathrm{d}t \int_{-\infty}^{\infty} \mathrm{d}t' \, e^{i\omega_1 t} e^{-i\omega_2 t'} D(t, t') \tag{7.16}$$

While we now have to deal with two additional "outer" time integrations in addition to the inner time integration in (7.14), the three time arguments can be treated on a similar footing, which allows for a more systematic evaluation.

Let us briefly establish the relation between these two Fourier transform variants in a more general way. Let $f(t - t')$ be a general function of the time difference and

$$f(\omega) = \int_{-\infty}^{\infty} \mathrm{d}(t-t') \, e^{i\omega(t-t')} f(t-t') \tag{7.17}$$

denote the Fourier transform with respect to $t - t'$. The separate Fourier transformations

$$f(\omega_1, \omega_2) = \int_{-\infty}^{\infty} \mathrm{d}t \int_{-\infty}^{\infty} \mathrm{d}t' \, e^{i\omega_1 t} e^{-i\omega_2 t'} f(t-t') \tag{7.18}$$

define a function of two variables ω_1 and ω_2. The two time integrations in Eq. (7.18) can readily be performed using the following transformation of the time variables:

$$t_1 = t - t', \quad t_2 = t + t' \tag{7.19}$$

or

$$t = \tfrac{1}{2}(t_1 + t_2), \quad t' = \tfrac{1}{2}(t_2 - t_1) \tag{7.20}$$

With the new variables, the integral (7.18) becomes separable:

$$
\begin{aligned}
f(\omega_1,\omega_2) &= \tfrac{1}{2}\int_{-\infty}^{\infty} \mathrm{d}t_1\, e^{i\frac{\omega_1+\omega_2}{2}t_1} f(t_1) \int_{-\infty}^{\infty} \mathrm{d}t_2\, e^{i\frac{\omega_1-\omega_2}{2}t_2} \\
&= f(\frac{\omega_1+\omega_2}{2}) \int_{-\infty}^{\infty} \frac{\mathrm{d}t_2}{2} e^{i\frac{\omega_1-\omega_2}{2}t_2} \\
&= 2\pi\delta(\omega_1-\omega_2) f(\omega_1)
\end{aligned} \tag{7.21}
$$

This establishes a simple relation between $f(\omega_1, \omega_2)$ and $f(\omega_1)$. The rule to obtain $f(\omega)$ from $f(\omega_1, \omega_2)$ is to discard the factor $2\pi\delta(\omega_1 - \omega_2)$, and replace both ω_1 and ω_2 by ω.

The two-variable Fourier transform for the first-order diagram (7.14) takes on the form

$$
D(\omega_1,\omega_2) = w_{pq} \int_{-\infty}^{\infty} \mathrm{d}t \int_{-\infty}^{\infty} \mathrm{d}t' \int_{-\infty}^{\infty} \mathrm{d}t_1\, e^{i\omega_1 t} e^{-i\omega_2 t'}\, G_p^0(t,t_1) G_q^0(t_1,t') \tag{7.22}
$$

There are $3! = 6$ possible orderings of the three time arguments t, t', t_1, e.g., $t > t_1 > t'$. This means that the original threefold time integral can be split into six distinct parts by performing the integration in the respective time-ordered ways. Obviously, this partitioning of the time integrations in the original Feynman diagram can be visualized by the **time-ordered diagrams** in Fig. 7.2. Here, a vertical time axis is assumed, larger times being placed above smaller ones.

The important aspect of the time-ordered (or Goldstone) diagrams is that the result of the time-ordered integration can be derived directly from the respective diagram. As a demonstration, let us perform the integration according to time-ordering (a):

$$
\begin{aligned}
D^{(a)}(\omega_1,\omega_2) = w_{pq} \int_{-\infty}^{\infty} \mathrm{d}t \int_{-\infty}^{\infty} \mathrm{d}t_1 \int_{-\infty}^{\infty} \mathrm{d}t'\, e^{i\omega_1 t}\, e^{-i\omega_2 t'} \\
G_p^0(t,t_1) G_q^0(t_1,t')\theta(t-t_1)\theta(t_1-t')
\end{aligned} \tag{7.23}
$$

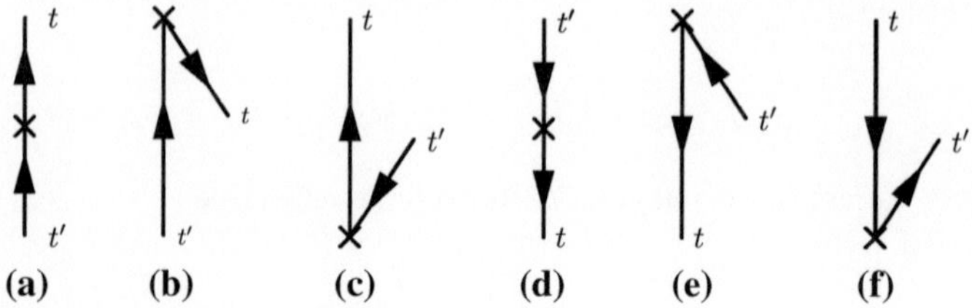

Fig. 7.2 Time orderings of the first-order one-particle diagram

Here, the time-ordering $t > t_1 > t'$ is assured by the product $\theta(t - t_1)\theta(t_1 - t')$ of step functions in the integrand. Recalling that $G^0_p(t, t_1)$ is given by

$$G^0_p(t, t_1) = -i\theta(t - t_1)e^{-i(\epsilon_p - i\eta)(t - t_1)}\bar{n}_p + i\theta(t_1 - t)e^{-i(\epsilon_p + i\eta)(t - t_1)}n_p$$

we see that the first part is compatible with time-ordering (a), whereas the second part is projected out according to $\theta(t_1 - t)\theta(t - t_1) \equiv 0$:

$$G^0_p(t, t_1) \rightarrow -i\theta(t - t_1)e^{-i(\epsilon_p - i\eta)(t - t_1)}\bar{n}_p$$

Analogously, the second part of $G^0_q(t_1, t')$ is to be discarded, that is,

$$G^0_q(t_1, t') \rightarrow -i\theta(t_1 - t')e^{-i(\epsilon_q - i\eta)(t_1 - t')}\bar{n}_q$$

This finding may readily be generalized. In a time-ordered diagram, thc direction (up or down) of a free fermion line has a distinct meaning: In upwards directed lines, the first time argument is larger than the second one, which means that only the particle part ($\bar{n}_r = 1$) of the free electron propagator comes into play; in downwards directed lines, the first time argument is smaller than the second one, and therefore, only the hole part ($n_r = 1$) survives. Accordingly, we will denote upwards and downwards directed lines in time-ordered diagrams as particle and hole lines, respectively.

We may now write Eq. (7.23) more explicitly as

$$\begin{aligned} D^{(a)}(\omega_1, \omega_2) =&(-i)^2\bar{n}_p\bar{n}_q w_{pq} \int_{-\infty}^{\infty} \mathrm{d}t e^{i(\omega_1 - \epsilon_p + i\eta)t} \\ & \int_{-\infty}^{t} \mathrm{d}t_1 e^{it_1(\epsilon_p - \epsilon_q)} \int_{-\infty}^{t_1} \mathrm{d}t' e^{-i(\omega_2 - \epsilon_q + i\eta)t'} \end{aligned} \tag{7.24}$$

where the various factors have been reordered according to their dependence on t, t_1, and t'. Note that the convergence factors for the t_1-integration have cancelled each other out. Now, the three integrations can be performed successively in the order t', t_1, and t. The t'-integration yields

$$\int_{-\infty}^{t_1} \mathrm{d}t' e^{-i(\omega_2 - \epsilon_q + i\eta)t'} = \frac{e^{it_1(\epsilon_q - \omega_2 - i\eta)}}{i(\epsilon_q - \omega_2 - i\eta)} \tag{7.25}$$

While the numerator enters the ensuing t_1-integration, the denominator constitutes a factorial part of the final result (see Eq. 7.29 below). As is stated more specifically in the diagram rules below, this factor can directly be related to a horizontal line ("cut") above the t'-vertex. The horizontal line crosses the particle line with the index q and an auxiliary ω-line connecting the outer vertices t and t', which allows one to specify the entries (ϵ_q, ω) and their signs in the denominator.

Taking the numerator of Eq. (7.25) into account, we see that the ϵ_q terms drop out of the t_1-integral, which is consistent with the fact that the ϵ_q particle line ends at the inner vertex t_1. The t_1-integration can readily be evaluated yielding

$$\int_{-\infty}^{t} \mathrm{d}t_1 e^{it_1(\epsilon_p-\omega_2-i\eta)} = \frac{e^{it(\epsilon_p-\omega_2-i\eta)}}{i(\epsilon_p-\omega_2-i\eta)} \tag{7.26}$$

Again, the denominator, being part of the final result, can be associated with a cut, here above the t_1-vertex. The t-dependent numerator enters the final t-integration, eliminating the ϵ_p term. Moreover, also the convergence factor $e^{-\eta t}$ is cancelled, so that the t-integration simply generates a delta function,

$$\int_{-\infty}^{\infty} \mathrm{d}t e^{it(\omega_1-\omega_2)} = 2\pi\delta(\omega_1-\omega_2) \tag{7.27}$$

as required by the two-variable Fourier transform. Combining the results (7.25)–(7.27) gives

$$D^{(a)}(\omega_1,\omega_2) = 2\pi\delta(\omega_1-\omega_2)w_{pq}\frac{\bar{n}_p}{(\omega_2-\epsilon_p+i\eta)}\frac{\bar{n}_q}{(\omega_2-\epsilon_q+i\eta)} \tag{7.28}$$

from which the final result

$$D^{(a)}(\omega) = w_{pq}\frac{\bar{n}_p}{(\omega-\epsilon_p+i\eta)}\frac{\bar{n}_q}{(\omega-\epsilon_q+i\eta)} \tag{7.29}$$

for the single-variable Fourier transform is obtained.

Let us take a brief look at the second time-ordering $t_1 > t > t'$ represented by diagram $D^{(b)}$:

$$D^{(b)}(\omega_1,\omega_2) = w_{pq}\int_{-\infty}^{\infty}\mathrm{d}t_1\int_{\infty}^{t_1}\mathrm{d}t\int_{\infty}^{t}\mathrm{d}t'$$
$$e^{i\omega_1 t}e^{-i\omega_2 t'}G_p^0(t,t_1)G_q^0(t_1,t')\theta(t_1-t)\theta(t-t') \tag{7.30}$$

As above, a product of θ-functions has been introduced in order to make the time-ordering $t_1 > t > t'$ explicit. Note that $\theta(t_1-t)\theta(t-t')$ implies $\theta(t_1-t')$, which makes apparent that $G_p^0(t,t_1)$ and $G_q^0(t_1,t')$ have to be replaced by their hole and particle part, respectively:

$$G_p^0(t,t_1) \rightarrow i\theta(t_1-t)e^{-i(\epsilon_p+i\eta)(t-t_1)}\, n_p$$
$$G_q^0(t_1,t') \rightarrow -i\theta(t_1-t')e^{-i(\epsilon_q-i\eta)(t_1-t')}\, \bar{n}_q$$

The three successive time integrations can easily be performed, yielding $D^{(b)}(\omega_1, \omega_2)$, from which the final result

$$D^{(b)}(\omega) = -w_{pq} \frac{\bar{n}_q}{(\omega - \epsilon_q + i\eta)} \frac{n_p}{(\epsilon_p - \epsilon_q)} \tag{7.31}$$

can be deduced. As above, $D^{(b)}(\omega)$ is given essentially by a product of two denominators, which can be related to cuts above the vertex t' and t. The latter cut does not cross the auxiliary ω-line, and accordingly, the second denominator is ω-independent. Note that $\epsilon_p \neq \epsilon_q$ because of the restriction $n_p\bar{n}_q = 1$. Accordingly, the infinitesimal in the ω-independent denominator (being originally of the form $(\epsilon_p - \epsilon_q - 2i\eta)^{-1}$) is dispensable and has been omitted in Eq. (7.31).

In a similar way, the remaining time-ordered diagrams $D^{(c)}, \dots, D^{(f)}$ can be evaluated (see Exercise 7.1). One may observe that whenever $t > t'$, such as in $D^{(a)}$, $D^{(b)}$, $D^{(c)}$, the ω-dependent denominators are of the form $(\omega \cdots + i\eta)^{-1}$ (representing poles in the lower complex ω-plane). By contrast, in the time-orderings $D^{(d)}$, $D^{(e)}$, $D^{(f)}$ with $t < t'$, the poles are located in the upper complex ω-plane.

It is interesting to contrast the compact result of Eq. (7.13) with the fragmented form associated with the use of time-ordered diagrams (see Exercise 7.3):

$$D(\omega) = D^{(a)}(\omega) + D^{(b)}(\omega) + \cdots + D^{(f)}(\omega) \tag{7.32}$$

Obviously, the product $G^0_p(\omega)G^0_q(\omega)$ can be expanded in a sum of four products of each two simple poles, two of which can readily be identified with $D^{(a)}(\omega)$ and $D^{(d)}(\omega)$. The other two products are of mixed type combining poles in the upper and lower complex plane. Here, the expansion into partial fractions generates each two contributions of the type $D^{(b,c)}(\omega)$ and $D^{(e,f)}(\omega)$.

Rules for Time-Ordered or Goldstone Diagrams

A general treatment of the inner time and Fourier integrations in the Feynman (or Abrikosov) diagrams is given in Appendix A.4, which establishes the following diagram rules:

(G1) A Feynman (or Abrikosov) diagram of nth order gives rise to $(n + 2)!$ time-ordered or Goldstone diagrams corresponding to the $(n + 2)!$ permutations of the two outer vertices t, t' and n inner vertices $t_1, \dots, t_n$. Draw an auxiliary line (ω-line) from vertex t to vertex t'.

(G2) In the time-ordered diagrams, the direction of the G^0-lines has the following meaning: upwards and downwards directed lines are associated with unoccupied (particle) and occupied (hole) orbitals, respectively. Introduce the corresponding restrictions in the one-particle indices.

(G3) Each (horizontal) 'cut' between two successive vertices gives rise to a denominator of the kind

$$(\sigma\omega + \varepsilon_k + \varepsilon_l + \cdots - \varepsilon_a - \varepsilon_b - \cdots + i\eta)^{-1}$$

Here, any G^0-line crossing the cut contributes an orbital energy: with positive sign for hole lines (occupied orbitals), $+\varepsilon_k, +\varepsilon_l, \dots$; with negative sign for particle lines (unoccupied orbitals), $-\varepsilon_a, -\varepsilon_b, \dots$. The energy variable ω arises, if the auxiliary line (ω-line) between the outer vertices ($t \to t'$) crosses the cut. Here, the sign $\sigma = 1$ applies, if the ω-line is directed downwards, and $\sigma = -1$, if the ω-line is directed upwards. Put $\sigma = 0$ if the ω-line does not cross the cut; note that in that case the infinitesimal $i\eta$ can be dropped.

(G4) Each hole line contributes a factor (-1). This leads to a total factor of $(-1)^{L+M}$, where L is the number of closed loops and M is the number of hole lines. The various i-factors total $+1$:

$$\left.\begin{array}{rl} -i & \text{from the definition of the GF} \\ (-i)^n & \text{from the } n\text{th-order perturbation theory} \\ i^{n+1} & \text{from the } n+1 \text{ cuts} \end{array}\right\} +1$$

Two important consequences of the Goldstone diagram analysis should be noted:

1. As confirmed by the above rules, the time integrations arising in a linked Feynman diagram can always be performed and lead to well-defined expressions. This means that the adiabatic switching functions $e^{-\epsilon|t_\nu|}$ originally accompanying the interaction terms are no longer needed to ensure convergence of the time integrations. This justifies the a priori limit $\epsilon \to 0$ supposed in Sect. 5.3.
2. The time-ordered diagrams associated with a given Feynman (or Abrikosov) diagram can be divided into two classes, I, II, according to the ordering of the external vertices t and t'. In the diagrams of class I ($t > t'$), any ω-dependent denominators are of the type $(\omega \cdots + i\eta)$. Obviously, the diagrams of class I are analytic in the upper complex ω-plane and, thus, contribute exclusively to the $\boldsymbol{G}^+$ part of the electron propagator. Conversely, the diagrams of class II ($t < t'$) are analytic in the lower complex plane and contribute exclusively to $\boldsymbol{G}^-$. In this way, the Goldstone diagrams establish independent diagrammatic PT expansions of the $\boldsymbol{G}^+$ and $\boldsymbol{G}^-$ parts.

The second-order Abrikosov diagram (Fig. 6.7) gives rise to $4! = 24$ time-ordered diagrams, as there are altogether four vertices, that is, each two outer and inner ones. Figure 7.3 shows the 12 diagrams of class I ($t > t'$); the 12 diagrams of class II ($t' > t$) are obtained by turning the class I diagrams upside down.

For a demonstration of the use of the Goldstone rules, we evaluate the second diagram in Fig. 7.3:

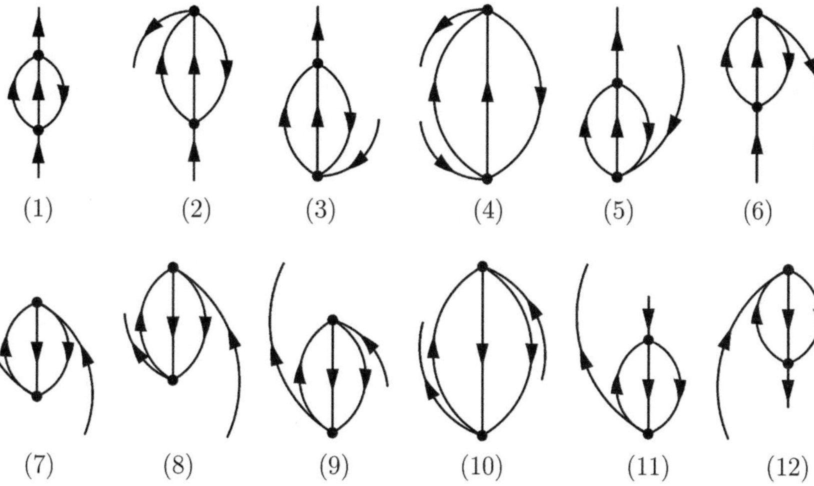

Fig. 7.3 Second-order Goldstone (time-ordered) diagrams in Abrikosov notation shown are the 12 diagrams with $t > t'$, contributing to $\boldsymbol{G}^+$

$A^{(2,2)} \equiv$

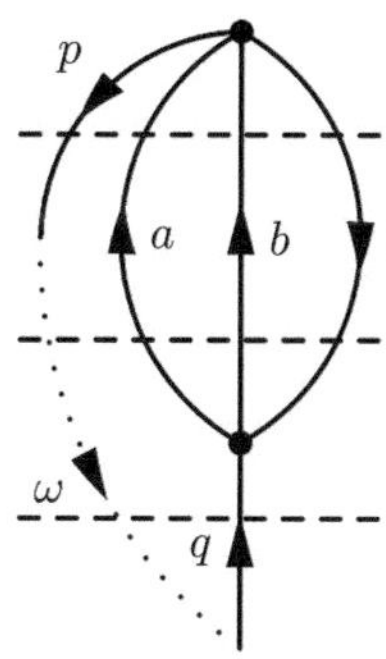

The three cuts between every two successive vertices have been depicted by dashed lines, while the directed dotted line represents the auxiliary ω-line, connecting the outer t, t' vertices. Moreover, one-particle indices have been assigned to the free fermion lines, where a, b and k denote particle and hole states, respectively. For the two outer fermion lines, denoted by the general (unspecific) indices p, q, the HF occupation numbers n_p and $\bar{n}_q$ are used to restrict p and q to hole and particle states, respectively. With the help of the Goldstone rules, diagram $A_{pq}^{(2,2)}$ can readily be expressed as follows:

$$A_{pq}^{(2,2)}(\omega) = (-1)^s \frac{n_p \bar{n}_q}{\omega - \epsilon_q + i\eta} \frac{1}{2} \sum_{a,b,k} \frac{V_{pk[ab]} V_{ab[qk]}}{(\epsilon_p + \epsilon_k - \epsilon_a - \epsilon_b)(\omega + \epsilon_k - \epsilon_a - \epsilon_b + i\eta)} \tag{7.33}$$

where the overall sign $(-1)^s$ still needs to be fixed. Note that the factor $\frac{1}{2}$ accounts for the pair of equivalent particle lines (a, b) according to the Abrikosov rule (A5). To determine the overall sign, one has to inspect one of the Feynman diagrams comprised in the Abrikosov diagram $A_{pq}^{(2,2)}$, for example:

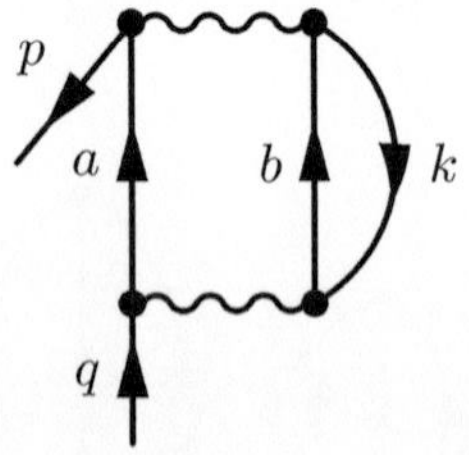

(7.34)

Here, we have two hole lines and one closed loop, so that the overall sign is $(-1)^3 = -1$. The overall sign in the analytical expression (7.33) has to be $(-1)^s = -1$, since with that choice the Feynman diagram (7.34) is correctly reproduced by (7.33).

Exercises

7.1 Evaluate the six time-ordered diagrams in Fig. 7.2 by performing the time integrations.

7.2 Adjust the Goldstone diagram rules to the case of a one-particle interaction diagrams and apply these rules to the first-order diagrams $(a), \ldots, (f)$ in Fig. 7.2.

7.3 Verify that the compact expression (7.13) is identical with the result deriving from the six first-order Goldstone diagrams (Exercise 7.2).

7.4 Supplementing Exercise 6.1, evaluate the energy representation of the diagrams $T1$ and $T3$ in Fig. 6.8

7.5 Evaluate the second-order Goldstone diagrams (7–10) shown in Fig. 7.3, which individually feature $1p/3p$-$2h$ interactions. Verify that in the sum of these diagrams, $S = (7) + (8) + (9) + (10)$ the $3p$-$2h$-denominators cancel out.

7.6 Compare the contributions of the compact second-order expression (7.11), (7.12) to $\boldsymbol{G}^+(\omega)$ with the 12 Goldstone diagrams in Fig. 7.3.

References

1. Goldstone J (1957) Proc R Soc A 239:267
2. Cederbaum LS (1973) Theoret Chim Acta 31:239
3. Cederbaum LS, Domcke W (1977) Adv Chem Phys 36:205

Part III
Approximations and Computational Schemes

Chapter 8
Self-Energy and the Dyson Equation

In the four preceding chapters, we have established the formalism of diagrammatic perturbation theory for the electron propagator, which allows one to derive successively higher-order contributions $\boldsymbol{G}^{(n)}(\omega)$. However, a finite perturbation expansion, e.g., through third order,

$$\boldsymbol{G}(\omega) = \boldsymbol{G}^{0}(\omega) + \boldsymbol{G}^{(2)}(\omega) + \boldsymbol{G}^{(3)}(\omega) + O(4) \tag{8.1}$$

does not result in a useful approximation scheme to determine the physical quantities of interest, that is, ionization energies, electron affinities, and the corresponding spectral factors. The reason is that the components $G_{pq}(\omega)$ are analytical functions, and a finite perturbation expansion does not recover the proper analytical structure (3.17), being a sum over simple poles, from which the desired information could be extracted. So the question is how to translate the diagrammatic perturbation expansion into a viable computational scheme. What is needed here is to sum the perturbation expansion, even if only partially, through infinite order. A possible path toward such infinite partial summations, recovering the proper analytical structure of the electron propagator, is provided by the Dyson equation, which we will address in this chapter.

8.1 Diagrammatic Approach to the Self-Energy

According to the diagram rules, the Feynman or Abrikosov diagrams of order $n \geq 1$ begin and end with a free fermion line. This suggests to write the perturbation expansion of the electron propagator in a form depicted graphically in Fig. 8.1. Here, the hatched symbol represents the quantity $\widetilde{\Sigma}_{pq}(t, t')$, referred to as **improper self-energy part**. $\widetilde{\Sigma}_{pq}(t, t')$ is defined as the sum ($n > 0$) of all diagrammatic contributions obtained by stripping off the respective two outer free fermion lines. The analytical expression corresponding to Fig. 8.1 reads

J. Schirmer, *Many-Body Methods for Atoms, Molecules and Clusters*, Lecture Notes in Chemistry 94, https://doi.org/10.1007/978-3-319-93602-4_8

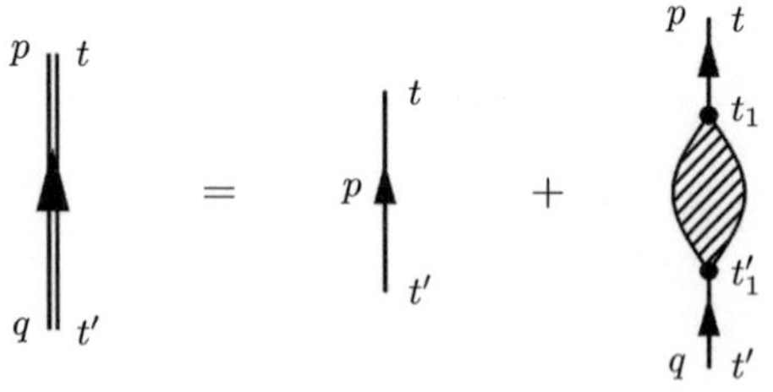

Fig. 8.1 Definition of the improper self-energy part

$$G_{pq}(t,t') = \delta_{pq} G_p^0(t,t') + \int \mathrm{d}t_1 \int \mathrm{d}t_1' \, G_p^0(t,t_1) \widetilde{\Sigma}_{pq}(t_1,t_1') G_q^0(t_1',t') \tag{8.2}$$

or, in ω-representation,

$$G_{pq}(\omega) = \delta_{pq} G_p^0(\omega) + G_p^0(\omega) \widetilde{\Sigma}_{pq}(\omega) G_q^0(\omega) \tag{8.3}$$

Here

$$\widetilde{\Sigma}_{pq}(\omega) = \int_{-\infty}^{\infty} \mathrm{d}(t-t') \, e^{i\omega(t-t')} \, \widetilde{\Sigma}_{pq}(t,t') \tag{8.4}$$

is the Fourier transform of $\widetilde{\Sigma}_{pq}(t,t')$. Using an obvious matrix notation, Eq. (8.3) can be written in the form

$$\boldsymbol{G}(\omega) = \boldsymbol{G}^0(\omega) + \boldsymbol{G}^0(\omega) \widetilde{\boldsymbol{\Sigma}}(\omega) \boldsymbol{G}^0(\omega) \tag{8.5}$$

To proceed to the definition of the less trivial **proper self-energy part**, let us consider the fourth-order diagram shown in Fig. 8.2. Obviously, this diagram can be separated into two parts by cutting a single free fermion line, namely the one connecting the two second-order fragments. In a similar way, any diagram can be characterized as being either *separable* or *non-separable* with respect to cutting a single free fermion line. This allows one to define a quantity $\Sigma_{pq}(t,t')$ in analogy to $\widetilde{\Sigma}_{pq}(t,t')$, but with the restriction to non-separable diagrams:

$$\Sigma_{pq}(t,t') \equiv \{\text{sum over non-separable contributions to } \widetilde{\Sigma}_{pq}(t,t')\} \tag{8.6}$$

$\Sigma_{pq}(t,t')$ is referred to as proper self-energy part, or simply, **self-energy**. Obviously, the improper self-energy part can be constructed from the proper one in the way shown graphically in Fig. 8.3, where the symbols designated Σ represent $\Sigma_{pq}(t,t')$. The geometrical-type series of "powers" of $\Sigma_{pq}(t,t')$ reflects the fact that diagrams can be multiply separable. The third term in the power series, for example, comprises all diagrams that decompose into three fragments upon cutting each two free fermion lines.

Fig. 8.2 A fourth-order Abrikosov diagram composed of two second-order diagrams

Fig. 8.3 Self-energy part $\Sigma_{pq}(t, t')$ as a non-separable contribution

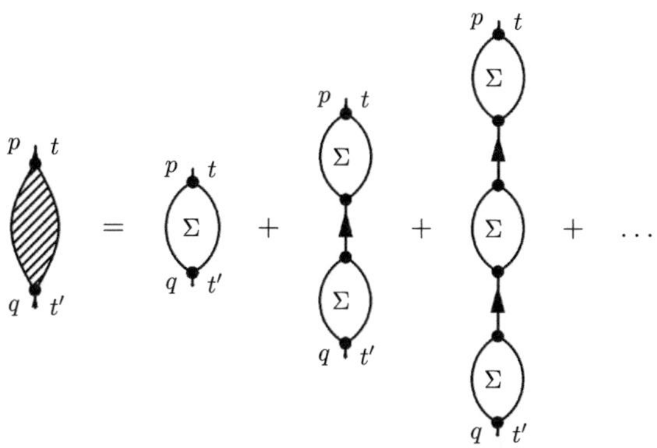

Fig. 8.4 Electron propagator as geometric series of powers of the self-energy part

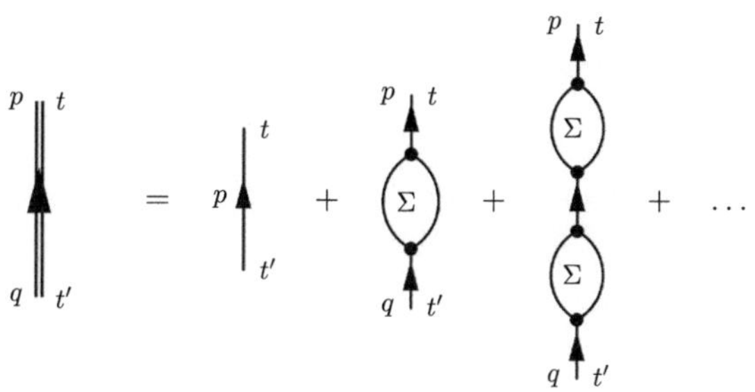

The geometrical series for $\widetilde{\Sigma}_{pq}(t,t')$ in Fig. 8.3 may be used in the representation of the electron propagator according to Fig. 8.1 or Eq. (8.2). The result, shown graphically in Fig. 8.4, can be written as follows:

$$G_{pq}(t,t') = G^0_{pq}(t,t') + \iint dt_1\, dt_1'\, G^0_p(t,t_1)\Sigma_{pq}(t_1,t_1')G^0_q(t_1',t') +$$
$$\sum_r \int \cdots \int dt_1\, dt_1'\, dt_2\, dt_2'\, G^0_p(t,t_1)\Sigma_{pr}(t_1,t_1')G^0_r(t_1',t_2)\Sigma_{rq}(t_2,t_2')G^0_q(t_2',t') + \cdots \tag{8.7}$$

The latter equation can be cast into a more compact, albeit implicit form:

$$G_{pq}(t,t') = G^0_{pq}(t,t') + \sum_r \iint dt_1\, dt_2\, G^0_p(t,t_1)\Sigma_{pr}(t_1,t_2)G_{rq}(t_2,t') \tag{8.8}$$

This is the famous Dyson equation [1], relating the electron propagator $G_{pq}(t,t')$ to the self-energy part $\Sigma_{pq}(t,t')$. Note that the explicit expansion of Eq. (8.7) results by solving the Dyson equation iteratively for $G_{pq}(t,t')$. A graphical representation of the Dyson equation is given in Fig. 8.5.

Using the energy representation,

$$\Sigma_{pq}(\omega) = \int_{-\infty}^{\infty} e^{i\omega(t-t')}\,\Sigma_{pq}(t,t')\,d(t-t') \tag{8.9}$$

and an obvious matrix notation, the Dyson equation can be cast in the compact form

$$\boldsymbol{G}(\omega) = \boldsymbol{G}^0(\omega) + \boldsymbol{G}^0(\omega)\boldsymbol{\Sigma}(\omega)\boldsymbol{G}(\omega) \tag{8.10}$$

The formal solution for $\boldsymbol{G}(\omega)$ is readily obtained, reading

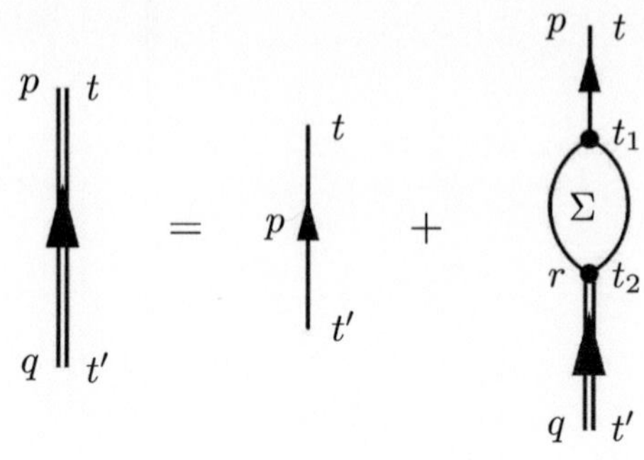

Fig. 8.5 Schematic representation of the Dyson equation

$$\boldsymbol{G}(\omega) = \left(\boldsymbol{G}^0(\omega)^{-1} - \boldsymbol{\Sigma}(\omega)\right)^{-1} \tag{8.11}$$

According to

$$\begin{aligned}\boldsymbol{G}(\omega) &= \left(\mathbf{1} - \boldsymbol{G}^0(\omega)\boldsymbol{\Sigma}(\omega)\right)^{-1} \boldsymbol{G}^0(\omega) \\ &= \boldsymbol{G}^0(\omega) + \boldsymbol{G}^0(\omega)\boldsymbol{\Sigma}(\omega)\boldsymbol{G}^0(\omega) + \ldots\end{aligned} \tag{8.12}$$

the right-hand side can be expanded as a geometrical series in powers of $\boldsymbol{\Sigma}(\omega)$, which is the energy representation analogue to Eq. (8.7).

The Dyson equation can be viewed as providing a formal definition of the self-energy part $\boldsymbol{\Sigma}(\omega)$, e.g., in the form

$$\boldsymbol{\Sigma}(\omega) = \boldsymbol{G}^0(\omega)^{-1} - \boldsymbol{G}(\omega)^{-1} \tag{8.13}$$

obtained from Eq. (8.10). Its usefulness, however, derives from the fact that there is a direct diagrammatic approach to $\boldsymbol{\Sigma}(\omega)$. An approximation to $\boldsymbol{\Sigma}(\omega)$ (obtained for example from a low-order diagrammatic expansion) will lead via the Dyson equation to an approximation for the electron propagator in the form of an infinite, if only partial, summation of terms in the original perturbation expansion. This may result in a viable computational scheme.

1. Diagram Rules for the Self-energy Part

The diagram rules for the self-energy part can readily be obtained by obvious modifications of the original Feynman diagram rules for the electron propagator:

Employ the original Feynman rules (F1)–(F4) or the Abrikosov rules (A1)–(A4) for the electron propagator $G_{pq}(t, t')$ with the following modifications:

- consider only diagrams that cannot be separated into two fragments by cutting one free fermion line;
- remove the two outer free fermion lines from the respective wiggly interaction line (interaction dots), while keeping their one-particle indices, p, q, in the interaction line (dot) expressions;
- the number of free fermion lines is $2n - 1$ rather than $2n + 1$. The overall phase i^n of rule (F4), however, still applies, because the two i-factors associated with the two outer free fermion lines persist.

The second-order Feynman diagram for $\Sigma_{pq}(t, t')$ shown in Fig. 8.6 derives from diagram (A) in Fig. 6.5. Its analytical expression reads

$$\Sigma_{pq}^{(2,A)}(t, t') = \sum_{r,u,v} V_{pruv} V_{uvqr} G_u^0(t, t') G_v^0(t, t') G_r^0(t', t) \tag{8.14}$$

Note the absence of time integrations, since there are only (two) external vertices in the second-order diagram. The second-order Abrikosov diagram for the self-energy, comprising both second-order Feynman diagrams, is shown in Fig. 8.7. The diagram can easily be translated into the following analytical expression:

$$\Sigma^{(2)}_{pq}(t,t') = \frac{1}{2}\sum_{r,u,v} V_{pr[uv]}V_{uv[qr]}G^0_u(t,t')G^0_v(t,t')G^0_r(t',t) \tag{8.15}$$

In the energy representation (Eqs. (7.8), (7.9)) of the second-order electron propagator, the G^0 functions associated with the two outer free fermion lines factorize, and $\Sigma^{(2)}_{pq}(\omega)$ is obtained by simply discarding those G^0-factors:

$$\Sigma^{(2)}_{pq}(\omega) = \frac{1}{2}\sum_{r,u,v} V_{pr[uv]}V_{uv[qr]}\int\frac{\mathrm{d}\omega_1}{2\pi}\int\frac{\mathrm{d}\omega_2}{2\pi}\,G^0_u(\omega_1)G^0_v(\omega_2)G^0_r(\omega_1+\omega_2-\omega) \tag{8.16}$$

The internal ω-integrations have been performed in Sect. 7.1 (cf. Eq. 7.10), and the resulting explicit expression for $\Sigma^{(2)}_{pq}(\omega)$ is given by Eq. (7.12). The rules presented in Sect. 7.1 for evaluating diagrams in the energy representation can easily be adapted to the case of the self-energy diagrams.

As discussed in Sect. 7.2, time-ordered or Goldstone diagrams can be used to derive directly explicit ω-dependent expressions for the electron propagator diagrams. The same technique can be applied to the self-energy diagrams. In an nth-order self-energy diagram, there are n vertices. Hence, the number of time-orderings is $n!$ rather than $(n+2)!$, which means a considerable reduction in comparison with the case of the electron propagator. The original diagram rule (G1) must be modified accordingly:

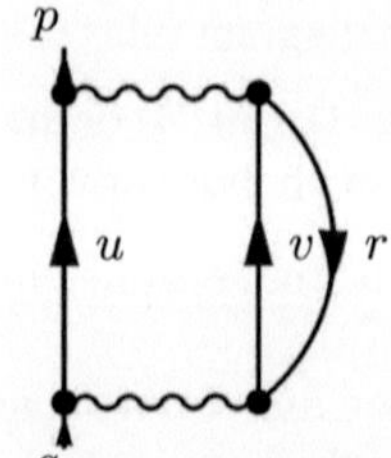

Fig. 8.6 Feynman diagram for the self-energy part in second order

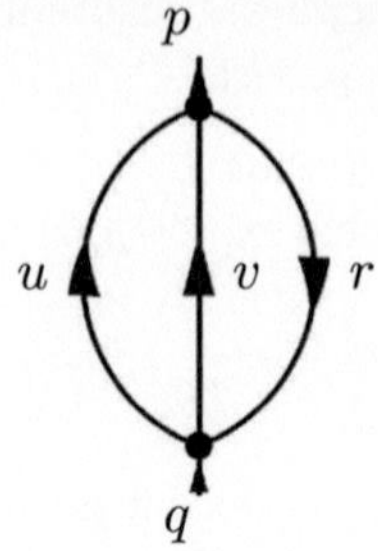

Fig. 8.7 Second-order Abrikosov diagram for the self-energy part

2. Time-Ordered Diagrams for the Self-energy Part

(G1') A Feynman (or Abrikosov) diagram of nth order gives rise to $n!$ time-ordered or Goldstone diagrams corresponding to the $n!$ permutations of the two outer vertices t, t' and $n-2$ inner vertices $t_1, \ldots, t_{n-2}$. Draw an auxiliary line (ω-line) from vertex t to vertex t'.

The other rules (G2)–(G4) apply in their original form. In particular, the overall phase factor remains to be $+1$, which comes about as follows:

$$
+1 \leftarrow \begin{cases} -i & \text{from the definition of the propagator} \\ (-i)^n & \text{from the } n\text{th-order perturbation theory} \\ i^{n-1} & \text{from the } N-1 \text{ cuts} \\ i^2 & \text{from the two external free fermion lines} \end{cases}
$$

As an example, we will apply the Goldstone analysis to evaluate $\Sigma^{(2)}_{pq}(\omega)$. The two time-ordered diagrams associated with the second-order Abrikosov diagram (Fig. 8.7) are shown in Fig. 8.8. Combining the analytical expressions deriving from the Goldstone rules yields the following result for the second-order self-energy:

$$
\Sigma^{(2)}_{pq}(\omega) = \frac{1}{2}\sum_{a,b,k} \frac{V_{pk[ab]}V_{ab[qk]}}{\omega + \epsilon_k - \epsilon_a - \epsilon_b + i\eta} + \frac{1}{2}\sum_{a,k,l} \frac{V_{pa[kl]}V_{kl[qa]}}{\omega + \epsilon_a - \epsilon_k - \epsilon_l - i\eta} \tag{8.17}
$$

where the indices a, b and k, l are restricted to particle and hole states, respectively. The latter expression is of course equivalent to the result in Eq. (7.12), as can be seen by renaming the summation indices.

The three third-order Abrikosov diagrams for the electron propagator shown in Fig. 6.8 give rise to the corresponding third-order self-energy diagrams in Fig. 8.9. These diagrams can readily be evaluated using the Goldstone analysis. Each diagram has six time-orderings, which is in striking contrast to the propagator case, where each third-order diagram gives rise to 120 Goldstone diagrams. The first three Goldstone diagrams belonging to $T1$ are shown in Fig. 8.10. Applying the above diagram rules,

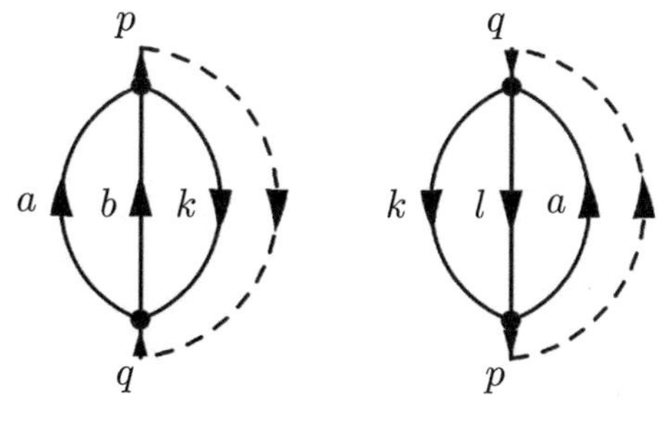

Fig. 8.8 Second-order Goldstone diagrams for the self-energy

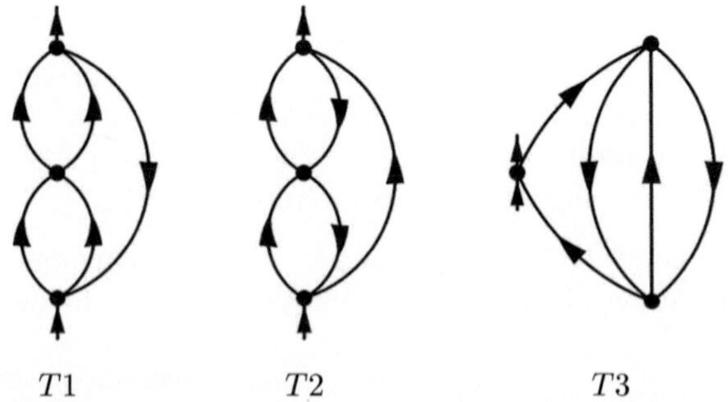

Fig. 8.9 Third-order Feynman/Abrikosov diagrams for the self-energy part

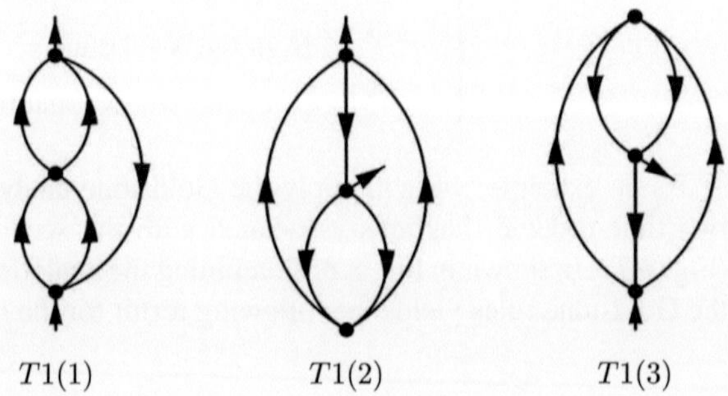

Fig. 8.10 The first three time-orderings (Goldstone diagrams) of the third-order self-energy diagram $T1$

the analytic expressions can easily be obtained (see Exercise 8.2); the results for $T1(1)$ and $T1(2)$ are given in Eqs. (9.13) and (9.16) of Sect. 9.1.

Both the $T1$ and $T2$ diagrams give rise to ω-dependent expressions. By contrast, the $T3$ diagram leads to constant (ω-independent) contributions. There is only one external vertex here, suspending the possibility of auxiliary ω-lines. In the corresponding $T3$ diagram for the electron propagator (Fig. 6.8), the two outer free fermion lines begin and end at the same vertex. The analytical expression is of the form given by Eq. (6.18). The self-energy contribution obtained by removing the two outer free fermion lines reads

$$\Sigma^{T3}_{pq}(t, t') = \delta(t - t')X_{pq}$$

in time representation and

$$\Sigma^{T3}_{pq}(\omega) = X_{pq}$$

in ω-representation, where X_{pq} is a constant given by Eq. (6.19). The finding that the third-order self-energy consists of ω-dependent and ω-independent (constant) contributions reveals a general analytical structure which will be addressed in the ensuing Sect. 8.2.

8.2 Analytical Properties of the Self-Energy

The self-energy can be written in the general form

$$\Sigma_{pq}(\omega) = \Sigma_{pq}(\infty) + M_{pq}(\omega) \tag{8.18}$$

Here, $M_{pq}(\omega)$ and $\Sigma_{pq}(\infty)$ are referred to as dynamical (ω-dependent) part and static (ω-independent) part, respectively. The notation $\Sigma_{pq}(\infty)$ used for the static self-energy part reflects the fact that $M_{pq}(\omega) \to 0$ for $\omega \to \infty$. Analogous to Eq. (3.17) for the electron propagator, there is a spectral representation for $M_{pq}(\omega)$,

$$M_{pq}(\omega) = M^+_{pq}(\omega) + M^-_{pq}(\omega) = \sum_{\nu} \frac{m_p^{(\nu)} m_q^{(\nu)*}}{\omega - \omega_\nu + i\eta} + \sum_{\mu} \frac{m_p^{(\mu)} m_q^{(\mu)*}}{\omega - \omega_\mu - i\eta} \tag{8.19}$$

The $M^{\pm}_{pq}(\omega)$ terms differ in the location of the poles, being in the lower and upper complex ω-plane, respectively. Like in the case of the electron propagator, $M^+_{pq}(\omega)$ and $M^-_{pq}(\omega)$ are referred to as $(N+1)$-particle (or affinity) and $(N-1)$-particle (or ionisation) part, respectively. Other than in the spectral representation of the electron propagator, the positions of the poles ω_λ and the amplitudes $m_p^{(\lambda)}$ cannot directly be related to physical quantities.

A general proof of Eq. (8.18) and the spectral representation (8.19) has been presented in Refs. [2, 3]. In principle, the analytical form of the self-energy part may be inferred from the spectral properties of $\boldsymbol{G}(\omega)$ via Eq. (8.13), which can be seen as a definition of $\boldsymbol{\Sigma}(\omega)$. To get an idea of the algebra here at work, one may consider a simple function modelled after the diagonal approximation to Eq. (8.13),

$$s(\omega) = \omega - \epsilon - g(\omega)^{-1} \tag{8.20}$$

where

$$g(\omega) = \sum_{k=1}^{m} \frac{p_k}{\omega - e_k}, \quad \sum p_k = 1 \tag{8.21}$$

serves as a surrogate for the electron propagator. As the reader may verify (see Exercise 8.3), the function $s(\omega)$ can be written in the desired form,

$$s(\omega) = x + \sum_{k=1}^{m-1} \frac{q_k}{\omega - \omega_k} \tag{8.22}$$

where $\omega_1, \ldots, \omega_{m-1}$ are the $m-1$ zeros of $g(\omega)$, and x is a constant.

Of course, a corresponding analysis for the $\boldsymbol{\Sigma}(\omega)$ and $\boldsymbol{G}(\omega)$ matrices is more demanding. We confine ourselves to the derivation of Eq. (8.18) and a specification of the constant self-energy part, $\boldsymbol{\Sigma}(\infty)$. To this end, we write the spectral representation (3.17) of the electron propagator in the generalizing form

$$G_{pq}(\omega) = \sum_n \frac{x_p^{(n)} x_q^{(n)*}}{\omega - e_n} \tag{8.23}$$

where the index n runs over both $(N+1)$- and $(N-1)$-particle states; the imaginary infinitesimals $\pm i\eta$ are not essential for the ensuing analysis and have been dropped. Using a compact matrix notation, the spectral representation can be written as

$$\boldsymbol{G}(\omega) = \boldsymbol{x}(\omega\mathbf{1} - \boldsymbol{E})^{-1}\boldsymbol{x}^\dagger \tag{8.24}$$

Here, $\boldsymbol{E}$ is a diagonal matrix of energies e_n, that is, negative electron affinities and ionization potentials:

$$e_n = \begin{cases} -A_n, & n \in \{N+1\} \\ -I_n, & n \in \{N-1\} \end{cases} \tag{8.25}$$

The matrix $\boldsymbol{x}$ is a rectangular matrix of amplitudes x_{pn}:

$$x_{pn} = x_p^{(n)} = \begin{cases} \langle \Psi_0 | c_p | \Psi_n^{N+1} \rangle, & n \in \{N+1\} \\ \langle \Psi_n^{N-1} | c_p | \Psi_0 \rangle, & n \in \{N-1\} \end{cases} \tag{8.26}$$

Note that

$$\boldsymbol{x}\boldsymbol{x}^\dagger = \mathbf{1} \tag{8.27}$$

which is the matrix form of Eq. (3.22).

Using Eq. (8.24) in (8.13) and $\boldsymbol{G}^0(\omega)^{-1} = \omega\mathbf{1} - \boldsymbol{\epsilon}$, where $\boldsymbol{\epsilon}$ denotes the diagonal matrix of HF orbital energies, the self-energy takes on the form

$$\boldsymbol{\Sigma}(\omega) = \omega\mathbf{1} - \boldsymbol{\epsilon} - \left[\boldsymbol{x}(\omega\mathbf{1} - \boldsymbol{E})^{-1}\boldsymbol{x}^\dagger\right]^{-1} \tag{8.28}$$

which now can be analyzed with respect to taking the limit $\omega \to \infty$. For this purpose, we expand the second part on the right-hand side in a power series in ω^{-1} using twice the geometric series:

$$\begin{aligned} \boldsymbol{G}(\omega)^{-1} &= \omega \left[\boldsymbol{x} \left(\mathbf{1} - \frac{\boldsymbol{E}}{\omega} \right)^{-1} \boldsymbol{x}^\dagger \right]^{-1} \\ &= \omega \left[\boldsymbol{x} \left(\mathbf{1} + \frac{\boldsymbol{E}}{\omega} + O(\omega^{-2}) \right) \boldsymbol{x}^\dagger \right]^{-1} \\ &= \omega \left[\mathbf{1} + \boldsymbol{x}\frac{\boldsymbol{E}}{\omega}\boldsymbol{x}^\dagger + O\left(\omega^{-2}\right) \right]^{-1} = \omega\mathbf{1} - \boldsymbol{x}\boldsymbol{E}\boldsymbol{x}^\dagger + O\left(\frac{1}{\omega}\right) \end{aligned}$$

Thus, the expansion of the self-energy becomes

$$\boldsymbol{\Sigma}(\omega) = -\boldsymbol{\epsilon} + \boldsymbol{x}\boldsymbol{E}\boldsymbol{x}^{\dagger} + O\left(\frac{1}{\omega}\right) \tag{8.29}$$

which allows us to identify the constant self-energy part as

$$\boldsymbol{\Sigma}(\infty) = -\boldsymbol{\epsilon} + \boldsymbol{x}\boldsymbol{E}\boldsymbol{x}^{\dagger} \tag{8.30}$$

This result, reading more explicitly

$$\begin{aligned}\Sigma_{pq}(\infty) = -\epsilon_p\delta_{pq} - &\sum_{n\in\{N+1\}} (E_0 - E_n^{N+1})\langle\Psi_0|c_p|\Psi_n^{N+1}\rangle\langle\Psi_n^{N+1}|c_q^{\dagger}|\Psi_0\rangle \\ - &\sum_{n\in\{N-1\}} (E_n^{N-1} - E_0)\langle\Psi_0|c_q^{\dagger}|\Psi_n^{N-1}\rangle\langle\Psi_n^{N-1}|c_p|\Psi_0\rangle\end{aligned} \tag{8.31}$$

can be seen as a sum rule for the energies (pole positions) and amplitudes of the electron propagator. The sum over states on the right-hand side can be replaced by the following closed-form expressions:

$$\begin{aligned}\Sigma_{pq}(\infty) &= -\epsilon_p\delta_{pq} + \langle\Psi_0|[c_p, \hat{H}]c_q^{\dagger}|\Psi_0\rangle + \langle\Psi_0|c_q^{\dagger}[c_p, \hat{H}]|\Psi_0\rangle \\ &= -\epsilon_p\delta_{pq} + \langle\Psi_0|\{c_q^{\dagger}, [c_p, \hat{H}]\}|\Psi_0\rangle\end{aligned} \tag{8.32}$$

The equivalence of Eqs. (8.31) and (8.32) can be seen by inserting the complete sets of $(N+1)$- and $(N-1)$-electron states in the commutator expectation values on the right-hand side of Eq. (8.32). The anticommutator/commutator $\{c_q^{\dagger}, [c_p, \hat{H}]\}$ on the right-hand side of Eq. (8.32) can readily be evaluated, upon which the static self-energy takes on the form

$$\Sigma_{pq}(\infty) = \sum_{u,v} V_{pu[qv]}\left(\langle\Psi_0|c_u^{\dagger}c_v|\Psi_0\rangle - \delta_{uv}n_u\right) \tag{8.33}$$

Here, the HF relations (4.6) have been supposed. Recalling the definition (3.26) of the density matrix, we may write

$$\Sigma_{pq}(\infty) = \sum_{u,v} V_{pu[qv]}\left(\rho_{vu} - \rho_{vu}^{(0)}\right) \tag{8.34}$$

where

$$\rho_{vu}^{(0)} = \langle\Phi_0|c_u^{\dagger}c_v|\Phi_0\rangle = \delta_{uv}n_u \tag{8.35}$$

denote HF density matrix elements. Using Eqs. (3.28) and (3.34), the density matrix elements can be replaced by electron propagator elements, which allows us to write the static self-energy elements in the form

$$\Sigma_{pq}(\infty) = \sum_{u,v} V_{pu[qv]}(-i)\left(G_{vu}(t,t^+) - G^0_{vu}(t,t^+)\right) \tag{8.36}$$

or

$$\Sigma_{pq}(\infty) = \sum_{u,v} V_{pu[qv]}\frac{1}{2\pi i}\oint \left(G_{vu}(\omega) - G^0_{vu}(\omega)\right) \mathrm{d}\omega \tag{8.37}$$

This result, establishing a relationship between the static self-energy and the electron propagator, can further be expanded by inserting the formal solution (8.11) of the Dyson equation for $\boldsymbol{G}(\omega)$ on the right-hand side:

$$\Sigma_{pq}(\infty) = \sum_{u,v} V_{pu[qv]}\left[\frac{1}{2\pi i}\oint \left(\boldsymbol{G}^0(\omega)^{-1} - \boldsymbol{\Sigma}(\infty) - \boldsymbol{M}(\omega)\right)^{-1}\Big|_{vu} \mathrm{d}\omega - \delta_{uv}n_u\right] \tag{8.38}$$

Here, the partitioning (8.18) of the self-energy has been used. As Eq. (8.38) shows, the static self-energy part is determined by the dynamical part: For a given $\boldsymbol{M}(\omega)$, it constitutes an implicit equation for $\boldsymbol{\Sigma}(\infty)$. This means that in devising approximation schemes for the self-energy one can focus on the dynamical part. Once a suitable approximation for $\boldsymbol{M}(\omega)$ has been devised, a consistent approximation for $\boldsymbol{\Sigma}(\infty)$ can be obtained via Eq. (8.38). A practical procedure for the evaluation of $\boldsymbol{\Sigma}(\infty)$ is described in Appendix A.5.

The closed-form expressions (8.36) and (8.37) can be obtained in a more intuitive way using diagrammatic analysis. The diagrams contributing to $\boldsymbol{\Sigma}(\infty)$ are of the form of the third-order diagram $T3$ in Fig. 8.9. Any nth-order diagram, $n > 0$, in the electron propagator expansion can be transformed into a corresponding $\boldsymbol{\Sigma}(\infty)$ diagram of order $n+1$ by joining the two outer free fermion lines in an interaction dot (external vertex). Symbolically, the sum of all diagrams contributing to $\boldsymbol{\Sigma}(\infty)$ can be depicted as follows:

$\Sigma^{\infty}_{pq}(t,t') \equiv$ 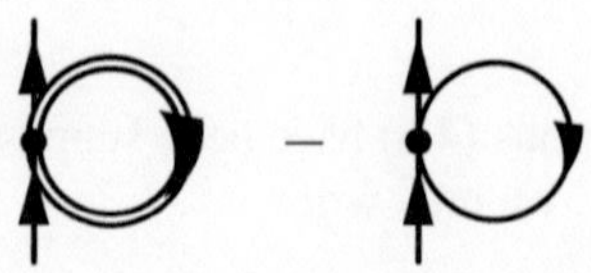 (8.39)

Here, as in Eq. (6.20), the double line represents the full electron propagator. That is, the first graph on the right-hand side comprises all diagrams contributing to $\boldsymbol{\Sigma}(\infty)$. However, there is one extra diagram, namely the first-order "tadpole" diagram (second graph on the right-hand side), associated with the zeroth-order diagram (free fermion line). In the HF representation supposed here, there is no first-order contribution to the electron propagator and, thus, to the self-energy, as was discussed in Sect. 6.2. Therefore, the tadpole contribution to the static self-energy must be subtracted from the first term on the right-hand side.

The graphical representation (8.39) of the static self-energy part can be translated into an analytical expression as follows

$$\Sigma^{\infty}_{pq}(t,t') = \delta(t-t')\Sigma_{pq}(\infty)$$
$$\Sigma_{pq}(\infty) = \lim_{t_1'\to t_1}\sum_{r,s} V_{pr[qs]}(-i)\left(G_{sr}(t_1,t_1') - G^0_{sr}(t_1,t_1')\right) \tag{8.40}$$

The overall phase factor $(-i)$ on the right-hand side comprises several distinct contributions, for example a factor (-1) arising from the fact that a closed loop is formed when the two outer free fermion lines are joined at the same entry of a wiggly interaction line.

It should be noted that equating the time arguments t_1 and t_1' on the right-hand side of Eq. (8.40) does not depend on the order of the time arguments:

$$\begin{aligned}(-i)G_{sr}(t,t^+) - (-i)G^0_{sr}(t,t^+) &= (-i)G_{sr}(t^+,t) - (-i)G^0_{sr}(t^+,t)\\ &= \langle\Psi_0|c_r^\dagger c_s|\Psi_0\rangle - n_r\delta_{rs}\end{aligned}$$

This can be seen by using the anticommutation relation $\{c_r^\dagger, c_s\} = \delta_{rs}$ both in the definitions of G_{sr} and G^0_{sr}.

Finally, we note that Eq. (8.34) allows us to establish a physical interpretation of the diagonal elements of the static self-energy part. We consider a diagonal element, $\Sigma_{k\alpha k\alpha}(\infty)$, where k is a spatial orbital, and retain only the Coulomb parts of the antisymmetrized two-particle integrals,

$$\Sigma_{k\alpha k\alpha}(\infty) \sim \sum_{u,v} 2V_{kukv}\left(\rho_{vu} - \rho^{(0)}_{vu}\right) \tag{8.41}$$

Here u, v denote spatial indices, and $\rho_{vu} \equiv \rho_{v\gamma,u\gamma}$. Using the electron density function associated with the given density matrix,

$$\rho(\boldsymbol{x}) = \sum 2\phi_u^*(\boldsymbol{x})\phi_v(\boldsymbol{x})\rho_{vu} \tag{8.42}$$

Eq. (8.41) can be written as

$$\Sigma_{k\alpha k\alpha}(\infty) \sim \int\int d\boldsymbol{x}d\boldsymbol{x}'\frac{|\phi_k(\boldsymbol{x})|^2}{|\boldsymbol{x}-\boldsymbol{x}'|}(\rho(\boldsymbol{x}') - \rho^{(0)}(\boldsymbol{x}')) \tag{8.43}$$

As this expression shows, $\Sigma_{k\alpha k\alpha}(\infty)$ accounts for the ground-state correlation effect, $\Delta\rho(\boldsymbol{x}) = \rho(\boldsymbol{x}) - \rho^{(0)}(\boldsymbol{x})$, in the Coulomb repulsion between the electron density and the charge, $|\phi_k(\boldsymbol{x})|^2$, of the electron in the HF orbital k. Note that the self-interaction error due to the neglect of the exchange integrals in Eq. (8.34) is not relevant in $\Delta\rho(\boldsymbol{x})$.

8.3 Solving the Dyson Equation

1. The Dyson Secular Matrix

Using the general form (8.18) for the self-energy and the matrix notation $\boldsymbol{G}^0(\omega)^{-1} = \omega\mathbf{1} - \boldsymbol{\epsilon}$ for the inverse of the free electron propagator, the formal solution (8.11) of the Dyson equation can be written more explicitly as

$$\boldsymbol{G}(\omega) = \left(\boldsymbol{G}^0(\omega)^{-1} - \boldsymbol{\Sigma}(\omega)\right)^{-1} = (\omega\mathbf{1} - \boldsymbol{\epsilon} - \boldsymbol{\Sigma}(\infty) - \boldsymbol{M}(\omega))^{-1} \tag{8.44}$$

Moreover, one may adopt the following matrix notation for the dynamical self-energy parts:

$$\boldsymbol{M}(\omega) = \boldsymbol{M}^+(\omega) + \boldsymbol{M}^-(\omega), \quad \boldsymbol{M}^\pm(\omega) = \boldsymbol{m}^{\pm\dagger}\left(\omega\mathbf{1} - \boldsymbol{\Omega}^\pm\right)^{-1}\boldsymbol{m}^\pm \tag{8.45}$$

Here, $\boldsymbol{\Omega}^\pm$ denote the diagonal matrices of the pole positions in the spectral representations (8.19) of the two $M^\pm_{pq}(\omega)$ parts,

$$\Omega^+_{\nu\nu} = \omega_\nu,\ \nu \in \{N+1\}, \quad \Omega^-_{\mu\mu} = \omega_\mu,\ \mu \in \{N-1\} \tag{8.46}$$

and $\boldsymbol{m}^\pm$ are the corresponding matrices of the amplitudes,

$$m^+_{\nu p} = m^{(\nu)*}_p,\ \nu \in \{N+1\}, \quad m^-_{\mu p} = m^{(\mu)*}_p,\ \mu \in \{N-1\} \tag{8.47}$$

As a result, the Dyson equation takes on the form

$$\boldsymbol{G}(\omega) = \left(\omega\mathbf{1} - \boldsymbol{\epsilon} - \boldsymbol{\Sigma}(\infty) - \boldsymbol{m}^{-\dagger}(\omega\mathbf{1} - \boldsymbol{\Omega}^-)^{-1}\boldsymbol{m}^- - \boldsymbol{m}^{+\dagger}(\omega\mathbf{1} - \boldsymbol{\Omega}^+)^{-1}\boldsymbol{m}^+\right)^{-1} \tag{8.48}$$

which has some semblance of the partitioning formulas reviewed in Appendix A.1. In fact, one may introduce the following Dyson secular matrix

$$\boldsymbol{A} = \begin{pmatrix} \boldsymbol{\epsilon} + \boldsymbol{\Sigma}(\infty) & \boldsymbol{m}^{-\dagger} & \boldsymbol{m}^{+\dagger} \\ \boldsymbol{m}^- & \boldsymbol{\Omega}^- & \mathbf{0} \\ \boldsymbol{m}^+ & \mathbf{0} & \boldsymbol{\Omega}^+ \end{pmatrix} \tag{8.49}$$

Using the partitioning formulas (A.1.26) and (A.1.27) for the upper left diagonal block of $\boldsymbol{A}$, that is, $\boldsymbol{A}_{11} = \boldsymbol{\epsilon} + \boldsymbol{\Sigma}(\infty)$, the right-hand side of Eq. (8.48) can readily be identified with the upper left diagonal block of the inverse of $(\omega\mathbf{1} - \boldsymbol{A})$:

$$\boldsymbol{G}(\omega) = (\omega\mathbf{1} - \boldsymbol{A})^{-1}\big|_{11} \tag{8.50}$$

Let us note that in the upper left diagonal block, denoted by $|_{11}$, the matrix indices are the HF one-particle indices, comprising both occupied (h) and unoccupied (p) orbitals.

The inversion of an ω-dependent matrix of the form $(\omega\mathbf{1} - \boldsymbol{A})$, where $\boldsymbol{A}$ is hermitian, is equivalent to solving the eigenvalue problem (see Appendix A.1)

$$\boldsymbol{AX} = \boldsymbol{XE}, \qquad \boldsymbol{X}^\dagger\boldsymbol{X} = \mathbf{1} \tag{8.51}$$

as

$$(\omega\mathbf{1} - \boldsymbol{A})^{-1} = \boldsymbol{X}(\omega\mathbf{1} - \boldsymbol{E})^{-1}\boldsymbol{X}^\dagger \tag{8.52}$$

Here, $\boldsymbol{E}$ and $\boldsymbol{X}$ denote the diagonal matrix of the eigenvalues and eigenvectors of $\boldsymbol{A}$, respectively. Accordingly, the solution of the Dyson equation can be written as

$$\boldsymbol{G}(\omega) = \boldsymbol{X}(\omega\mathbf{1} - \boldsymbol{E})^{-1}\boldsymbol{X}^\dagger\big|_{11} \tag{8.53}$$

or, more explicitly,

$$G_{pq}(\omega) = \sum_n \frac{x_p^{(n)} x_q^{(n)*}}{\omega - e_n} \tag{8.54}$$

where $e_n = E_{nn}$ and $x_p^{(n)} = X_{pn}$, and n runs over both the $(N+1)$- and $(N-1)$-electron states. As the comparison with Eqs. (8.23)–(8.26) shows, this is just the spectral representation of the electron propagator. This means that the ionization and electron attachment energies are obtained as the eigenvalues of the Dyson matrix $\boldsymbol{A}$, while the corresponding spectroscopic factors are given by the h or p orbital components of the eigenvectors.

It should be noted that the sum rules (3.22) and (8.30), (8.31) can directly be inferred from the 11-block of the eigenvalue equations (8.51):

$$\sum_n X_{pn}X_{qn}^* = \delta_{pq}, \quad \sum_n E_n X_{pn}X_{qn}^* = A_{pq} = \epsilon_p\delta_{pq} + \Sigma_{pq}(\infty) \tag{8.55}$$

In conclusion, we have seen that the Dyson equation can be formulated as the eigenvalue problem of a hermitian matrix, referred to as Dyson secular matrix, where the entries derive from the spectral representation of the dynamic self-energy part (amplitudes and pole positions), the matrix elements of the static self-energy part, and the HF orbital energies.

2. Dyson Equation as an Effective One-Particle Eigenvalue Equation

Rather than dealing with the full eigenvalue problem (8.51) of the Dyson secular matrix, one may resort to the partitioning of the original eigenvalue problem as exemplified in Eq. (A1.31) of Appendix A.1. Contracting the eigenvalue problem to the (11)-block of $\boldsymbol{A}$, the Dyson equation takes on the form

$$(\boldsymbol{\epsilon} + \boldsymbol{\Sigma}(e_n))\,\boldsymbol{x}_n = e_n\boldsymbol{x}_n \tag{8.56}$$

where $e_n = E_{nn}$ are the eigenvalues of $\boldsymbol{A}$ and $\boldsymbol{x}_n$ is the vector of the p/h-components, $x_p^{(n)} = X_{pn}$, of the full eigenvectors $\boldsymbol{X}_n$. This is an energy-dependent (pseudo-) eigenvalue equation for a one-particle problem, in which the self-energy matrix $\boldsymbol{\Sigma}(\omega)$ can be seen as orbital representation of an energy-dependent, non-local one-particle operator $\hat{\Sigma}(\boldsymbol{x}, \boldsymbol{x}'; \omega)$. The enormous reduction of the dimensionality with respect to Eq. (8.51) comes at a price, as the solutions of the pseudo-eigenvalue problem have to be sought using iterative techniques, which are prone to computational problems. It should be noted that the usual orthonormalization features do not apply to the pseudo-eigenvectors $\boldsymbol{x}_n$ (see Appendix A.1).

3. Dyson Orbitals

As a tool for visualizing many-body effects in ionization (or electron attachment) the so-called Dyson orbitals have proven useful. Consider a final ionic state, say of $N-1$ electrons, $|\Psi_n^{N-1}\rangle$. A **Dyson orbital** $\Phi_n(\boldsymbol{\xi})$ can be assigned to this state as follows:

$$\begin{aligned}\Phi_n(\boldsymbol{\xi}) &= \langle\Psi_n^{N-1}|\hat{\psi}(\boldsymbol{\xi})|\Psi_0\rangle \\ &= \sum_r \langle\Psi_n^{N-1}|c_r|\Psi_0\rangle\psi_r(\boldsymbol{\xi}) = \sum_r x_r^{(n)}\psi_r(\boldsymbol{\xi})\end{aligned} \tag{8.57}$$

As in Eq. (1.3), $\boldsymbol{\xi} \equiv \boldsymbol{x}\sigma$ combines the spatial and spin variables; $\psi_r(\boldsymbol{\xi})$ denote the spin orbitals associated with the fermion operators, $c_r^\dagger$, c_r, and

$$\hat{\psi}(\boldsymbol{\xi}) = \sum_r c_r\psi_r(\boldsymbol{\xi}) \tag{8.58}$$

is the field operator according to Eq. (2.45). Alternatively, the Dyson orbital can be obtained directly from the ground- and ionic-state wave functions,

$$\Phi_n(\boldsymbol{\xi}) = \sqrt{N!}\int \Psi_n^{N-1}(\boldsymbol{\xi}_2, \ldots, \boldsymbol{\xi}_N)^*\Psi_0(\boldsymbol{\xi}, \boldsymbol{\xi}_2, \ldots, \boldsymbol{\xi}_N)\, \mathrm{d}\boldsymbol{\xi}_2 \ldots \mathrm{d}\boldsymbol{\xi}_N \tag{8.59}$$

where as in Eq. (1.5) the integration over $\boldsymbol{\xi}_i$ comprises the summation over spin variables. To show that both definitions are equivalent (see Exercise 8.4) one may insert in Eq. (8.57) the resolution of the identity in terms of the coordinate eigenstates $|\boldsymbol{\xi}_1 \ldots \boldsymbol{\xi}_N\rangle$ (respecting here Eq. 1.24) and evaluate $\hat{\psi}(\boldsymbol{\xi})|\boldsymbol{\xi}_1 \ldots \boldsymbol{\xi}_N\rangle$ according to Eq. (2.9).

Like spin orbitals, the Dyson orbitals $\Phi_n(\boldsymbol{\xi})$ can be written as products of a spatial orbital and a spin function:

$$\Phi_{n\gamma}(\boldsymbol{x}, \sigma) = \Phi_n(\boldsymbol{x})\chi_{\overline{\gamma}}(\sigma) \tag{8.60}$$

Here, $\gamma = \alpha, \beta$ specifies the z-component of the spin in the final state, $n \equiv n\gamma$, and $\overline{\gamma}$ denotes the spin quantum number complementary to γ, e.g., $\overline{\alpha} = \beta$. The spatial

Dyson orbital is given by

$$\Phi_n(\boldsymbol{x}) = \langle \Psi_{n\gamma}^{N-1} | \hat{\psi}_{\overline{\gamma}}(\boldsymbol{x}) | \Psi_0 \rangle = \sum_r x_r^{(n)} \phi_r(\boldsymbol{x}) \tag{8.61}$$

where $\hat{\psi}_\gamma$ is the mixed representation (2.49) of the field operator with $\phi_r(\boldsymbol{x})$ denoting the spatial functions in $\psi_r(\boldsymbol{\xi})$. The alignment of γ and $\overline{\gamma}$ is of course a consequence of the spin symmetry in the above matrix element. In the second equation, the amplitudes $x_r^{(n)}$ can be assumed to be spin-free as

$$x_r^{(n)} = x_{r\gamma}^{(n\overline{\gamma})} = x_{r\overline{\gamma}}^{(n\gamma)} \tag{8.62}$$

We note again that, in general, the Dyson orbitals are not orthonormal,

$$\int \Phi_n^*(\boldsymbol{x}) \Phi_m(\boldsymbol{x})\, \mathrm{d}\boldsymbol{x} = \sum_r x_r^{(n)*} x_r^{(m)} \neq \delta_{nm} \tag{8.63}$$

because the amplitudes $x_r^{(n)}$ do not constitute the full final-state eigenvector.

Note that for an uncorrelated ground state, $|\Phi_0\rangle$, and a corresponding single-hole ionic state, $c_{p\gamma}|\Phi_0\rangle$, the Dyson orbital is simply given by the HF orbital $\phi_p(\boldsymbol{x})$.

4. Second-Order Approximation to the Self-energy

The construction of a Dyson secular matrix and the ensuing solution of the secular equations applies not only to the exact self-energy but also to suitable approximations, more specifically, approximations in which the dynamic self-energy part is given in the form of the spectral representation (8.19). As an illustrative example, we will consider the second-order approximation to the self-energy part in the following.

Obviously, the second-order self-energy, given by Eq. (8.17), is of the same analytic form as the spectral representation. Since there is no static second-order contribution, we may write

$$\Sigma_{pq}^{(2)}(\omega) = M_{pq}^{(2)}(\omega) = M_{pq}^{(2)+}(\omega) + M_{pq}^{(2)-}(\omega) \tag{8.64}$$

where

$$\begin{aligned} M_{pq}^{(2)+}(\omega) &= \sum_{a<b,k} \frac{V_{pk[ab]} V_{ab[qk]}}{\omega + \epsilon_k - \epsilon_a - \epsilon_b + i\eta} \\ M_{pq}^{(2)-}(\omega) &= \sum_{a,k<l} \frac{V_{pa[kl]} V_{kl[qa]}}{\omega + \epsilon_a - \epsilon_k - \epsilon_l - i\eta} \end{aligned} \tag{8.65}$$

The poles in the $(N+1)$- and $(N-1)$-parts are labeled by $2p$-$1h$ and $2h$-$1p$ index triples. Note the $2p$-$1h$ and $2h$-$1p$ indices are restricted according to $k, a < b$ and $a, k < l$, respectively, which eliminates the factors $\frac{1}{2}$ in the original expression (8.17).

The matrix elements of the second-order Dyson secular matrix, $\boldsymbol{A}(2)$, can directly be taken from Eq. (8.65):

$$
\begin{aligned}
\Omega^{-}_{akl,akl} &= -\,\epsilon_a + \epsilon_k + \epsilon_l \\
\Omega^{+}_{abk,abk} &= -\,\epsilon_k + \epsilon_a + \epsilon_b \\
m^{-}_{akl,q} &= V_{qa[kl]}, \qquad m^{+}_{abk,q} = V_{ab[qk]}
\end{aligned}
\tag{8.66}
$$

The structure of the $\boldsymbol{A}(2)$ matrix, as given below,

$$
\begin{array}{c|c|c|c|}
 & 1h/1p & 2h\text{-}1p & 2p\text{-}1h \\ \hline
 & \ddots \;\; \epsilon_q \;\; \ddots & \vdots \;\; \cdots V_{kl[qa]} \cdots \;\; \vdots & \vdots \;\; \cdots V_{ab[qk]} \cdots \;\; \vdots \\ \hline
\boldsymbol{A}(2) \equiv & \vdots \;\; \cdots V_{qa[kl]} \cdots \;\; \vdots & \ddots \;\; -\varepsilon_a + \varepsilon_k + \varepsilon_l \;\; \ddots & \mathbf{0} \\ \hline
 & \vdots \;\; \cdots V_{qk[ab]} \cdots \;\; \vdots & \mathbf{0} & \ddots \;\; -\varepsilon_k + \varepsilon_a + \varepsilon_b \;\; \ddots \\ \hline
\end{array}
$$

reflects the partitioning of the secular expansion manifold into three subsets corresponding to HF orbitals (or $1h$ and $1p$ states), the $2h$-$1p$, and $2p$-$1h$ configurations.

It is instructive to compare the second-order Dyson secular matrix to the secular matrices arising in the context of a wave-function approach, that is, separate CI expansions for the $(N-1)$-particle and $(N+1)$-particle states (see Sect. 12.3). Let us consider the $1h$ and $2h$-$1p$ CI configurations,

$$
\begin{aligned}
|\Phi^{N-1}_{j}\rangle &= c_j|\Phi_0\rangle \\
|\Phi^{N-1}_{akl}\rangle &= c^{\dagger}_a c_k c_l|\Phi_0\rangle, \; k < l
\end{aligned}
$$

of $N-1$ particles. The corresponding matrix elements of the CI secular matrix, taken with respect to the subtracted hamiltonian, $\hat{H}' = \hat{H} - E_0(1)$, where $E_0(1)$ is the HF (first-order) ground-state energy, read

$$
\begin{aligned}
H'_{ij} &= -\epsilon_i \delta_{ij} \\
H'_{j,akl} &= V_{kl[ja]} \\
H'_{akl,a'k'l'} &= (\epsilon_a - \epsilon_k - \epsilon_l)\delta_{aa'}\delta_{kk'}\delta_{ll'} + O(1)
\end{aligned}
\tag{8.67}
$$

Here, the matrix elements of the $2h$-$1p$ diagonal block are specified only through zeroth order. Obviously, the secular matrix given by Eq. (8.67) is (up to a sign) part of the second-order Dyson matrix. In a similar way, the analogous CI secular matrix for $N+1$ particles is retrieved within $\boldsymbol{A}(2)$. What makes $\boldsymbol{A}(2)$ peculiar is the fact that there is a coupling of the $(N-1)$- and $(N+1)$-particle parts. How can such a coupling, which is not feasible within a wave-function approach, be rationalized? An analysis of the results through second order of perturbation theory can give some clue.

5. Analysis of the Dyson Equation Using Second-Order Perturbation Theory

Consider the ionization energy I_k of the state deriving from the one-hole configuration $|\Phi_k\rangle = c_k|\Phi_0\rangle$. Through second order, straightforward matrix perturbation theory for the $\boldsymbol{A}(2)$ secular problem gives

$$I_k(2) = -\epsilon_k - \sum_{a,j<l} \frac{|V_{ka[jl]}|^2}{\epsilon_k + \epsilon_a - \epsilon_j - \epsilon_l} - \sum_{j,b<c} \frac{|V_{kj[bc]}|^2}{\epsilon_k + \epsilon_j - \epsilon_b - \epsilon_c} \tag{8.68}$$

where the first and second sum on the right-hand side correspond to the coupling of the single-hole configuration k with $2h$-$1p$ and $2p$-$1h$ configurations, respectively. In accordance with Koopmans' theorem, the ionization energy through first order is given by the negative orbital energy, $I_k(1) = -\epsilon_k$. How does the expansion (8.68) compare with the exact ionization energy $I_k = E_k^{N-1} - E_0$? The formal perturbation expansion of I_k through second order can be written as

$$I_k = -\epsilon_k + W_k^{(2)}(2h\text{-}1p) + W_k^{(2)}(3h\text{-}2p) - E_0^{(2)} + O(3) \tag{8.69}$$

Here, $W_h^{(2)}(2h\text{-}1p)$ denotes the second-order energy arising from the coupling of the $1h$ configuration with $2h$-$1p$ configurations and

$$E_0^{(2)} = -\sum_{b<c,i<j} \frac{|V_{bc[ij]}|^2}{\epsilon_b + \epsilon_c - \epsilon_i - \epsilon_j} \tag{8.70}$$

is the second-order contribution to the ground-state energy, as specified in Eq. (4.65). The explicit expression for $W_k^{(2)}(2h\text{-}1p)$ reads

$$W_k^{(2)}(2h\text{-}1p) = -\sum_{a,j<l} \frac{|V_{ka[jl]}|^2}{\epsilon_k + \epsilon_a - \epsilon_j - \epsilon_l} \tag{8.71}$$

which is just the second term on the right-hand side of the Dyson result (8.68). The other second-order term in Eq. (8.69),

$$W_k^{(2)}(3h\text{-}2p) = -\sum_{\substack{b<c \\ i<j\neq k}} \frac{|V_{bc[ij]}|^2}{\epsilon_b + \epsilon_c - \epsilon_i - \epsilon_j} \tag{8.72}$$

results from the coupling of the $1h$ configuration with $3h$-$2p$ configurations (see Exercise 8.6). Apart from the restrictions $i < j \neq k$ in the summation indices, $W_k^{(2)}(3h\text{-}2p)$ is of the same form as $E_0^{(2)}$. As can readily be seen, the difference $W_h^{(2)}(3h\text{-}2p) - E_0^{(2)}$ of the $3h$-$2p$ contributions in the ionic energy and the $2p$-$2h$ contributions in the ground-state energy can be identified with the second sum in the Dyson result (8.68):

$$W_k^{(2)}(3h\text{-}2p) - E_0^{(2)} = \sum_{j,b<c} \frac{|V_{kj[bc]}|^2}{\epsilon_b + \epsilon_c - \epsilon_j - \epsilon_k} \tag{8.73}$$

This shows that the strange coupling of $(N-1)$- and $(N+1)$-particle states in the Dyson secular problem can be seen as a means to account for the admixture of $3h$-$2p$ configurations in the ionic state and the second-order correlation energy in the ground state. A CI treatment of the $1h$ states at comparable accuracy would require CI expansions comprising $1h$, $2h$-$1p$, and $3h$-$2p$ configurations. In the second-order Dyson approach, by contrast, the secular matrix is formed by the manifold of $1h$, $1p$, $2h$-$1p$, and $2p$-$1h$ configurations.

Besides the $1h$ main states, the $\boldsymbol{A}(2)$ secular matrix accounts for so-called satellite states deriving from $2h$-$1p$ configurations, treated however only in zeroth order:

$$I_{akl} = \epsilon_a - \epsilon_k - \epsilon_l + O(1) \tag{8.74}$$

In a completely analogous way, the $1p$ and $2p$-$1h$ states of $N+1$ particles can be analyzed.

The second-order Dyson approximation, based on using the second-order self-energy in the Dyson equation, provides a very simple computational approach to ionization energies and electron affinities of closed-shell atoms and molecules. However, the accuracy hereby afforded is rather modest. Typically, the error in the ionization energies of outer valence $1h$ main states is in the order of 1–2 eV. An extension to higher order is by no means straightforward. Already at the third-order level, the dynamic self-energy deviates from the form of a sum over simple poles as here products of poles come into play, which means that contributions $\boldsymbol{M}^{(n)}(\omega)$ for $n \geq 3$ cannot readily be incorporated in the Dyson secular matrix.

6. Graphical Solution of the Dyson Equation in the Diagonal Approximation

For a given expression of the self-energy, the Dyson equation (8.11) can also be solved in a more descriptive way, where the desired information, that is, pole positions and residues of the electron propagator, is extracted as the zero points and the respective slopes of the inverse of the electron propagator,

$$\boldsymbol{G}(\omega)^{-1} = \omega\mathbf{1} - \boldsymbol{\epsilon} - \boldsymbol{\Sigma}(\omega) \tag{8.75}$$

In particular, such a procedure applies to the diagonal approximation in which the non-diagonal elements of the self-energy part are neglected. As a consequence, there is an individual equation for each diagonal element of the electron propagator:

$$G_{pp}(\omega) = \left(\omega - \epsilon_p - \Sigma_{pp}(\omega)\right)^{-1} \tag{8.76}$$

Assuming that $\Sigma_{pp}(\omega)$ is of the form

$$\Sigma_{pp}(\omega) = \Sigma_{pp}(\infty) + \sum_n \frac{|m_{pn}|^2}{\omega - \Omega_n} \tag{8.77}$$

the zero points of $G_{pp}(\omega)$ are given by the solutions of the equation

$$\omega - \epsilon_p - \Sigma_{pp}(\infty) = \sum_n \frac{|m_{pn}|^2}{\omega - \Omega_n} \tag{8.78}$$

As depicted in Fig. 8.11, the function on the left-hand side is a straight line with slope 1 crossing the ω-axis at $\epsilon_p + \Sigma_{pp}(\infty)$, while the right-hand side is a sum over simple poles located at the positions Ω_n. Obviously, the straight line cuts the pole function once between two successive pole positions, so that for each pair of successive poles there is a single zero point of $G_{pp}(\omega)^{-1}$. This amounts to a "graphical solution" of the one-component Dyson equation in the diagonal approximation. Of course, the zero points between successive pole positions can also be determined numerically, using, e.g., a Newton–Raphson-type procedure. The residues (or pole strengths) of the poles of $G_{pp}(\omega)$ are determined by the slopes of the pole function at the intersection points. Let ω_0 be a pole position resulting from the single-component Dyson equation for $G_{pp}(\omega)$ and P_0 denote the residue to be determined. Expanding $G_{pp}^{-1}(\omega)$ in a Taylor series about ω_0 yields

$$G_{pp}^{-1}(\omega) = \left(\omega - \epsilon_p - \Sigma_{pp}(\omega)\right) = \frac{1}{P_0}(\omega - \omega_0) + \alpha(\omega - \omega_0)^2 + \cdots \tag{8.79}$$

where $1/P_0$ is given by

$$\frac{1}{P_0} = 1 - \Sigma'_{pp}(\omega_0) \tag{8.80}$$

The slope of the pole function is always negative so that the pole strength,

$$P_0 = (1 - \Sigma'_{pp}(\omega_0))^{-1} \tag{8.81}$$

is a positive number smaller 1, $P_0 \leq 1$.

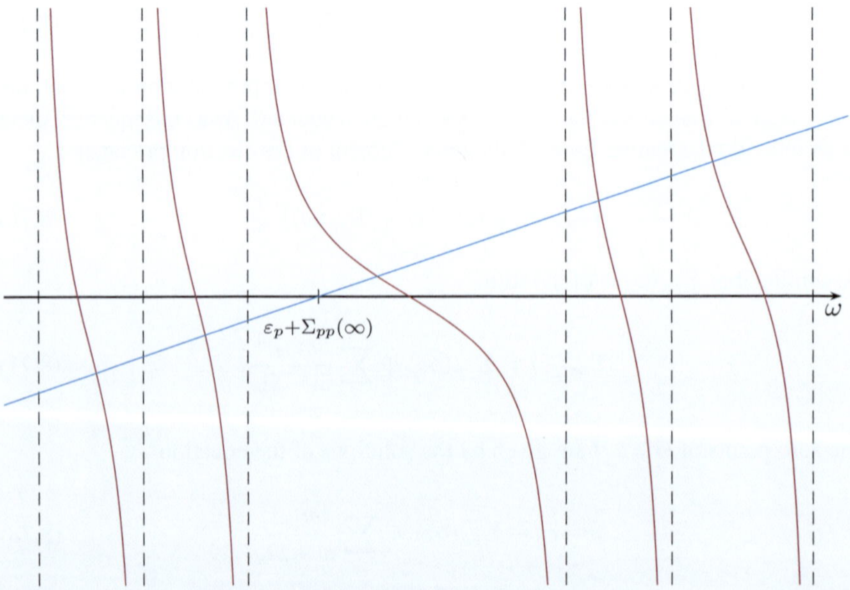

Fig. 8.11 Graphical solution of the diagonal form of the Dyson equation

Figure 8.11 shows a typical arrangement of the poles. On the left side, there are the self-energy poles associated with the $(N-1)$-particle (ionization) part, being separated from the $(N+1)$-particle (electron affinity) poles on the right-hand side by an energy gap of the order $3(\epsilon_{LUMO} - \epsilon_{HOMO})$, where the acronyms HOMO and LUMO refer to highest occupied and lowest unoccupied molecular (HF) orbital, respectively. As assumed in Fig. 8.11, the orbital energy ϵ_p, say, of an occupied orbital in the outer valence region, is located within the gap between the $(N-1)$- and the $(N+1)$-particle poles. Accordingly, there is one solution within the gap with a relatively large residue, since the modulus of the slope at the interaction point tends to be small. This solution corresponds to the "single-hole" main state. In addition, there are solutions within successive pole positions on the left- and right-hand side, which however, have much smaller residues since here the pole function have steep slopes at the crossing points. Those solutions correspond to "secondary" or "shake-up" states.

7. Beyond Second Order: Outer Valence Green's Function (OVGF) Method

As already mentioned, the expansion of a diagonal self-energy element $\Sigma_{pp}(\omega)$ through third order,

$$\Sigma_{pp}(\omega) = \Sigma_{pp}^{(2)}(\omega) + \Sigma_{pp}^{(3)}(\omega) + O(4) \tag{8.82}$$

is not of the simple analytic form (8.77) presumed in the graphical solution, as the third-order diagrams introduce products of simple poles. Nevertheless, in the energy region within the ionization/electron-affinity gap, sufficiently far from the two outermost poles, $\Sigma_{pp}(\omega)$ is a smooth function of ω, and the third-order expansion (8.82) may represent that region quite reasonably, irrespective of its inadequacy in the pole regions. Accordingly, for outer valence main ($1h$) states, where the graphical solutions of the respective diagonal Dyson equations lie in the gap region, a third-order self-energy expansion may afford a good approximation. An approach based on the third-order expansion of the self-energy is the widely used and utmost successful outer valence Green's function (OVGF) method [4–6]. A related scheme is the P3 method basing on a partial third-order self-energy expression [7].

The two main ingredients of the OVGF method are:

1. Diagonal approximation for the electron propagator: $G_{pq}(\omega) \approx \delta_{pq} G_{pp}(\omega)$
2. A modified third-order expansion of the diagonal self-energy matrix elements:

$$\Sigma_{pp}(\omega) \approx \Sigma^{(2)}_{pp}(\omega) + \frac{1}{1 - A_p} \Sigma^{(3)}_{pp}(\omega) \tag{8.83}$$

where $A_p < 1$ is a predefined correction coefficient.

In the OVGF expression for the self-energy, a correction factor is attached to the strict third-order contribution $\Sigma^{(3)}_{pp}(\omega)$, the purpose of which is to extrapolate the expansion to higher orders. The usual recipe is

$$A_p = \frac{\sum_{i=2}^{5} \left(T1(i)|_{\omega=\epsilon_p} + T2(i)|_{\omega=\epsilon_p} \right)}{M^{(2)}_{pp}(\epsilon_p)} \tag{8.84}$$

Here, $T1$ and $T2$ are the two ω-dependent third-order diagrams (see Fig. 8.9), and $i = 2, \ldots, 5$ label the time-orderings that have only one ω-denominator. Obviously, A_p somehow measures the ratio of the third- and second-order self-energy contributions (for the pth diagonal element), and, thus, the factor $(1 - A_p)^{-1} = 1 + A_p + A_p^2 + \ldots$ in (8.83) can be understood as sort of an extrapolation to higher-order contributions. It should be noted that there are two more extrapolation schemes to be applied in specific cases (see Refs. [4, 6] for details).

Exercises

8.1 Consider a general one-particle system with a hamiltonian of the form $\hat{h} = \hat{h}_0 + \hat{v}$, where $\hat{h}_0$ is a "free" hamiltonian and $\hat{v}$ a perturbation. Accordingly, there are resolvent operators $\hat{g}(\omega) = (\omega - \hat{h})^{-1}$ and $\hat{g}_0(\omega) = (\omega - \hat{h}_0)^{-1}$ corresponding to the full and free hamiltonian, respectively. Derive a Dyson-like equation for $\hat{g}(\omega)$ and another equation analogous to Eq. (8.2). What are the analogues to the improper and proper self-energy parts?

8.2 (a) Evaluate the three time-orderings contributing to $\mathbf{\Sigma}^{-}(\omega)$ ($t' > t$) of the diagram $T1$ in Fig. 8.9.
(b) Draw and evaluate the six time-orderings of diagram $T3$ in Fig. 8.9.

8.3 (a) Consider the model functions (8.20) and (8.21) in the simple two-pole case ($m = 2$) and verify the corresponding Eq. (8.22).
(b) Perform the corresponding analysis for $m = 3$. (Use partial fraction decomposition without explicitly solving the cubic or quadratic equations).
(c) Consider the general case (m arbitrary) and establish that $s(\omega)$ can be written in the form of Eq. (8.22).

8.4 Adapt the matrix representation of Abrikosov diagrams (Sect. 6.3) to the diagrams for the dynamical self-energy part $\boldsymbol{M}(\omega)$ and test that concept for the second- and third-order diagrams.

8.5 Show that the definitions (8.57) and (8.59) for the Dyson orbitals are equivalent. (Proceed here as indicated in Sect. 8.3.)

8.6 Use (formal) RSPT for the $(N-1)$-electron state deriving from $c_k|\Phi_0\rangle$ and verify the second-order expressions (8.71)–(8.73).

References

1. Dyson FJ (1949) Phys Rev 75:1736
2. Ethofer S, Schuck P (1969) Z Phys 228:264
3. Winter J (1972) Nucl Phys A 194:535
4. Cederbaum LS (1975) J Phys B 8:290
5. Cederbaum LS, Domcke W (1977) Adv Chem Phys 36:205
6. von Niessen W, Schirmer J, Cederbaum L (1984) Comput Phys Rep 1:57
7. Ortiz JV, Zakrzewski VG, Dolgounitcheva O (1997) In: Kryachko E (ed) Conceptual trends in quantum chemistry. Kluwer, Dordrecht, p 3

Chapter 9
Algebraic–Diagrammatic Construction (ADC)

As was discussed in the beginning of Chap. 8, a finite perturbation expansion of the electron propagator does not reproduce its correct analytical form. While this applies also to the self-energy part beyond second order, here finite perturbation expansions together with the Dyson equation may provide viable approximations to the electron propagator, as shown by the third-order OVGF approach discussed in Sect. 8.3. The applicability of the OVGF approximation, however, is restricted to the energy region above the highest pole of $\boldsymbol{M}^-(\omega)$ and below the first pole of $\boldsymbol{M}^+(\omega)$. For the treatment of ionization energies and electron affinities outside that outer valence regime, the behavior of the self-energy part near its poles matters. This means that one has to recover the proper analytical form (8.19) of the dynamical self-energy part. Approximations of that type are obtained as a result of performing infinite summations of a certain class of diagrams. An example of such an infinite partial summation of diagrams in the case of the self-energy part is Hedin's GW approximation [1], which, in turn, is based on the famous random-phase approximation (RPA), being itself an infinite partial summation of diagrams of the polarization propagator (see Chap. 15). An alternative way of generating infinite partial summations in a diagrammatic perturbation expansion is the algebraic-diagrammatic construction (ADC) [2, 3]. In the following we will discuss how the ADC procedure can be performed in the case of the dynamic self-energy part. A direct ADC approach to the $\boldsymbol{G}^-(\omega)$ and $\boldsymbol{G}^+(\omega)$ parts of the electron propagator is presented in Chap. 10.

9.1 ADC Formulation of the Dynamic Self-Energy Part

According to Eq. (8.19), the dynamical self-energy consists of two parts, $\boldsymbol{M}^+(\omega)$ and $\boldsymbol{M}^-(\omega)$, referred to as affinity and ionization part, respectively. The ADC formulation applies independently to either of them. To be specific, we will consider the affinity part $\boldsymbol{M}^+(\omega)$ in the following and drop the superscripts + for notational ease. The treatment of $\boldsymbol{M}^-(\omega)$ is completely analogous.

J. Schirmer, *Many-Body Methods for Atoms, Molecules and Clusters*, Lecture Notes in Chemistry 94, https://doi.org/10.1007/978-3-319-93602-4_9

The starting point is the spectral representation according to Eq. (8.19). In matrix notation (see Eq. 8.45), the matrix element $M_{pq}(\omega)$ can be written as

$$M_{pq}(\omega) = \boldsymbol{m}_p^\dagger(\omega\mathbf{1} - \boldsymbol{\Omega})^{-1}\boldsymbol{m}_q \tag{9.1}$$

Here, $\boldsymbol{\Omega}$ is the diagonal matrix of self-energy pole locations ω_ν (affinity part), and $\boldsymbol{m}_p$ is a vector of Dyson amplitudes $m_{\nu p} = m_p^{(\nu)*}$. Now, the diagonal spectral representation can be replaced by the more general non-diagonal **ADC form**

$$M_{pq}(\omega) = \boldsymbol{U}_p^\dagger(\omega\mathbf{1} - \boldsymbol{K} - \boldsymbol{C})^{-1}\boldsymbol{U}_q \tag{9.2}$$

where $\boldsymbol{K} + \boldsymbol{C}$ is a constant hermitian matrix, referred to as ADC secular matrix, and $\boldsymbol{U}_p$ is a constant vector of 'effective coupling' matrix elements. The latter form results from applying a general unitary transformation $\boldsymbol{Q}$ to Eq. (9.1), where

$$\begin{aligned} \boldsymbol{U}_p &= \boldsymbol{Q}\boldsymbol{m}_p \\ \boldsymbol{K} + \boldsymbol{C} &= \boldsymbol{Q}\boldsymbol{\Omega}\boldsymbol{Q}^\dagger \end{aligned} \tag{9.3}$$

relate the Dyson amplitudes and self-energy poles to the corresponding quantities of the ADC representation.

At this point, the matrix elements of $\boldsymbol{K} + \boldsymbol{C}$ and $\boldsymbol{U}_p$ are still unspecified. However, we may suppose that the following perturbation expansions apply:

$$\boldsymbol{C} = \boldsymbol{C}^{(1)} + \boldsymbol{C}^{(2)} + \cdots \tag{9.4}$$

$$\boldsymbol{U}_p = \boldsymbol{U}_p^{(1)} + \boldsymbol{U}_p^{(2)} + \cdots \tag{9.5}$$

Note that both expansions begin at first order. In the case of $\boldsymbol{U}_p$, this reflects the fact that the perturbation expansion of $M_{pq}(\omega)$ begins at second order. The matrix $\boldsymbol{C}$ begins at first order by definition, as the zeroth-order contribution to the ADC secular matrix is represented by $\boldsymbol{K}$. Obviously, $\boldsymbol{K}$ can be identified with $\boldsymbol{\Omega}^{(0)}$, which allows us to specify $\boldsymbol{K}$ as the diagonal matrix of the HF energies of $2p$-$1h$, $3p$-$2h$, ... configurations of $(N+1)$ electrons:

$$\begin{aligned} K_{jab,jab} &= -\epsilon_j + \epsilon_a + \epsilon_b \\ K_{ijabc,ijabc} &= -\epsilon_i - \epsilon_j + \epsilon_a + \epsilon_b + \epsilon_c \\ &\vdots \end{aligned} \tag{9.6}$$

Here, $i, j, \ldots$ label occupied HF orbitals, while $a, b, \ldots$ refer to unoccupied ones. The HF configurations allow one to designate also the expansion manifold underlying the effective interaction matrix (or ADC secular matrix) and the $\boldsymbol{U}$ vectors. Let $\{I, J, \ldots\} \equiv \{jab, a < b; ijabc, i < j, a < b < c; \ldots\}$ denote $(N+1)$-electron HF configurations of successive $2p$-$1h$, $3p$-$2h$, ... excitation classes (excluding the $1p$ class), then the secular matrix elements and $\boldsymbol{U}$ vector elements may be written as C_{IJ} and $U_{J,p}$, respectively.

The ADC form (9.2) can be expanded in a formal perturbation series. For this purpose, $(\omega\mathbf{1} - \boldsymbol{K} - \boldsymbol{C})^{-1}$ is replaced with its geometric series,

$$\begin{aligned} M_{pq}(\omega) &= \boldsymbol{U}_p^\dagger(\omega\mathbf{1} - \boldsymbol{K} - \boldsymbol{C})^{-1}\boldsymbol{U}_q \\ &= \boldsymbol{U}_p^\dagger(\omega\mathbf{1} - \boldsymbol{K})^{-1}\sum_{n=0}^{\infty}\left\{\boldsymbol{C}(\omega\mathbf{1} - \boldsymbol{K})^{-1}\right\}^n \boldsymbol{U}_q \end{aligned} \tag{9.7}$$

and the perturbation expansions (9.4), (9.5) are used for $\boldsymbol{C}$ and $\boldsymbol{U}_p$ ($\boldsymbol{U}_q$).

Now, we can formulate the **ADC procedure** as follows:

Compare the formal perturbation expansion of the ADC *form* (9.7) *to the original diagrammatic perturbation expansion for the self-energy part* $M^+_{pq}(\omega)$ *through a given order n of perturbation theory. Beginning at second order and proceeding to higher order, this comparison allows one to determine successively the terms in the expansions* (9.4) *and* (9.5) *of* $\boldsymbol{C}$ *and* $\boldsymbol{U}_p$*, respectively.*

The procedure is best explained by actually performing it. Let us first consider the trivial **second-order** case. The second-order ADC form, referred to as ADC(2), reads

$$M_{pq}(\omega) = \boldsymbol{U}_p^{(1)\dagger}(\omega\mathbf{1} - \boldsymbol{K})^{-1}\boldsymbol{U}_q^{(1)} + O(3) \tag{9.8}$$

which is to be compared with the second-order diagram for $M^{(2)+}_{pq}(\omega)$ shown in Fig. 8.8 (time-ordering $t > t'$). The corresponding analytical expression, given by the first term of Eq. (8.17), can be written as follows

$$M^{(2)+}_{pq}(\omega) = \sum_{j,a<b} \frac{V_{pj[ab]}V_{ab[qj]}}{\omega + \epsilon_j - \epsilon_a - \epsilon_b + i\eta} \tag{9.9}$$

Comparing Eqs. (9.8) and (9.9) allows one to determine the first-order contribution to $\boldsymbol{U}_q$,

$$U^{(1)}_{jab,q} = V_{ab[qj]} \tag{9.10}$$

and confirm that

$$K_{jab,jab} = -\epsilon_j + \epsilon_a + \epsilon_b \tag{9.11}$$

Note that the infinitesimal $+i\eta$ is not essential here; any pole in the time-orderings contributing to $\boldsymbol{M}^+(\omega)$ is of the type $(\omega \cdots + i\eta)$. Obviously, there are no contributions to $\boldsymbol{C}$ at the ADC(2) level.

The **third-order** or ADC(3) level is more interesting, as here the self-energy no longer is a sum over simple poles. The ADC expansion through third order reads

$$
\begin{aligned}
M_{pq}(\omega) = {} & \boldsymbol{U}_p^{(1)\dagger}(\omega\mathbf{1} - \boldsymbol{K})^{-1}\boldsymbol{U}_q^{(1)} \\
& + \boldsymbol{U}_p^{(2)\dagger}(\omega\mathbf{1} - \boldsymbol{K})^{-1}\boldsymbol{U}_q^{(1)} + \boldsymbol{U}_p^{(1)\dagger}(\omega\mathbf{1} - \boldsymbol{K})^{-1}\boldsymbol{U}_q^{(2)} \\
& + \boldsymbol{U}_p^{(1)\dagger}(\omega\mathbf{1} - \boldsymbol{K})^{-1}\boldsymbol{C}^{(1)}(\omega\mathbf{1} - \boldsymbol{K})^{-1}\boldsymbol{U}_q^{(1)} + O(4)
\end{aligned} \tag{9.12}
$$

The new quantities arising here, namely $\boldsymbol{U}_p^{(2)}$ and $\boldsymbol{C}^{(1)}$, are to be determined by comparing the second and third line with the diagrams for $M_{pq}^{(3)+}(\omega)$. Before inspecting the diagrams, let us note that, as in second order, the $\boldsymbol{K}$ denominators are restricted to those of $2p$-$1h$ type. Accordingly, only the $2p$-$1h$ matrix elements of $\boldsymbol{U}_p^{(2)}$ and $\boldsymbol{C}^{(1)}$ come into play at third order.

The two third-order diagrams $T1$ and $T2$, constituting $\boldsymbol{M}^{(3)}(\omega)$, are shown in Fig. 8.9. Each three Goldstone diagrams (with $t > t'$), shown in Fig. 8.10 for $T1$, contribute to the affinity part, $\boldsymbol{M}^{(3)+}(\omega)$. The diagrams $T1(1)$ and $T2(1)$ can directly be compared to the third line in the ADC expansion (9.12). Let us consider the $T1(1)$ contribution,

$$
T1(1)_{pq} = \sum_{j,a<b}\sum_{j',a'<b'} \frac{V_{pj[ab]}}{\omega + \epsilon_j - \epsilon_a - \epsilon_b}\delta_{jj'}V_{ab[a'b']}\frac{V_{qj'[a'b']}}{\omega + \epsilon_j' - \epsilon_a' - \epsilon_b'} \tag{9.13}
$$

where a superfluous $\sum_{j'}\delta_{jj'}$ summation has been inserted for clarity. The comparison with the ADC expression (third line of Eq. 9.12) reproduces the $\boldsymbol{U}^{(1)}$ expressions already determined and yields the following contribution to $\boldsymbol{C}^{(1)}$:

$$
C_{jab,j'a'b'}^{(1)} \leftarrow \delta_{jj'}V_{ab[a'b']} \tag{9.14}
$$

In a similar way, the contribution of the $T2(1)$ diagram can be taken into account. The final result can be written in the form

$$
C_{jab,j'a'b'}^{(1)} = \delta_{jj'}V_{ab[a'b']} - \left(\delta_{aa'}V_{j'b[jb']} + \delta_{bb'}V_{j'a[ja']}\right) + \left(a' \leftrightarrow b'\right) \tag{9.15}
$$

Here, the $2p$-$1h$ configurations (jab) and $(j'a'b')$ are restricted by requiring $a < b$ and $a' < b'$; together with these restrictions, the form (9.15), being anti-symmetrized with respect to the index pairs (ab) and $(a'b')$, is consistent with the diagrammatic expressions.

To determine $U_{jab,p}^{(2)}$, we may inspect the diagrams $T1(2)$ and $T2(2)$ conforming to the third term in Eq. (9.12); diagrams $T1(3)$ and $T2(3)$ simply reproduce the hermitian conjugate expression (second term of Eq. 9.12).
The analytic expression for $T1(2)$ reads

$$
T1(2)_{pq} = \sum_{j,a<b} V_{pj[ab]}\frac{1}{\omega + \epsilon_j - \epsilon_a - \epsilon_b}\frac{1}{2}\sum_{kl}\frac{V_{ab[kl]}V_{kl[qj]}}{\epsilon_k + \epsilon_l - \epsilon_a - \epsilon_b} \tag{9.16}
$$

from which the contribution to $U_{jab,q}^{(2)}$ is readily derived:

$$U^{(2)}_{jab,q} \leftarrow \tfrac{1}{2}\sum_{kl} \frac{V_{ab[kl]}V_{kl[qj]}}{\epsilon_k + \epsilon_l - \epsilon_a - \epsilon_b} \tag{9.17}$$

Taking also T2(2) into account the full second-order contribution is obtained:

$$\begin{aligned} U^{(2)}_{jab,q} =& \tfrac{1}{2}\sum_{kl} \frac{V_{ab[kl]}V_{kl[qj]}}{\epsilon_k + \epsilon_l - \epsilon_a - \epsilon_b} \\ &+ \left(\sum_{ck} \frac{V_{ac[kj]}V_{kb[qc]}}{\epsilon_a + \epsilon_c - \epsilon_j - \epsilon_k}\right) - (a \leftrightarrow b) \end{aligned} \tag{9.18}$$

To summarize, the ADC(3) secular matrix and $\boldsymbol{U}$ vectors read

$$(K + C)_{jab,j'a'b'} = (-\epsilon_j + \epsilon_a + \epsilon_b)\delta_{jj'}\delta_{aa'}\delta_{bb'} + C^{(1)}_{jab,j'a'b'} \tag{9.19}$$

$$U_{jab,q} = U^{(1)}_{jab,q} + U^{(2)}_{jab,q} \tag{9.20}$$

where $C^{(1)}_{jab,j'a'b'}$ is given by Eq. (9.15), and the first- and second-order contributions to $U_{jab,q}$ by Eqs. (9.10), (9.18). The explicit ADC(3) expressions for both $\boldsymbol{M}^{+}(\omega)$ and $\boldsymbol{M}^{-}(\omega)$ are listed in Appendix A.9.

At the **fourth-order** level, the ADC procedure is already somewhat elaborate. A comprehensive presentation has been given in Ref. [3], and we may confine us here to a brief sketch. There are 10 fourth-order Abrikosov diagrams shown in Fig. 9.1. Each fourth-order diagram entails 24 time-orderings or Goldstone diagrams, so that there

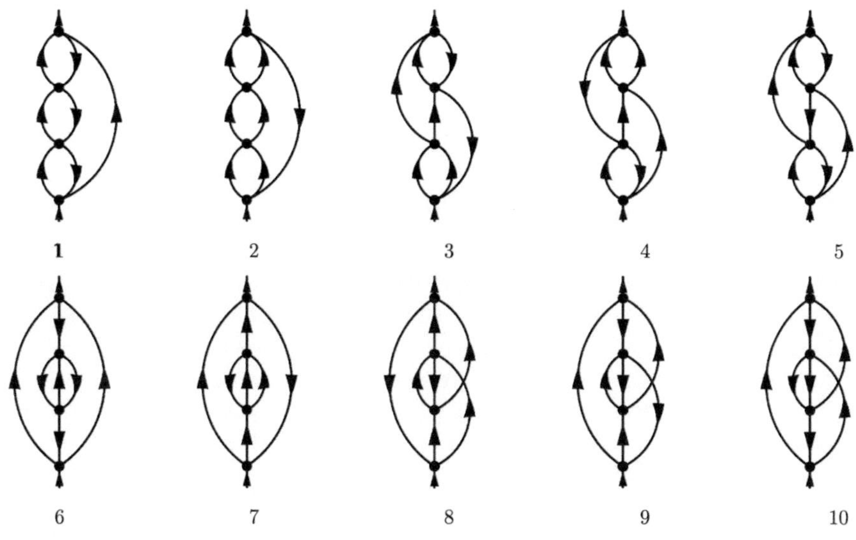

Fig. 9.1 Fourth-order Feynman diagrams (in Abrikosov form) for the dynamic self-energy part

are altogether 240 Goldstone diagrams, of which one half (120 diagrams) contributes to $M^{(4)+}$, the other half to $M^{(4)-}$. For the ADC procedure, many of these diagrams are redundant as they only recover ingredients already determined at the second- and third-order level. To determine those ADC(4) contributions which arise for the first time, it suffices to inspect a subset of certain key diagrams.

While the ADC(2) and ADC(3) configuration space for $\boldsymbol{M}^+(\omega)$ was spanned by the $2p$-$1h$ excitations, the next higher excitation class, $3p$-$2h$, comes explicitly into play at the fourth-order level. The ADC(4) expansions are of the structure

$$C_{jab,j'a'b'} = C^{(1)}_{jab,j'a'b'} + \underline{C^{(2)}_{jab,j'a'b'}} \tag{9.21}$$

$$C_{jab,i'j'a'b'c'} = \underline{C^{(1)}_{jab,i'j'a'b'c'}} \tag{9.22}$$

$$C_{ijabc,i'j'a'b'c'} = 0 \tag{9.23}$$

and

$$U_{jab,q} = U^{(1)}_{jab,q} + U^{(2)}_{jab,q} + \underline{U^{(3)}_{jab,q}} \tag{9.24}$$

$$U_{ijabc,q} = \underline{U^{(2)}_{ijabc,q}} \tag{9.25}$$

Here, the underlined contributions are those to be determined at the fourth-order level. Note that the $3p$-$2h$ components of the $\boldsymbol{U}$ vectors are of second order. There is a first-order $2p$-$1h$/$3p$-$2h$ coupling block of the $\boldsymbol{C}$ matrix, while the $3p$-$2h$ diagonal block of $\boldsymbol{C}$ vanishes. The $3p$-$2h$ diagonal block of the ADC(4) secular matrix, being of zeroth order, is given by

$$K_{ijabc,i'j'a'b'c'} = (-\epsilon_i - \epsilon_j + \epsilon_a + \epsilon_b + \epsilon_c)\delta_{ii'}\delta_{jj'}\delta_{aa'}\delta_{bb'}\delta_{cc'} \tag{9.26}$$

For the explicit ADC(4) expressions, the reader is referred to Ref. [3].

9.2 Dyson-ADC Secular Equations

In the preceding section, we have seen how the ADC procedure can be used to derive in a systematic way higher-order approximations to the ADC secular matrix $\boldsymbol{K} + \boldsymbol{C}$ and the matrix of $\boldsymbol{U}$ vectors. To generate an explicit representation of the dynamic self-energy part to be employed in the Dyson equation, the respective ADC(n) secular problem has to be solved. Alternatively, the ADC secular quantities can directly be incorporated into a common Dyson-ADC secular matrix. This will be discussed in the following.

For a given ADC secular matrix $\boldsymbol{K} + \boldsymbol{C}$ and matrix of $\boldsymbol{U}$ vectors, the dynamic self-energy in the form of Eq. (9.1) is obtained via diagonalization according to

$$(\boldsymbol{K}+\boldsymbol{C})\,\boldsymbol{Y}=\boldsymbol{Y}\boldsymbol{\Omega},\quad \boldsymbol{Y}^{\dagger}\boldsymbol{Y}=\mathbf{1} \tag{9.27}$$

Here, $\boldsymbol{\Omega}$ is the diagonal matrix of the poles of $\boldsymbol{M}(\omega)$, and $\boldsymbol{Y}$ is the matrix of eigenvectors $\boldsymbol{Y}_n$. The Dyson amplitudes are given by

$$m_{np}=\boldsymbol{Y}_n^{\dagger}\boldsymbol{U}_p \tag{9.28}$$

which may be written in a compact matrix notation as

$$\boldsymbol{m}=\boldsymbol{Y}^{\dagger}\boldsymbol{U} \tag{9.29}$$

As the comparison with Eq. (9.3) shows, the eigenvector matrix $\boldsymbol{Y}$ can be identified with the unitary transformation, $\boldsymbol{Q}=\boldsymbol{Y}$, that relates the diagonal (spectral representation) form (9.1) to the non-diagonal ADC form (9.2).

The Dyson equation can then be solved as described in Sect. 8.3 by diagonalizing the Dyson secular matrix (8.49)

$$\boldsymbol{A}=\begin{pmatrix}\boldsymbol{\epsilon}+\boldsymbol{\Sigma}(\infty) & \boldsymbol{m}^{-\dagger} & \boldsymbol{m}^{+\dagger}\\ \boldsymbol{m}^{-} & \boldsymbol{\Omega}^{-} & \mathbf{0}\\ \boldsymbol{m}^{+} & \mathbf{0} & \boldsymbol{\Omega}^{+}\end{pmatrix} \tag{9.30}$$

and expressing the electron propagator according to Eq. (8.50),

$$\boldsymbol{G}(\omega)=(\omega\mathbf{1}-\boldsymbol{A})^{-1}|_{11} \tag{9.31}$$

Here, the energies $\Omega_n^{\pm}$ and Dyson amplitudes $\boldsymbol{m}_p^{\pm}$ are obtained from the respective ADC approximation for $\boldsymbol{M}^{\pm}(\omega)$.

Rather than using the two-step approach described above, the ADC and Dyson eigenvalue problems can be combined into the eigenvalue problem of a common Dyson-ADC secular matrix $\boldsymbol{B}$, being of the form

$$\boldsymbol{B}=\begin{pmatrix}\boldsymbol{\epsilon}+\boldsymbol{\Sigma}(\infty) & \boldsymbol{U}^{-\dagger} & \boldsymbol{U}^{+\dagger}\\ \boldsymbol{U}^{-} & \boldsymbol{K}^{-}+\boldsymbol{C}^{-} & \mathbf{0}\\ \boldsymbol{U}^{+} & \mathbf{0} & \boldsymbol{K}^{+}+\boldsymbol{C}^{+}\end{pmatrix} \tag{9.32}$$

The electron propagator is given by the 11-block of $(\omega\mathbf{1}-\boldsymbol{B})^{-1}$,

$$\boldsymbol{G}(\omega)=(\omega\mathbf{1}-\boldsymbol{B})^{-1}|_{11} \tag{9.33}$$

Again, this can be seen by comparing Eqs. (8.44), (9.2) with the partitioning formulas (A1.26), (A1.27). Accordingly, Eq. (9.33) can be written as

$$\boldsymbol{G}(\omega)=\boldsymbol{X}(\omega\mathbf{1}-\boldsymbol{E})^{-1}\boldsymbol{X}^{\dagger}|_{11} \tag{9.34}$$

in terms of the eigenvalue and eigenvector matrices of $\boldsymbol{B}$:

$$\boldsymbol{B}\boldsymbol{X} = \boldsymbol{X}\boldsymbol{E}, \qquad \boldsymbol{X}^{\dagger}\boldsymbol{X} = \mathbf{1} \tag{9.35}$$

Here, $\boldsymbol{E}$ and $\boldsymbol{X}$ denote the diagonal matrix of eigenvalues and the eigenvector matrix, respectively. The obvious advantage of the Dyson-ADC secular matrix $\boldsymbol{B}$ is that one can directly compute selected ionization energies, $I_n = -E_n$, or electron affinities, $A_m = -E_m$, using iterative matrix diagonalization routines such as the Davidson [4] or Lanczos [5, 6] methods. It should be noted, however, that the desired roots, e.g., for the energetically lowest ionization energies, lie in the middle of the eigenvalue spectrum of $\boldsymbol{B}$, which complicates the use of iterative diagonalization techniques.

As we have argued in Sect. 8.2, the static part $\boldsymbol{\Sigma}(\infty)$ of the self-energy entering the Dyson-ADC secular matrix in the $1p/1h$ block should be chosen to be consistent with the respective approximation scheme employed for the dynamical self-energy part $\boldsymbol{M}(\omega)$. A practical method to determine $\boldsymbol{\Sigma}(\infty)$ for an ADC representation of $\boldsymbol{M}(\omega)$ is presented in Appendix A.5.

The structure of the $\boldsymbol{B}$ matrix at the ADC(3) level is shown in Fig. 9.2. The explicit configuration space comprises the $1h$ and $2h$-$1p$ configurations of $N-1$ particles, and the $1p$ and $2p$-$1h$ configurations of $N+1$ particles. The perturbation expansions of the $\boldsymbol{U}^{\pm}$ vector elements extend to second order, $\boldsymbol{U}^{(1,2)} = \boldsymbol{U}^{(1)} + \boldsymbol{U}^{(2)}$, while the $\boldsymbol{C}^{\pm}$ matrix blocks are confined to first order. At the fourth-order level, the explicit configuration spaces extend to the $3h$-$2p$ and $3p$-$2h$ configurations. The required perturbation expansions of the secular matrix elements are of the form of Eqs. (9.21)–(9.25).

Fig. 9.2 Structure of the Dyson secular matrix $\boldsymbol{B}$ at the ADC(3) level

	$1h/1p$	$2h$-$1p$	$2p$-$1h$
$1h/1p$	$\epsilon + \Sigma(\infty)$	$U^{-(1,2)\dagger}$	$U^{+(1,2)\dagger}$
$2h$-$1p$	$U^{-(1,2)}$	$K^{-} + C^{-(1)}$	0
$2p$-$1h$	$U^{+(1,2)}$	0	$K^{+} + C^{+(1)}$

Let us state some **general properties** of the Dyson-ADC approximation schemes for the electron propagator.

1. **Infinite partial summation**: The ADC(n) approximation (in the form of the Dyson-ADC secular problem) represents an infinite partial summation of the diagrammatic perturbation expansion, being *complete* through order n of perturbation theory. The ionization energies (electron affinities) of $1h$ ($1p$) main states are treated consistently through order n.
2. **Diagonalization and perturbation theory**: The resulting computational method combines the eigenvalue problem (*diagonalisation*) of a hermitian secular matrix with (finite) *perturbation expansions* of the secular matrix elements.
3. **Regularity**: With respect to convergence, the PT expansions of the secular matrix elements behave essentially like the RSPT expansion of the ground-state energy. The energy denominators are of the type

$$\nu \times c_{virt} - \nu \times \epsilon_{occ} \geq \nu \, \Delta\epsilon, \;\; \nu = 1, 2, \ldots \tag{9.36}$$

where $\Delta\epsilon$ is the energy gap between occupied and virtual HF orbitals. There are no "dangerous" denominators (with small or even zero absolute values) provided the energy gap is sufficiently large. This is referred to as the *regularity* of the PT expansions of the secular matrix elements.
4. **Compact configuration spaces**: At the second-order level, the explicit configuration space of the Dyson-ADC secular matrix comprises the $1h/1p$ and $2h$-$1p/2p$-$1h$ configurations. At each even order $n = 4, 6, \ldots$, the explicit configuration space grows by the next higher class of $(N-1)$- and $(N+1)$-electron configurations. Accordingly, for the order levels $n = 2m$ and $n = 2m + 1$, $m = 1, 2, \ldots$, the explicit configuration space comprises the classes $1h, \ldots, (m+1)h$-mp and $1p, \ldots, (m+1)p$-mh.
5. **Size-consistency**: The Dyson-ADC approach is *size-consistent*. For a system S consisting of two separated fragments A and B, the results obtained for a "local" ionization, say, on fragment A, do not depend on whether the method is applied to S or to fragment A. As is well known, this does not apply to the CI treatment based on restricted (as opposed to full) CI expansions (see Sect. 12.3).

As a consequence of the compactness property, the ADC(n) configuration spaces are smaller (more *compact*) than those of comparable CI expansions. The consistent treatment of $1h$ main states through second (and third order) requires the CI expansions to extend through the $3h$-$2p$ configurations. By contrast, the ADC(2) (and ADC(3)) configuration spaces are restricted to the $2h$-$1p$ configurations of $N-1$ particles, but also comprise the $2h$-$1p$ configurations of $N+1$ particles. With increasing order n, the CI expansions required for a consistent treatment grow twice as fast as their ADC counterparts: At each even order two further excitation classes rather than one have to be taken into account (see Table 9.1). For example, at the fourth-order level, the $1h$, $2h$-$1p, \ldots, 5h$-$4p$ CI expansion is contrasted with the ADC(4) configuration space spanned by the $1h/1p$, $2h$-$1p$/$2p$-$1h$, and $3h$-$2p$/$3p$-$2h$ configurations.

Table 9.1 ADC and CI configuration spaces required for a consistent treatment of $1h$ main state ionization energies through increasing order n

n	ADC	CI
0, 1	1h/1p	1h
2, 3	1h, 2h-1p/1p, 2p-1h	1h, 2h-1p, 3h-2p
4, 5	1h, 2h-1p, 3h-2p/1p, 2p-1h, 3p-2h	1h, 2h-1p,...,5h-4p
⋮	⋮	⋮

An alternative way of formulating the compactness property is based on the inspection of *truncation error orders* (TEO). When the ADC configuration space is truncated after the excitation class m, that is, the mh-$(m-1)p$ and mp-$(m-1)h$ configurations of $N-1$ and $N+1$ particles, respectively, the error in the $1h$ (or $1p$) main state ionization energies is of the order $2m$. For example, truncation after the $2h$-$1p$/$2p$-$1h$ configurations ($m = 2$) results in a fourth-order error in the $1h$ (and $1p$) main states. The corresponding TEO in the CI expansions is only 2, reflecting the fact that the $1h$/$3h$-$2p$ coupling is of first order, which means their omission causes a second-order error in the $1h$ main state ionization energies.

So far the compactness property has only been stated but not justified. At the diagrammatic level, a rationalization is as follows. According to the ADC analysis, the second-order diagram for $\boldsymbol{M}(\omega)$ (Fig. 8.7) corresponds to the $2h$-$1p$ and $2p$-$1h$ configurations (cf. the energy denominators in the analytical expressions). The next higher class, that is, $3h$-$2p$ (and $3p$-$2h$) configurations can appear upon a (1–3)-branching of one of the three free fermion lines in the second-order diagram (see the paragraph 'Systematic construction of Abrikosov diagrams' in Sect. 6.3). In order to obtain a valid diagram, the (1–3)-branching must be accompanied by a corresponding (3–1)-junction. Since both the branching and junction are first-order constituents (having one interaction dot each), real $3h$-$2p$/$3p$-$2h$ denominators can occur only in fourth- and higher-order diagrams. Obviously, this finding can be generalized: It takes two additional orders to install the next higher configuration class in the diagrams. Stated differently, mh-$(m-1)p$ and mp-$(m-1)p$ configurations enter the explicit ADC configuration space at the level of order $2m - 2$.

A more general justification of the compactness property can be inferred from the spectral representation (3.17) of the electron propagator. Let us consider the $(N-1)$-electron part,

$$G^{-}_{pq}(\omega) = \sum_n \frac{\langle\Psi_0|c_q^\dagger|\Psi_n^{N-1}\rangle\langle\Psi_n^{N-1}|c_p|\Psi_0\rangle}{\omega + E_n^{N-1} - E_0 - i\eta} \tag{9.37}$$

The exact $(N-1)$-electron states $|\Psi_n^{N-1}\rangle$ can be classified according to their descent from CI configuration classes, mh-$(m-1)p$, $m = 1, \ldots$. We will denote the excitation class of $|\Psi_n^{N-1}\rangle$ by $[n]$, that is, $[n] = \mu$ if $|\Psi_n^{N-1}\rangle$ derives from a μh-$(\mu - 1)p$ CI configuration. The amplitude $\langle\Psi_n^{N-1}|c_p|\Psi_0\rangle$ fulfills the following remarkable order relation [3, 7, 8]:

$$\langle\Psi_n^{N-1}|c_p|\Psi_0\rangle \sim O([n]-1) \tag{9.38}$$

This means that the lowest non-vanishing term in the perturbation expansion of the amplitude is of the order $[n]-1$. Now, we may ask at which order states of class $[n]$ will appear in the (diagrammatic) perturbation expansion of the electron propagator. Using the order relation (9.38), and the fact that the energy denominators on the right-hand side of (9.37) always begin at zeroth order, that is, in the form $\omega + \epsilon_a + \cdots - \epsilon_k \ldots$, the answer is $2[n]-2$, which is in agreement with the diagrammatic argument outlined above. A completely analogous reasoning applies to the $(N+1)$-electron states.

The size-consistency of the ADC(n) approximations is a consequence of the diagrammatic perturbation theory, more specifically, the linked-cluster theorem. The Feynman diagrams constituting the electron propagator or the self-energy part are **locally correct**. Let us again inspect the case of separate fragments. If a diagram begins with a local free fermion line, associated, for example, with a one-particle state of fragment A, then the entire diagram pertains to fragment A, because any part of the diagram is ultimately linked to the initial free fermion line and the interaction dots (or interaction lines) can have only fragment-A entries due to the separate fragment model (which implies that interaction points with both fragment-A and fragment-B entries vanish). The separation of the diagrams into subsets of A- and B-type is reflected in a corresponding separation of the ADC secular matrices. We will come back to the size-consistency of the ADC approximations in Chap. 12.

Exercises

9.1 Apply the systematic construction of Abrikosov diagrams discussed in Sect. 6.3 to the self-energy diagrams and verify that Fig. 9.1 shows all fourth-order diagrams for $\boldsymbol{M}(\omega)$.

9.2 Use the matrix representation of self-energy diagrams (Exercise 8.4) to generate the fourth-order diagrams for $\boldsymbol{M}(\omega)$.

9.3 Redraw the diagrams 3–5 in Fig. 9.1 with the order of the two inner vertices inverted. Apply the same procedure to diagram 8. Discuss the finding for diagram 8 with regard to the rule (A5) for Abrikosov diagrams (Sect. 6.3).

9.4 Use RSPT for the $(N-1)$-electron state $|\Psi_{abjkl}\rangle$ deriving from $c_a^\dagger c_b^\dagger c_j c_k c_l|\Phi_0\rangle$ and verify that the matrix element $\langle\Psi_{abjkl}|c_p|\Psi_0\rangle$ vanishes in first order.

References

1. Hedin L (1965) Phys Rev A 139:796
2. Schirmer J (1982) Phys Rev A 26:2395
3. Schirmer J, Cederbaum LS, Walter O (1983) Phys Rev A 28:1237
4. Davidson ER (1975) J Comput Phys 17:87
5. Parlett BN (1980) The symmetric eigenvalue problem. Prentice Hall, Englewood Cliffs
6. Cullum JK, Willoughby RA (1985) Lanczos algorithm for large symmetric eigenvalue computations. Birkhäuser, Boston
7. Schirmer J (1991) Phys Rev A 43:4647
8. Schirmer J, Mertins F (2009) Theor Chem Acc 125:145

Chapter 10
Direct ADC Procedure for the Electron Propagator

In the preceding Chap. 9, the algebraic–diagrammatic construction (ADC) was established as a procedure to derive systematically higher-order approximations to the $\boldsymbol{M}^+(\omega)$ and $\boldsymbol{M}^-(\omega)$ parts of the dynamical self-energy $\boldsymbol{M}(\omega)$. The respective ADC matrices could then be incorporated within a common Dyson secular matrix allowing one to solve the Dyson equation in the form of a hermitian eigenvalue problem. As a characteristic feature, the Dyson approach combines the $(N-1)$- and $(N+1)$-particle problems as parts of a common computational scheme. The ADC procedure, however, is quite general and can be applied to the electron propagator as well, more precisely, to the $\boldsymbol{G}^-(\omega)$ or $\boldsymbol{G}^+(\omega)$ parts. Such a direct ADC approach, described in this chapter, avoids the Dyson equation altogether and leads to separate ADC schemes for the $(N-1)$- and $(N+1)$-particle problems. The obvious advantage of these *direct* or *non-Dyson* schemes is the smaller size of the secular problem, being roughly half the size of a comparable Dyson formulation. Moreover, the lowest ionization energies (or electron affinities) are at the edge of the eigenvalue spectrum (and not in the middle of the joint $(N \pm 1)$-particle energy spectrum). The price to be paid here is a higher complexity in the PT expansions of the secular matrix elements, which, however, is workable through the third-order ADC(3) level.

10.1 ADC Representation of $G^-(\omega)$

The direct ADC procedures for the $(N \pm 1)$-parts of the electron propagator are completely analogous, and we shall confine ourselves to the $\boldsymbol{G}^-(\omega)$ part in the following. As discussed in Sect. 3.1, the $(N-1)$-particle part of the electron propagator can be written in the form of Eq. (3.25),

$$G^-_{pq}(\omega) = \left\langle \Psi_0 | c^\dagger_q (\omega + \hat{H} - E_0 - i\eta)^{-1} c_p | \Psi_0 \right\rangle \tag{10.1}$$

J. Schirmer, *Many-Body Methods for Atoms, Molecules and Clusters*, Lecture Notes in Chemistry 94, https://doi.org/10.1007/978-3-319-93602-4_10

that is, as matrix elements of the generalized resolvent operator $(\omega + \hat{H} - E_0 - i\eta)^{-1}$. Inserting the resolution of the identity in terms of the exact $(N-1)$-electron states,

$$\hat{\mathbb{1}} = \sum |\Psi_n^{N-1}\rangle\langle\Psi_n^{N-1}| \tag{10.2}$$

before and after the resolvent operator on the right-hand side yields the spectral representation

$$G^-_{pq}(\omega) = \sum_n \frac{x_p^{(n)} x_q^{(n)*}}{\omega - \omega_n - i\eta} \tag{10.3}$$

where

$$\omega_n = -I_n = E_0 - E_n^{N-1} \tag{10.4}$$

and

$$x_p^{(n)} = \langle\Psi_n^{N-1}|c_p|\Psi_0\rangle \tag{10.5}$$

are the negative ionization energies and spectroscopic factors, respectively. The infinitesimal $-i\eta$ in the denominator no longer is essential and will be omitted in the following. Using a compact matrix notation, the spectral representation (10.3) can be written as

$$\boldsymbol{G}^-(\omega) = \boldsymbol{x}(\omega\mathbf{1} - \boldsymbol{\Omega})^{-1}\boldsymbol{x}^\dagger \tag{10.6}$$

where $\boldsymbol{\Omega}$ is the diagonal matrix of negative ionization energies, and $\boldsymbol{x}$ is the matrix of elements $x_{pn} = x_p^{(n)}$.

For a convenient notation, we introduce the transposed matrix

$$\tilde{\boldsymbol{G}}(\omega) = \boldsymbol{G}^-(\omega)^t = \tilde{\boldsymbol{x}}^\dagger(\omega\mathbf{1} - \boldsymbol{\Omega})^{-1}\tilde{\boldsymbol{x}} \tag{10.7}$$

where $\tilde{\boldsymbol{x}}$ is the transpose of $\boldsymbol{x}$, that is, $\tilde{x}_{np} = x_p^{(n)}$. In a similar way as in Sect. 9.1, the diagonal spectral representation (10.7) can be transformed into the non-diagonal **ADC representation**

$$\tilde{\boldsymbol{G}}(\omega) = \boldsymbol{f}^\dagger(\omega - \boldsymbol{K} - \boldsymbol{C})^{-1}\boldsymbol{f} \tag{10.8}$$

by means of a general unitary transformation yet to be determined (see Eq. 9.3). In the present case, the transformation can be made physically more explicit by supposing a complete set of so-called **intermediate states**, $|\tilde{\Psi}_I^{N-1}\rangle$, mediating between the exact energy eigenstates and the $(N-1)$-electron HF (or CI) configurations. Indicative of the latter connection, the capital letter indices refer to HF configurations,

$$\{I, J, \dots\} = \{k;\ akl, k < l;\ abjkl, a < b, j < k < l; \dots\} \tag{10.9}$$

classified as $1h$, $2h$-$1p$, $3h$-$2p, \dots,$ configurations. As before, the indices $a, b, c, \dots$ and $i, j, k, \dots$ denote unoccupied (virtual) and occupied orbitals, respectively, with regard to the HF ground state. In the following, the designation $N-1$ in the notation of the intermediate states will be skipped, as the affiliation with $(N-1)$-electron states can be inferred from the configuration indices $I, J, \ldots$. In Chap. 11, a direct construction of the intermediate states will be presented. For the present purpose, though, it suffices to suppose just the existence of such states, dispensing with the need for further specification.

Analogously to the spectral representation, the ADC form (10.8) can be obtained by using in Eq. (10.1) the resolution of the identity

$$\hat{\mathbb{1}} = \sum |\tilde{\Psi}_I\rangle\langle\tilde{\Psi}_I| \tag{10.10}$$

in terms of the intermediate states. This allows us to define the ADC secular matrix $\boldsymbol{K}+\boldsymbol{C}$ and the matrix $\boldsymbol{f}$ of "effective" transition amplitudes as representations with respect to the (so far hypothetical) intermediate states:

$$\begin{aligned}(\boldsymbol{K}+\boldsymbol{C})_{IJ} &= -\langle\tilde{\Psi}_I \hat{H} - E_0 \tilde{\Psi}_J\rangle \\ f_{Iq} &= \langle\tilde{\Psi}_I|c_q|\Psi_0\rangle\end{aligned} \tag{10.11}$$

In the ensuing Sect. 10.2, we discuss how the ADC matrices $\boldsymbol{K}+\boldsymbol{C}$ and $\boldsymbol{f}$ can successively be derived from the diagrammatic perturbation expansion for the electron propagator part $\boldsymbol{G}^{-}(\omega)$.

Let us assume that the ADC procedure has provided approximate (or exact) expressions for $\boldsymbol{K}+\boldsymbol{C}$ and $\boldsymbol{f}$. Then, the ADC **secular equations**,

$$(\boldsymbol{K}+\boldsymbol{C})\boldsymbol{X} = \boldsymbol{X}\boldsymbol{\Omega}, \qquad \boldsymbol{X}^{\dagger}\boldsymbol{X} = \mathbf{1} \tag{10.12}$$

allow one to derive the physical information of interest. Here $\boldsymbol{X}$ denotes the matrix of eigenvectors and $\boldsymbol{\Omega}$ is the diagonal matrix of eigenvalues ω_n, to be identified with the negative ionization energies, $\omega_n = -I_n$. The spectroscopic factors, $x_p^{(n)}$, are obtained from the scalar product of the nth eigenvector and the respective columns of the matrix $\boldsymbol{f}$:

$$x_p^{(n)} = \sum_J X_{Jn}^{*} f_{Jp} \tag{10.13}$$

The eigenvector components can be viewed as the expansion coefficients

$$X_{Jn} = \langle\tilde{\Psi}_J|\Psi_n^{N-1}\rangle \tag{10.14}$$

of the exact (or approximate) energy eigenstates written as linear combinations of the hypothetical intermediate states.

10.2 Explicit ADC Procedure Through Second Order

So far the ADC form (10.8) is merely a formal construct yet to be related to the diagrammatic perturbation expansion of $\boldsymbol{G}^-(\omega)$. To this end, one may expand the ADC form in a (still formal) perturbation series supposing the same MP partitioning of the hamiltonian as in the diagrammatic expansion of the electron propagator. Both the ADC secular matrix and the $\boldsymbol{f}$ matrix are subject to perturbation expansions

$$\boldsymbol{K} + \boldsymbol{C} = \boldsymbol{K} + \boldsymbol{C}^{(1)} + \boldsymbol{C}^{(2)} + \boldsymbol{C}^{(3)} + \dots \tag{10.15}$$

$$\boldsymbol{f} = \boldsymbol{f}^{(0)} + \boldsymbol{f}^{(1)} + \boldsymbol{f}^{(2)} + \dots \tag{10.16}$$

Here $\boldsymbol{K}$ denotes the zeroth-order part of the secular matrix, that is, the diagonal matrix of (negative) HF ionization energies,

$$\begin{aligned} K_{kk} &= \epsilon_k \\ K_{akl,akl} &= \epsilon_k + \epsilon_l - \epsilon_a \\ &\vdots \end{aligned} \tag{10.17}$$

The partitioning of the ADC matrix into a zeroth-order part $\boldsymbol{K}$ and a remainder $\boldsymbol{C}$ allows one to expand the resolvent matrix $(\omega - \boldsymbol{K} - \boldsymbol{C})^{-1}$ as a geometric series,

$$\tilde{\boldsymbol{G}}(\omega) = \boldsymbol{f}^\dagger(\omega - \boldsymbol{K} - \boldsymbol{C})^{-1}\boldsymbol{f} = \boldsymbol{f}^\dagger\,(\omega - \boldsymbol{K})^{-1}\sum_{\nu=0}^{\infty}\left(\frac{\boldsymbol{C}}{\omega - \boldsymbol{K}}\right)^{\nu}\boldsymbol{f} \tag{10.18}$$

Together with the PT expansions for $\boldsymbol{C}$ and $\boldsymbol{f}$, this establishes the formal ADC series for $\tilde{\boldsymbol{G}}(\omega)$.

The ADC expansion can further be specified by considering the block structure of the ADC matrices associated with the partitioning of the configurations J into classes $\mu = 1, 2, 3, \dots$ of $1h$, $2h$-$1p$, $3h$-$2p$, ... excitations. The respective matrix blocks will be denoted by $\boldsymbol{K}_\mu$, $\boldsymbol{C}_{\mu\mu'}$, and $\boldsymbol{f}_\mu$, where μ denotes the class of μh-$(\mu - 1)p$ excitations. The $\boldsymbol{f}_\mu$ matrix blocks may be distinguished further according to the second matrix index, f_{Iq}, as particle ($\overline{n}_q = 1$) or hole terms ($n_q = 1$).

The ADC expansion (10.18) is to be compared with the diagrammatic PT expansion for $\boldsymbol{G}^-(\omega)$. According to Chap. 7, the latter is constituted directly from the class of time-ordered (Goldstone) diagrams with $t < t'$. As an instructive demonstration, let us perform the ADC procedure for $\boldsymbol{G}^-(\omega)$ through second order of PT.

Zeroth Order:
The zeroth-order ADC matrix

$$\tilde{\boldsymbol{G}}^{(0)}(\omega) = \boldsymbol{f}^{(0)\dagger}(\omega - \boldsymbol{K})^{-1}\boldsymbol{f}^{(0)} \tag{10.19}$$

is to be compared with the explicit zeroth-order expression

$$\tilde{G}_{pq}^{(0)}(\omega) = \delta_{pq}(\omega - \epsilon_p)^{-1} n_p$$

This shows that all blocks $\boldsymbol{f}_\mu^{(0)}$ vanish except for $\mu = 1$,

$$\boldsymbol{f}_\mu^{(0)} = 0, \text{ for } \mu > 1 \tag{10.20}$$

and $\boldsymbol{f}_1^{(0)}$ is given by

$$f_{pq}^{(0)} = \delta_{pq} n_p \tag{10.21}$$

First Order:

Since $\boldsymbol{f}_1^{(0)}$ is the only non-vanishing zeroth-order block of $\boldsymbol{f}$, the first-order ADC form can be written more specifically as

$$\begin{aligned}\tilde{\boldsymbol{G}}^{(1)}(\omega) = \boldsymbol{f}_1^{(1)\dagger}(\omega - \boldsymbol{K}_1)^{-1}\boldsymbol{f}_1^{(0)} + h.c.\\ +\boldsymbol{f}_1^{(0)\dagger}(\omega - \boldsymbol{K}_1)^{-1}\boldsymbol{C}_{11}^{(1)}(\omega - \boldsymbol{K}_1)^{-1}\boldsymbol{f}_1^{(0)}\end{aligned} \tag{10.22}$$

displaying the matrix blocks $\boldsymbol{f}_1^{(1)}$ and $\boldsymbol{C}_{11}^{(1)}$ as new constituents. The formal ADC expression is to be compared to the (vanishing) first-order propagator contribution, $\boldsymbol{G}^{(1)-}(\omega) = 0$, (see Sect. 6.2), which gives

$$\begin{aligned}\boldsymbol{C}_{11}^{(1)} &= \boldsymbol{0}\\ \boldsymbol{f}_1^{(1)} &= \boldsymbol{0}\end{aligned} \tag{10.23}$$

Second Order:

In second order, the ADC expansion takes on the form

$$\begin{aligned}\tilde{\boldsymbol{G}}^{(2)}(\omega) =& \boldsymbol{f}_1^{(2)\dagger}(\omega - \boldsymbol{K}_1)^{-1}\boldsymbol{f}_1^{(0)} + h.c. && (a)\\ &+\boldsymbol{f}_1^{(0)\dagger}(\omega - \boldsymbol{K}_1)^{-1}\boldsymbol{C}_{11}^{(2)}(\omega - \boldsymbol{K}_1)^{-1}\boldsymbol{f}_1^{(0)} && (b)\\ &+\boldsymbol{f}_2^{(1)\dagger}(\omega - \boldsymbol{K}_2)^{-1}\boldsymbol{f}_2^{(1)} + \dots && (c)\\ &+\boldsymbol{f}_1^{(0)\dagger}(\omega - \boldsymbol{K}_1)^{-1}\boldsymbol{C}_{12}^{(1)}(\omega - \boldsymbol{K}_2)^{-1}\boldsymbol{C}_{21}^{(1)}(\omega - \boldsymbol{K}_1)^{-1}\boldsymbol{f}_1^{(0)} + \dots && (d)\\ &+\boldsymbol{f}_1^{(0)\dagger}(\omega - \boldsymbol{K}_1)^{-1}\boldsymbol{C}_{12}^{(1)}(\omega - \boldsymbol{K}_2)^{-1}\boldsymbol{f}_2^{(1)} + \dots + h.c. && (e)\end{aligned} \tag{10.24}$$

Here the findings (10.20), (10.23) from the zeroth- and first-order levels have already been taken into account. The dots in lines (*c*), (*d*), and (*e*) indicate terms with higher block indices μ, e.g., $\boldsymbol{f}_\mu^{(1)\dagger}(\omega - \boldsymbol{K}_\mu)^{-1}\boldsymbol{f}_\mu^{(1)}$, $\mu > 2$, in line (*c*). Anticipating the result of the comparison with the second-order diagrams, none of the latter terms, having $(\omega - \boldsymbol{K}_\mu)^{-1}$ denominators, $\mu > 2$, are retrieved in the diagrams, which means that the quantities $\boldsymbol{f}_\mu^{(1)}$ vanish for $\mu > 2$,

$$\boldsymbol{f}_{\mu}^{(1)} = \mathbf{0}, \ \boldsymbol{C}_{1\mu}^{(1)} = \mathbf{0} \ \text{for} \ \mu > 2 \tag{10.25}$$

and need no longer be considered in second and higher order.

The non-vanishing ADC terms, to be determined by the comparison with the second-order diagrams, are $\boldsymbol{f}_1^{(2)}$, $\boldsymbol{f}_2^{(1)}$, $\boldsymbol{C}_{12}^{(1)}$, and $\boldsymbol{C}_{11}^{(2)}$. As discussed in Chap. 7, there are each 12 second-order Goldstone diagrams contributing to $\boldsymbol{G}^+(\omega)$ and $\boldsymbol{G}^-(\omega)$, respectively. The diagrams for $\boldsymbol{G}^+(\omega)$ are shown in Fig. 7.3; the diagrams for $\boldsymbol{G}^-(\omega)$ are obtained by simply turning those of Fig. (7.3) upside down; corresponding to the numbering (1)–(12) in Fig. 7.3, we use the denotations $(\overline{1}), \ldots, (\overline{12})$ for the upside-down diagrams. Obviously, the diagrams $(\overline{1})-(\overline{6})$, $(\overline{11})$, and $(\overline{12})$ can directly be assigned to individual terms in the ADC expansion (10.24), namely $(\overline{1})$ to (d); $(\overline{2})$, $(\overline{3})$ to (e); $(\overline{4})$ to (c); $(\overline{5})$, $(\overline{6})$, $(\overline{11})$, $(\overline{12})$ to (a). Here, the derivation of the ADC terms from the diagrammatic expressions is straightforward. As an example, let us consider diagram $(\overline{3})$. This diagram can directly be identified with the ADC contribution (e), while the hermitian conjugate term relates to diagram $(\overline{2})$. Exchanging the indices p and q to be consistent with $\tilde{G}_{pq}$ the diagrammatic expression for $(\overline{3})$ reads

$$\tilde{A}_{pq}^{(2,3)}(\omega) = -\frac{1}{\omega - \epsilon_p} \sum_{a,i<j} \frac{V_{ij[pa]}}{\omega - \epsilon_i - \epsilon_j + \epsilon_a} \frac{V_{qa[ij]}}{\epsilon_q + \epsilon_a - \epsilon_i - \epsilon_j} n_p \bar{n}_q \tag{10.26}$$

Here, again, the infinitesimal $-i\eta$ in the ω-denominators has been omitted. The comparison of Eq. (10.26) and the ADC expression (e) of Eq. (10.24) allows us to determine the first-order contributions to the ADC matrix blocks $\boldsymbol{C}_{12}$ and $\boldsymbol{f}_2$:

$$C_{p,akl}^{(1)} = -V_{kl[pa]} n_p \tag{10.27}$$

$$f_{akl,q}^{(1)} = \frac{V_{qa[kl]}}{\epsilon_q + \epsilon_a - \epsilon_k - \epsilon_l} \bar{n}_q \tag{10.28}$$

It should be noted that only the relative sign of the $\boldsymbol{C}_{12}^{(1)}$ and $\boldsymbol{f}_2^{(1)}$ matrix elements can be extracted from diagram $(\overline{3})$; the absolute signs can be chosen consistently, herewith fixing the absolute sign of the entire $\boldsymbol{C}_{12}$ and $\boldsymbol{f}_2$ matrix blocks. Obviously, $\boldsymbol{C}_{12}^{(1)}$ and $\boldsymbol{f}_2^{(1)}$ could have been derived as well by equating the diagrams $(\overline{1})$ and $(\overline{4})$ with the terms (d) and (c), respectively. This redundancy can, of course, be used for checking the results in various ways.

Note that the $\boldsymbol{f}_1^{(2)}$ contributions deriving from comparing term (a) with the diagrams $(\overline{11})$, $(\overline{12})$ (or the $h.c.$ part with $(\overline{5})$, $(\overline{6})$) are of particle type, $\bar{n}_q = 1$.

The remaining four diagrams $(\overline{7})-(\overline{10})$ need special consideration since they do not individually match any of the ADC terms. Moreover, each diagram introduces ω_5-denominators, that is, denominators of the type $(\omega + \epsilon_a + \epsilon_b - \epsilon_i - \epsilon_j - \epsilon_k)^{-1}$ as there are cuts between successive vertices crossed by two particle and three hole lines (plus the ω-line). This seems to bring into play the $3h$-$2p$ configurations of class $\mu = 3$. However, when all four diagrams are combined, the ω_5-denominators cancel, and the resulting expressions can be assigned to the ADC terms (a) and (b), as will be shown in the following.

The diagrams $(\bar{7})$–$(\overline{10})$ differ only in their denominator products so that they can be combined according to

$$(\bar{7}) + (\bar{8}) + (\bar{9}) + (\overline{10})\Big|_{pq} = -\tfrac{1}{2}\sum_{a,b,j} V_{ab[pj]}V_{qj[ab]}X_{(pqabj)}n_p n_q \tag{10.29}$$

where $X_{(pqabj)}$ is the sum of the four denominator products arising in the diagrams $(\bar{7})$–$(\overline{10})$:

$$\begin{aligned} X_{(pqabj)} =& \omega_5^{-1}(\omega - \epsilon_p)^{-1}(\omega - \epsilon_q)^{-1} + \\ & \omega_5^{-1}(\omega - \epsilon_q)^{-1}(\epsilon_a + \epsilon_b - \epsilon_q - \epsilon_j)^{-1} + \\ & \omega_5^{-1}(\omega - \epsilon_p)^{-1}(\epsilon_a + \epsilon_b - \epsilon_p - \epsilon_j)^{-1} + \\ & \omega_5^{-1}(\epsilon_a + \epsilon_b - \epsilon_p - \epsilon_j)^{-1}(\epsilon_a + \epsilon_b - \epsilon_q - \epsilon_j)^{-1} \end{aligned}$$

Here, we use the abbreviation

$$\omega_5 = \omega - \epsilon_p - \epsilon_q - \epsilon_j + \epsilon_a + \epsilon_b$$

for the common $(3h\text{-}2p)$-type denominator. As the following small calculation shows, ω_5 is cancelled in the denominator and rather re-appears in the numerator:

$$\begin{aligned} X_{(pqabj)} &= \omega_5^{-1}\left[(\omega - \epsilon_p)^{-1} + (\epsilon_a + \epsilon_b - \epsilon_q - \epsilon_j)^{-1}\right]\left[(\omega - \epsilon_q)^{-1} + (\epsilon_a + \epsilon_b - \epsilon_p - \epsilon_j)^{-1}\right] \\ &= \omega_5\,(\omega - \epsilon_p)^{-1}(\omega - \epsilon_q)^{-1}(\epsilon_a + \epsilon_b - \epsilon_p - \epsilon_j)^{-1}(\epsilon_a + \epsilon_b - \epsilon_q - \epsilon_j)^{-1} \end{aligned}$$

Now, we may slightly rewrite ω_5,

$$\omega_5 = \tfrac{1}{2}(\omega - \epsilon_p) + \tfrac{1}{2}(\omega - \epsilon_q) + (\epsilon_a + \epsilon_b - \epsilon_j - \tfrac{1}{2}\epsilon_p - \tfrac{1}{2}\epsilon_q)$$

so that $X_{(pqabj)}$ splits into three terms,

$$\begin{aligned} X_{(pqabj)} =& \frac{1}{\omega - \epsilon_p}\,\frac{1}{\omega - \epsilon_q}\,\frac{\epsilon_a + \epsilon_b - \epsilon_j - \frac{1}{2}(\epsilon_p + \epsilon_q)}{(\epsilon_a + \epsilon_b - \epsilon_p - \epsilon_j)(\epsilon_a + \epsilon_b - \epsilon_q - \epsilon_j)} \\ & + \left(\frac{1}{\omega - \epsilon_p} + \frac{1}{\omega - \epsilon_q}\right)\tfrac{1}{2}(\epsilon_a + \epsilon_b - \epsilon_p - \epsilon_j)^{-1}(\epsilon_a + \epsilon_b - \epsilon_q - \epsilon_j)^{-1} \end{aligned}$$

Using this result in the full diagrammatic expression, (10.29) leads to a corresponding tripartite form fitting naturally in the ADC expansion (10.24). Here, the contribution featuring a product of poles,

$$\frac{n_p}{\omega - \epsilon_p} \left\{ -\tfrac{1}{2} \sum_{a,b,j} V_{ab[pj]} V_{qj[ab]} \frac{\epsilon_a + \epsilon_b - \epsilon_j - \frac{1}{2}(\epsilon_p + \epsilon_q)}{(\epsilon_a + \epsilon_b - \epsilon_p - \epsilon_j)(\epsilon_a + \epsilon_b - \epsilon_q - \epsilon_j)} \right\} \frac{n_q}{\omega - \epsilon_q}$$

corresponds to the ADC term (b), allowing us to derive the matrix elements of $\boldsymbol{C}_{11}^{(2)}$:

$$C_{pq}^{(2)} = \tfrac{1}{2} \sum_{a,b,j} v_{abpj} v_{qjab} (\epsilon_a + \epsilon_b - \epsilon_j - \tfrac{1}{2}\epsilon_p - \tfrac{1}{2}\epsilon_q), \; n_p = n_q = 1 \tag{10.30}$$

where the short notation

$$v_{rsuw} = \frac{V_{rs[uw]}}{\epsilon_r + \epsilon_s - \epsilon_u - \epsilon_w} \tag{10.31}$$

is used here and in the following.

Similarly, the two one-pole contributions match the ADC term (a) and its hermitian conjugate, supplying the matrix elements of $\boldsymbol{f}_1^{(2)}$:

$$f_{pq}^{(2)} = \frac{1}{4} \sum_{a,b,j} v_{abpj} v_{qjab} \, n_q \tag{10.32}$$

According to the restriction $n_q = 1$, this is a contribution to the hole part of $\boldsymbol{f}_1^{(2)}$.

Here a remark is appropriate. While the allocation of contributions to $\boldsymbol{C}_{11}^{(2)}$ and $\boldsymbol{f}_1^{(2)}$ follows from the diagrammatic expressions in a natural way, it can be modified, to a certain extent, by obvious algebraic manipulations. Consider a contribution to the ADC term (b) of the form

$$\frac{1}{\omega - \epsilon_p} z_{pq} \frac{1}{\omega - \epsilon_q}, \quad n_p = n_q = 1, \, p \neq q$$

where z_{pq} can be seen as contributing to $C_{pq}^{(2)}$. Now, we may apply partial fraction decomposition to the pole product yielding

$$\frac{1}{\omega - \epsilon_p} z_{pq} \frac{1}{\omega - \epsilon_q} = \frac{1}{\omega - \epsilon_p} \frac{z_{pq}}{\epsilon_p - \epsilon_q} + \frac{1}{\omega - \epsilon_q} \frac{z_{pq}}{\epsilon_q - \epsilon_p}$$

The form on the right-hand side matches the ADC terms (a) and the hermitian conjugate, and, as a consequence, now a contribution

$$\tilde{f}_{pq} = \frac{z_{pq}}{\epsilon_p - \epsilon_q}$$

can be allocated to the hole part of the $\boldsymbol{f}_1$ matrix. Note that the new $\boldsymbol{f}_{1h}$ contribution is anti-hermitian, $\tilde{f}_{pq} = -\tilde{f}_{qp}^*$. This shows that the off-diagonal matrix elements of

$\boldsymbol{C}_{11}$ (or parts thereof) could be transferred into $\boldsymbol{f}_{1h}$, resulting ultimately in a diagonal $\boldsymbol{C}_{11}$ matrix. While such a transfer from $\boldsymbol{C}$ contributions to $\boldsymbol{f}$ is formally possible, it does not lead to viable computational schemes, as is indicated by the "dangerous" denominator $(\epsilon_p - \epsilon_q)^{-1}$ in $\tilde{f}_{pq}$. Inversely, any anti-hermitian contributions in f_{pq} can be transferred into the off-diagonal matrix elements C_{pq} and C_{qp} of $\boldsymbol{C}_{11}$. A unique definition of both $\boldsymbol{C}_{11}$ and $\boldsymbol{f}_{1h}$ can be obtained by requiring that $\boldsymbol{f}_{1h}$ is hermitian.

At the **third-order** ADC level, the three diagrams $T1$, $T2$, and $T3$ shown in Fig. 6.8 have to be considered. Each diagram gives rise to 120 time-ordered (Goldstone) diagrams of which 60 contribute to $\boldsymbol{G}^-(\omega)$. The full ADC analysis has been given in Ref. [1], and we may confine ourselves to a few remarks. Due to their topological similarity, the treatment of the $T1$ and $T2$ Goldstone manifolds is largely analogous. Many of the 60 time-orderings are redundant for the ADC procedure as they repeat only terms already established at second and first order. Accordingly, it suffices to focus on certain key diagrams. The $T3$ diagrams can be treated in a closed analytical way, without the need to spell out the 60 time-orderings. The ADC contributions coming into play at the third-order level are $\boldsymbol{C}_{11}^{(3)}$, $\boldsymbol{C}_{12}^{(2)}$, $\boldsymbol{C}_{22}^{(1)}$ in the secular matrix, and $\boldsymbol{f}_1^{(3)}$, $\boldsymbol{f}_2^{(2)}$ in the spectroscopic amplitudes. The explicit ADC(3) expressions are listed in Appendix A.9.

10.3 Properties of the non-Dyson ADC Schemes

The ADC procedure for the ionization part $\boldsymbol{G}^-(\omega)$ of the electron propagator as outlined in the previous section leads to direct (or non-Dyson) ADC schemes for the computation of ionization energies and spectroscopic amplitudes. In Fig. 10.1, the structure of the second- and third-order ADC matrices is displayed. In both cases, the configuration space is spanned by the $1h$ and $2h$-$1p$ configurations. The secular matrix elements read

$$\begin{aligned}
(K+C)_{kl} &= \epsilon_k \delta_{kl} + \sum_{a<b,j} v_{abkj} v_{ljab} (\epsilon_a + \epsilon_b - \epsilon_j - \tfrac{1}{2}\epsilon_k - \tfrac{1}{2}\epsilon_l) \; + [\, C_{kl}^{(3)} \,] \\
C_{k,ak'l'} &= -V_{k'l'[ka]} \; + [\, C_{k,ak'l'}^{(2)} \,] \\
(K+C)_{akl,a'k'l'} &= (-\epsilon_a + \epsilon_k + \epsilon_l) \delta_{aa'} \delta_{kk'} \delta_{ll'} \; + [\, C_{akl,a'k'l'}^{(1)} \,]
\end{aligned} \tag{10.33}$$

where the ADC(2) scheme is constituted by the explicit expressions, while the terms in brackets are surrogates for the additions needed at the ADC(3) level. In a similar way, the ADC expressions for the transition amplitudes can be written as

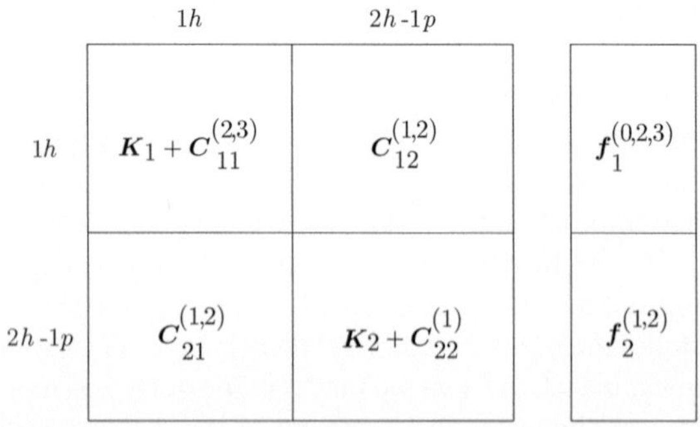

Fig. 10.1 Structure of the ADC matrices $\boldsymbol{K}+\boldsymbol{C}$ and $\boldsymbol{f}$ in second and third order

$$f_{kj} = \delta_{kj} + \tfrac{1}{2}\sum_{a<b,i} v_{abki}v_{jiab} \; + [\, f_{kj}^{(3)} \,]$$

$$f_{ka} = \frac{1}{\epsilon_a - \epsilon_k}\Big(\sum_{b<c,j} v_{bckj}V_{aj[bc]} - \sum_{b,i<j} v_{ab[ij]}V_{ij[kb]}\Big) \; + [\, f_{ka}^{(3)} \,]$$

$$f_{akl,q} = v_{qakl}\,\overline{n}_q \; + [\, f_{akl,q}^{(2)}\overline{n}_q \,] \tag{10.34}$$

Note that we here distinguish the hole part (first line) and the particle part (second line) of $\boldsymbol{f}_1$. In the $\boldsymbol{f}_2$ block (third line), there are particle contributions only.

The general properties attributed to the Dyson-ADC approach in Sect. 9.2 also characterize the direct ADC schemes established in this chapter. Let us recall the essential features:

- Via the ADC form (10.8), one can derive in a systematic way higher-order approximations (ADC(n) schemes) representing *infinite partial summations* of the diagrammatic PT expansion of $\boldsymbol{G}^-(\omega)$ (or likewise $\boldsymbol{G}^+(\omega)$) being *consistent through order* n.
- In the resulting computational scheme (cf. Eq. 10.12), the ionization energies are obtained as the negative eigenvalues of a hermitian secular ADC matrix, while the spectroscopic factors derive from the corresponding eigenvectors together with the ADC transition amplitudes (cf. Eq. 10.13).
- The direct ADC computational schemes combine *diagonalization* of a hermitian secular matrix with *regular perturbation expansions* of the secular matrix elements. At the ADC(n) level, the explicit configuration space comprises the excitations of $N-1$ electrons ($N+1$ electrons) through the class $\mu = n/2+1$ for even n and $\mu = (n+1)/2$ for odd n. This *compactness* of the explicit configuration spaces is reflected by a correspondingly high order of the *truncation error*: For the $1h$ ionization energies (or $1p$ attachment energies), the error due to the truncation of the configuration space at class μ is of the PT order 2μ. In the ADC(2) and ADC(3)

schemes, for example, the error caused by omitting the $3h$-$2p$ and higher excited configurations is of fourth order.
- Finally, the direct ADC(n) schemes are *size-consistent*, which is an indispensable requirement in applications to larger molecules.

The justification of both the compactness and size-consistency properties, as given in Sect. 9.2 for the Dyson-ADC approach, was based on diagrammatic arguments. While the latter apply here as well, the non-Dyson ADC variant allows for an alternative, more stringent formulation. Here, the key is that the direct ADC secular matrix can be identified as the representation of the (subtracted) hamiltonian with respect to a specific set of intermediate states presented in the following Chap. 11. The basic features of that intermediate state representation (ISR), ensuring the compactness and size-consistency of the ISR-ADC(n) computational schemes, will be addressed in Chap. 12.

It is instructive to compare the direct ADC scheme to the Dyson-ADC version presented in Chap. 9 at the lowest non-trivial, that is, second-order level. The matrix $\boldsymbol{K} + \boldsymbol{C}$ of the direct scheme (Eq. 10.33) can be largely retrieved in the $(N-1)$-electron parts of the Dyson secular matrix $\boldsymbol{B}$ (Eq. 9.32). For example, the coupling block $\boldsymbol{C}_{21}^{(1)}$ can be identified with the corresponding matrix elements of $\boldsymbol{U}^{(1)-}$. The notable difference arises in the $1h$ diagonal block, where the $\boldsymbol{C}_{11}^{(2)}$ second-order contributions are absent in $\boldsymbol{B}_{11}$ (the PT expansion of $\boldsymbol{\Sigma}(\infty)$ begins in third order). The counterpart to the $\boldsymbol{C}_{11}^{(2)}$ contribution is effected in the Dyson scheme by the coupling of the $1h$ configurations via $\boldsymbol{U}^{+}$ to the $2p$-$1h$ configurations of $N+1$ electrons. This can be put on a more rigorous foundation by applying a QDPT-type procedure (quasi-degenerate perturbation theory [2]) to the $1h$/$2p$-$1h$ coupling in $\boldsymbol{B}$.

An important indicator of the efficiency of a computational method is the *scaling* of the computing effort with the size of the system or, more specifically, with the number m of the one-particle basis states (molecular orbitals) entering the computation. At the ADC(2) level for example, the construction of the $\boldsymbol{C}_{11}$ block scales as m^5, as there are m^2 matrix elements, each requiring m^3 integral multiplications. (Obviously, a more refined scaling expression could have been obtained by distinguishing occupied and virtual orbitals.) The matrix-times-vector step in an iterative diagonalization procedure scales as m^4, as there are m^4 non-vanishing secular matrix elements (arising in the $\boldsymbol{C}_{12}$ block). This would mean that the overall scaling behaviour of the ADC(2) scheme is m^5. Actually, the overall scaling can be reduced to m^4 by using the so-called *direct diagonalization* procedure, in which, rather than computing and storing the secular matrix, the matrix elements are computed (and re-computed) "on the fly" as needed in the matrix-vector product cycles. This allows one to form suitable intermediates, thereby reducing the computational cost (see Exercise 10.4). The corresponding scaling of the ADC(3) scheme is m^5.

The computational performance of the direct third-order ADC scheme has been examined in model applications [3] to a series of small molecules, allowing for the comparison with full CI results. As expected, there is good mutual agreement between the results obtained using the direct and the Dyson-ADC variants. For the ionization energies of outer valence $1h$ states, the discrepancies of the ADC(3) and

FCI results are typically in the order of $\pm 0.2\,\text{eV}$. This level of accuracy is consistent with what has been found in numerous studies from the comparison of ADC(3) ionization energies with experimental results.

ADC Expressions for Ground-State Expectation Values

As discussed in Sects. 4.2 and 4.3, certain ground-state expectation values can be related to the electron propagator, or more specifically, to its $(N-1)$-electron part. For example, Eq. (3.34)

$$\boldsymbol{\rho} = \frac{1}{2\pi i} \oint \boldsymbol{G}^{-}(\omega)\, d\omega \tag{10.35}$$

allows one to derive the one-particle density matrix (3.26) from $\boldsymbol{G}^{-}(\omega)$. Using the ADC form (10.8), the contour integration for $\tilde{\boldsymbol{G}}(\omega)$ can readily be performed yielding

$$\frac{1}{2\pi i} \oint \tilde{\boldsymbol{G}}(\omega)\, d\omega = \frac{1}{2\pi i} \oint \boldsymbol{f}^{\dagger}(\omega - \boldsymbol{K} - \boldsymbol{C})^{-1} \boldsymbol{f}\, d\omega = \boldsymbol{f}^{\dagger}\boldsymbol{f} \tag{10.36}$$

As a result, the density matrix can be written as

$$\boldsymbol{\rho} = (\boldsymbol{f}^{\dagger}\boldsymbol{f})^{t} \tag{10.37}$$

in terms of the transition amplitudes comprised in the ADC matrix $\boldsymbol{f}$.

According to Eq. (10.16), the ADC expression (10.37) establishes a PT expansion for the density matrix. At a given nth-order ADC(n) scheme, this results in a truncated expansion which, though, is consistent through order n. At the ADC(2) level, Eq. (10.37) describes the density matrix consistently through second order. Here, for example, the h-h diagonal block of $\boldsymbol{\rho}$ can be written more explicitly as

$$\boldsymbol{\rho}_{hh} = \boldsymbol{1}_{hh} + \boldsymbol{f}_{1h}^{(2)} + \boldsymbol{f}_{1h}^{(2)*} + O(3) \tag{10.38}$$

in terms of the ADC(2) results for the $\boldsymbol{f}_{1h}$ sub-block of the $\boldsymbol{f}$ matrix.

As an immediate consequence of Eq. (10.37), the ground-state expectation value (3.35) of a general one-particle operator $\boldsymbol{A}$ can be written as

$$\langle \Psi_0 | \hat{A} | \Psi_0 \rangle = \mathrm{Tr}(\boldsymbol{A}\boldsymbol{\rho}) = \mathrm{Tr}(\boldsymbol{A}^{t} \boldsymbol{f}^{\dagger}\boldsymbol{f}) \tag{10.39}$$

Using the ADC(n) expressions for $\boldsymbol{f}$, one obtains a finite PT series for $\langle \Psi_0 | \hat{A} | \Psi_0 \rangle$, being consistent through order n.

A particular one-particle operator is the particle number operator $\hat{N} = \sum_q c_q^{\dagger} c_q$. Here, the (exact) expectation value is

$$N = \langle \Psi_0 | \hat{N} | \Psi_0 \rangle = \mathrm{Tr}(\boldsymbol{f}^{\dagger}\boldsymbol{f}) \tag{10.40}$$

This means that $\mathrm{Tr}(\boldsymbol{f}^\dagger \boldsymbol{f}) = N + O(n+1)$ at the ADC(n) level, which can be used as a check for the correctness of the ADC(n) expressions for $\boldsymbol{f}$.

The ADC form (10.8) can also be applied to Eq. (3.44), relating the ground-state energy to the electron propagator, which allows us to write E_0 in the form

$$E_0 = \frac{1}{2}\mathrm{Tr}(\boldsymbol{f}^\dagger(\boldsymbol{K}+\boldsymbol{C})\boldsymbol{f} + \boldsymbol{T}^t \boldsymbol{f}^\dagger \boldsymbol{f}) \qquad (10.41)$$

As above, this establishes a PT expansion of E_0 which, at the ADC(n) level, recovers the original RSPT series through order n.

Exercises

10.1 Evaluate the time-orderings $(\overline{6})$ and $(\overline{12})$ obtained by turning diagrams (6) and (12) in Fig. 7.3 upside down and derive therefrom the ADC(2) expression for f_{ka}.

10.2 Apply the direct ADC(2) scheme to ionization in the 2E-2O model (Exercise 2.4).

10.3 (a) Use Eq. (10.37) and the ADC(2) expressions (10.34) to expand the h-h and p-p matrix elements of the density matrix, $\rho_{kk'}$, $\rho_{aa'}$, through second order.
(b) Verify that $\mathrm{Tr}(\boldsymbol{\rho}) = N + O(3)$.

10.4 (a) Analyze the matrix-times-vector product for the ADC(2) secular matrix and establish that the direct diagonalization can be designed to scale as m^4.
(b) Perform a similar analysis for the four third-order contributions to the ADC(3) secular matrix listed in Appendix A.9.

References

1. Schirmer J, Trofimov AB, Stelter G (1998) J Chem Phys 109:4734
2. Durand S, Malrieu J-P (1987) Adv Chem Phys 67:321
3. Trofimov AB, Stelter G, Schirmer J (2002) J Chem Phys 117:6402

Chapter 11
Intermediate-State Representation (ISR)

The direct ADC procedure for the electron propagator part $\boldsymbol{G}^{-}(\omega)$ considered in the preceding chapter allows one to construct successively an in principle exact hermitian secular matrix, $\boldsymbol{K} + \boldsymbol{C}$, where (*i*) the eigenvalues are the negative ionization energies $-I_n$; (*ii*) the matrix index labels are given by the HF configurations (10.9) of $N-1$ electrons; (*iii*) the secular matrix elements can be expanded in regular PT expansions. As already anticipated in Chap. 10, these features suggest that $\boldsymbol{K} + \boldsymbol{C}$ is essentially a representation of the hamiltonian, or, more specifically, of $\hat{H} - E_0$, deriving from a set of $(N-1)$-electron "intermediate" basis states. But what actually are these presumed intermediate states underlying the ADC secular matrix? There is a surprisingly simple solution to this issue, establishing an alternative closed-form version of the ADC secular equations, completely independent of the original diagrammatic derivation [1, 2]. Being a wave-function approach, the new formulation overcomes certain limitations inherent to the propagator concept. Of the three sections of this chapter, Sect. 11.1 presents the general procedure for constructing the intermediate states and the consequent intermediate-state representation (ISR); Sect. 11.2 demonstrates the explicit derivation of the second-order ISR equations; and, finally, Sect. 11.3 discusses how the ISR concept can be applied to general operators.

11.1 Correlated Excited States and Excitation Class Orthogonalization

The starting point for the construction of the desired intermediate states are the so-called **correlated excited** (CE) **states** (here of $N-1$ electrons),

$$|\Psi_J^0\rangle = \hat{C}_J|\Psi_0\rangle \tag{11.1}$$

J. Schirmer, *Many-Body Methods for Atoms, Molecules and Clusters*, Lecture Notes in Chemistry 94, https://doi.org/10.1007/978-3-319-93602-4_11

Here, $\hat{C}_J$ denote physical excitation (ionization) operators of the manifold

$$\{\hat{C}_J\} = \left\{c_k;\ c_a^\dagger c_k c_l, k < l;\ c_a^\dagger c_b^\dagger c_j c_k c_l,\ a < b, j < k < l;\ \ldots\right\} \tag{11.2}$$

where the index notations conform to those used in Eq. (10.9). As in Sect. 10.2, the successive $1h$, $2h$-$1p$, $3h$-$2p, \ldots$ excitation classes are numbered $\mu = 1, 2, 3, \ldots$. To specify the class of a given configuration, J, we will use the symbol $[J]$, that is, $[J] = \mu$ if J designates a μh-$(\mu - 1)p$ configuration. The operators $\hat{C}_J$ are called physical, since, when acting on the HF ground state, they create the HF configurations

$$|\Phi_J\rangle = \hat{C}_J|\Phi_0\rangle \tag{11.3}$$

forming the CI expansion manifold for the $(N-1)$-electron system.

In contrast to the HF configurations, the CE states are not orthonormal:

$$S_{IJ} = \langle\Psi_I^0|\Psi_J^0\rangle = \langle\Psi_0|\hat{C}_I^\dagger\hat{C}_J|\Psi_0\rangle \neq \delta_{I,J} \tag{11.4}$$

Here, S_{IJ} denote the matrix elements of the CE-state overlap matrix $\boldsymbol{S}$, which also can be seen as a generalized density matrix. The CE states form a complete set of $(N-1)$-electron states [3, 4]. This suggests to generate proper basis states by applying a suitable orthonormalization procedure to the CE states. However, the most obvious choice, namely symmetric orthonormalization, according to

$$|\overline{\Psi}_J\rangle = \sum_I |\Psi_I^0\rangle(\boldsymbol{S}^{-1/2})_{IJ} \tag{11.5}$$

has to be discarded because the resulting $|\overline{\Psi}_J\rangle$ states lead to a secular problem that is neither compact nor size-consistent, as explained in Sects. 12.1 and 12.2. By contrast, the ADC features are recovered by Gram–Schmidt orthogonalization with respect to successively higher CE-state excitation classes μ, augmented by symmetric orthonormalization within each class.

For illustration, we shall construct the intermediate states in the two lowest excitation classes, $\mu = 1$ and 2. In case of the $1h$ states, being the lowest class, only symmetric orthonormalization is needed, and the intermediate states are obtained from the CE states, $|\Psi_k^0\rangle = c_k|\Psi_0\rangle$, by symmetric orthonormalization,

$$|\tilde{\Psi}_k\rangle = \sum_i c_i|\Psi_0\rangle(\boldsymbol{S}_1^{-1/2})_{ik} \tag{11.6}$$

where

$$(\boldsymbol{S}_1)_{ij} = \langle\Psi_0|c_i^\dagger c_j|\Psi_0\rangle \tag{11.7}$$

defines the overlap matrix $\boldsymbol{S}_1$ of the CE states of class 1. Note that $\boldsymbol{S}_1$ is the transpose of the h-h block of the one-particle density matrix as defined in Eq. (3.26).

Now, we may turn to the $2h$-$1p$ excitations. As the first step, the CE states $c_a^\dagger c_k c_l|\Psi_0\rangle$ have to be orthogonalized to the intermediate states $|\tilde{\Psi}_i\rangle$ of class 1. Applying the Gram–Schmidt procedure,

$$|\Psi^{\#}_{akl}\rangle = c_a^\dagger c_k c_l|\Psi_0\rangle - \sum_i |\tilde{\Psi}_i\rangle\langle\tilde{\Psi}_i|c_a^\dagger c_k c_l|\Psi_0\rangle \tag{11.8}$$

we obtain "precursor" states being orthogonal with respect to the intermediate states of class 1 but not yet orthonormal. Thus, in a second step, we apply symmetric orthonormalization to the precursor states. To that end, we introduce the overlap matrix

$$\boldsymbol{S}_2 \equiv \left(\langle\Psi^{\#}_{akl}|\Psi^{\#}_{a'k'l'}\rangle\right) \tag{11.9}$$

of the precursor states of class 2, and the final $2h$-$1p$ intermediate states are obtained according to

$$|\tilde{\Psi}_{akl}\rangle = \sum_{a',k',l'} |\Psi^{\#}_{a'k'l'}\rangle(\boldsymbol{S}_2^{-1/2})_{a'k'l',akl} \tag{11.10}$$

The extension to the higher classes, $\mu = 3, 4, \ldots$, is obvious, and we may formulate the general **excitation class orthogonalization** (ECO) procedure for intermediate states as follows:

(1) Assume that the intermediate states $|\tilde{\Psi}_K\rangle$ of the classes $1, \ldots, \nu - 1$ have been constructed. Then, orthogonalize the CE states $|\Psi^0_J\rangle$ of class ν with respect to the intermediate states of class $1, \ldots, \nu - 1$ according to

$$|\Psi^{\#}_J\rangle = |\Psi^0_J\rangle - \sum_{[K]<\nu} |\tilde{\Psi}_K\rangle\langle\tilde{\Psi}_K|\Psi^0_J\rangle, \quad [J] = \nu \tag{11.11}$$

(2) The "precursor states" $|\Psi^{\#}_J\rangle$ of class ν may then be orthonormalized symmetrically, yielding

$$|\tilde{\Psi}_J\rangle = \sum_{[I]=\nu} |\Psi^{\#}_I\rangle(\boldsymbol{S}_\nu^{-1/2})_{IJ} \tag{11.12}$$

where $\boldsymbol{S}_\nu$ is the overlap matrix of the precursor states of class ν,

$$(\boldsymbol{S}_\nu)_{IJ} = \langle\Psi^{\#}_I|\Psi^{\#}_J\rangle, \quad [I] = [J] = \nu \tag{11.13}$$

It should be noted that the N-electron ground state $|\Psi_0\rangle$ underlying the ECO-IS construction needs not be normalized to unity. A convenient choice is intermediate normalization, $\langle\Psi_0|\Phi_0\rangle = 1$ (see Appendix A.1), which will be supposed in the following.

The ECO construction allows one to establish a PT expansion of the intermediate states,

$$|\tilde{\Psi}_I\rangle = |\tilde{\Psi}^{(0)}_I\rangle + |\tilde{\Psi}^{(1)}_I\rangle + |\tilde{\Psi}^{(2)}_I\rangle + \ldots \tag{11.14}$$

deriving completely from the PT expansion of the N-electron ground state $|\Psi_0\rangle$. Obviously, in zeroth order the intermediate states coincide with the HF states:

$$|\tilde{\Psi}_I^{(0)}\rangle = |\Phi_I\rangle \tag{11.15}$$

In this sense, the intermediate states "mediate" between the HF states and the exact $(N-1)$-electron energy eigenstates.

Rather than expanding the intermediate states, it is more advantageous to deal directly with the PT expansions of the matrix elements of the **intermediate-state representations** (ISR). The ECO intermediate states $|\tilde{\Psi}_I\rangle$, forming a complete basis of the $(N-1)$-particle states, establish the ISR secular matrix $\boldsymbol{M}$,

$$M_{IJ} = \langle\tilde{\Psi}_I|\hat{H} - E_0|\tilde{\Psi}_J\rangle \tag{11.16}$$

representing the (shifted) hamiltonian $\hat{H} - E_0$ in terms of the intermediate states. The exact eigenstates $|\Psi_n^{N-1}\rangle$ can be expanded according to

$$|\Psi_n^{N-1}\rangle = \sum_I X_{In}|\tilde{\Psi}_I\rangle \tag{11.17}$$

where X_{In} are the components of the nth eigenvector of $\boldsymbol{M}$. In matrix notation, the ISR secular equations take on the form

$$\boldsymbol{MX} = \boldsymbol{X\Omega},\ \boldsymbol{X}^\dagger\boldsymbol{X} = \boldsymbol{1} \tag{11.18}$$

where $\boldsymbol{\Omega}$ denotes the diagonal matrix of eigenvalues ω_n, and $\boldsymbol{X}$ is the matrix of eigenvectors. According to Eq. (11.16), the eigenvalues can be identified as the ionization energies,

$$\omega_n = E_n^{N-1} - E_0 \tag{11.19}$$

Moreover, introducing the matrix $\tilde{\boldsymbol{f}}$ of ISR transition amplitudes,

$$\tilde{f}_{Ip} = \langle\tilde{\Psi}_I|c_p|\Psi_0\rangle \tag{11.20}$$

the spectroscopic factors (10.5) can be obtained according to

$$x_p^{(n)} = \sum_I X_{In}^* \tilde{f}_{Ip} \tag{11.21}$$

from the respective eigenvectors.

The PT expansions (11.14) of the intermediate states, based on the PT expansion of the N-electron ground state $|\Psi_0\rangle$, translate into PT expansions of the ISR secular matrix elements,

$$M_{IJ} = M_{IJ}^{(0)} + M_{IJ}^{(1)} + M_{IJ}^{(2)} + \ldots \tag{11.22}$$

where, of course, the PT expansion of the ground-state energy E_0 comes into play as well. These expansions can actually be put in practice, as will be demonstrated in the ensuing section. Here, it may already be noted that zeroth-order contributions arise only in the diagonal elements of $\boldsymbol{M}$ since

$$M_{IJ}^{(0)} = \langle\Phi_0|\hat{C}_I^\dagger(\hat{H}_0 - E_0^{(0)})\hat{C}_J|\Phi_0\rangle = -\delta_{IJ}K_I \tag{11.23}$$

Here, K_I denote the zeroth-order (HF) ionization energies (10.17).

In a similar way, PT expansions

$$\tilde{f}_{Ip} = \tilde{f}_{Ip}^{(0)} + \tilde{f}_{Ip}^{(1)} + \tilde{f}_{Ip}^{(2)} + \dots \tag{11.24}$$

can be established for the ISR transition amplitudes. Due to the orthogonality properties of the intermediate states, the transition amplitudes of the hole part,

$$\tilde{f}_{Jk} = \langle\tilde{\Psi}_J|c_k|\Psi_0\rangle = 0, \;\; n_k = 1, \;\; [J] > 1 \tag{11.25}$$

vanish for configurations J from excitation classes $\mu > 1$.

So far, the ECO-ISR formalism has been introduced as an autonomous approach, completely independent of the direct ADC approach deriving from the diagrammatic PT expansion of the propagator parts. Notwithstanding the distinct derivations, the resulting secular equations turn out to be essentially equivalent. As suggested by Eqs. (10.11) and (11.16), the ADC and ECO-ISR secular matrices are to be identified according to

$$\boldsymbol{K} + \boldsymbol{C} \equiv -\boldsymbol{M} \tag{11.26}$$

$$\boldsymbol{f} \equiv \tilde{\boldsymbol{f}} \tag{11.27}$$

With regard to the first line, it should be noted that the signs of off-diagonal secular matrix elements are to a certain extent conventional.

But how, actually, can these equivalencies be justified? Firstly, the explicit ECO-ISR equations through second order, as derived in the following Sect. 12.2, are identical with those of the direct ADC(2) scheme. Beyond second order, the derivation of explicit ECO-ISR expressions becomes rather unwieldy, and the complementing derivation of the third-order ADC(3) scheme via the ECO-ISR route has not been given yet. Anyway, demonstrating the equivalence of the direct ADC(n) and ECO-ISR(n) schemes for some low n, while a strong indication of correctness, is not a substitute for a proof. Rather, the essential argument underpinning the equivalence of the ADC and ECO-ISR secular equations is that both versions share two constituting features: Firstly, the secular matrix fulfills distinguished PT order relations, referred to as *canonical order relations*; secondly, it is *separable* with respect to a system of non-interacting fragments. The merit of the ECO-ISR approach is that these defining properties can be formulated and proven in a stringent manner, as will be discussed at length in Chap. 12.

It should be noted though that equivalence not necessarily means identity here. In fact, we have seen in Chap. 10 that the ADC procedure does not completely determine the secular matrix, but leaves some room for algebraic transformations. By contrast, for a given ground state and one-particle basis, the ECO-ISR procedure as defined by Eqs. (11.11)–(11.13) results in a unique IS representation. Here, however, the second step, that is, the symmetric orthonormalization of the respective precursor states, is a mere convention, not necessary for constituting the two basic features. In fact, any orthonormalization scheme could be used here. So there is some flexibility in the ECO-ISR concept, which may be seen as the counterpart to the residual non-uniqueness in the ADC expressions. Recognizing the general equivalence, we will use the terms ECO-ISR and ISR-ADC synonymously in the following.

11.2 Explicit ISR Procedure Through Second Order

In the following, we shall derive the explicit PT expressions of the ISR secular matrix $\boldsymbol{M}$ as needed for a consistent treatment of the $1h$ (main) ionic states through second order, to be referred to as ISR(2) scheme.

The explicit configuration space required at the ISR(2) level is spanned by the $1h$ and $2h$-$1p$ intermediate states. The coupling of $1h$ and $3h$-$2p$ states is already of second order (see Sect. 12.1). The PT expansions extend through second order in the $1h$ diagonal block $\boldsymbol{M}_{11}$,

$$M_{kl} = -\epsilon_k \delta_{kl} + M_{kl}^{(1)} + M_{kl}^{(2)} \tag{11.28}$$

through first-order in the $\boldsymbol{M}_{12}$ block,

$$M_{k,a'k'l'} = M_{k,a'k'l'}^{(1)} \tag{11.29}$$

while only the zeroth-order contributions

$$M_{akl,a'k'l'} = (\epsilon_a - \epsilon_k - \epsilon_l)\delta_{aa'}\delta_{kk'}\delta_{ll'} \tag{11.30}$$

are needed in the $\boldsymbol{M}_{22}$ block.

At the ISR(2) level, the PT expansion of the N-electron ground state underlying the ISR construction is needed through second order:

$$|\Psi_0\rangle = |\Phi_0\rangle + |\Psi_0^{(1)}\rangle + |\Psi_0^{(2)}\rangle + O(3) \tag{11.31}$$

The first-order term,

$$|\Psi_0^{(1)}\rangle = \sum_{a<b,k<l} v_{abkl} |\Phi_{abkl}\rangle \tag{11.32}$$

simply reflects the admixture of double excitations, $|\Phi_{abkl}\rangle = c_a^\dagger c_b^\dagger c_k c_l|\Phi_0\rangle$. Here, v_{abkl} is defined according to Eq. (10.31). The second-order wave function, comprising single, double, triple, and quadruple excitations, is already rather involved. Fortunately, though, an explicit specification of the various contributions is not needed in the following derivations.

As mentioned above, we may suppose intermediate normalization of $|\Psi_0\rangle$. This means that there is a PT expansion of the normalization integral,

$$I_0 = \langle\Psi_0|\Psi_0\rangle = 1 + I_0^{(2)} + O(3) \tag{11.33}$$

where the deviation from unity begins at second order:

$$I_0^{(2)} = \langle\Psi_0^{(1)}|\Psi_0^{(1)}\rangle = \sum_{a<b,i<j} |v_{abij}|^2 \tag{11.34}$$

The **first-order** expansions of the intermediate states are straightforward as orthonormalization does not come into play before second order:

$$|\tilde{\Psi}_k\rangle = |\Phi_k\rangle + c_k|\Psi_0^{(1)}\rangle + O(2) \tag{11.35}$$

$$|\tilde{\Psi}_{akl}\rangle = |\Phi_{akl}\rangle + c_a^\dagger c_k c_l|\Psi_0^{(1)}\rangle + O(2) \tag{11.36}$$

Herewith, the first-order matrix elements of $\boldsymbol{M}$ can easily be established. Let us consider the $\boldsymbol{M}_{12}$ block. The first-order expression just reproduces the familiar CI coupling matrix element:

$$M_{k,ak'l'}^{(1)} = \langle\Phi_k|\hat{H}_I|\Phi_{ak'l'}\rangle = V_{k'l'[ka]} \tag{11.37}$$

Note that the other potential first-order contributions, such as $\langle\tilde{\Psi}_k^{(1)}|\hat{H}_0 - E_0^{(0)}|\Phi_{a'k'l'}\rangle$, vanish. In the $\boldsymbol{M}_{11}$ block, the first-order contributions are given by

$$M_{kl}^{(1)} = \langle\Phi_k|\hat{H}_I - E_0^{(1)}|\Phi_l\rangle \tag{11.38}$$

which, for HF orbitals supposed here, vanishes according to Eq. (4.6).

A more demanding task is to evaluate the **second-order** contributions in

$$M_{kl} = \langle\tilde{\Psi}_k|\hat{H} - E_0|\tilde{\Psi}_l\rangle \tag{11.39}$$

The $1h$ intermediate states are given by

$$|\tilde{\Psi}_k\rangle = \sum_i c_i|\Psi_0\rangle(\boldsymbol{s}^{-1/2})_{ik} \tag{11.40}$$

where $\boldsymbol{S}_1 = \boldsymbol{s}$ is used for the CE overlap matrix to simplify the notation. Note that $\boldsymbol{s}$ is a hermitian and positive definite matrix so that $\boldsymbol{s}^{1/2}$ is well defined. The PT expansion of $\boldsymbol{s}$ through second order reads

$$s_{kl} = \langle\Psi_0|c_k^\dagger c_l|\Psi_0\rangle = \delta_{kl} + s_{kl}^{(2)} + O(3)$$

as the potential first-order contributions are seen to vanish,

$$s_{kl}^{(1)} = \langle\Psi_0^{(1)}|c_k^\dagger c_l|\Phi_0\rangle + \langle\Phi_0|c_k^\dagger c_l|\Psi_0^{(1)}\rangle = 0$$

Accordingly, $\boldsymbol{s}$ has the structure

$$\boldsymbol{s} = \mathbf{1} + \boldsymbol{s}^{(2)} + O(3) \tag{11.41}$$

where $\mathbf{1}$ denotes the unit matrix. This, in turn, entails the PT structure

$$\boldsymbol{s}^{-1/2} = \mathbf{1} - \tfrac{1}{2}\boldsymbol{s}^{(2)} + O(3) \tag{11.42}$$

for the inverse of $\boldsymbol{s}^{1/2}$. As a preparatory step, let us evaluate the second-order contribution to $\boldsymbol{s}$:

$$s_{kl}^{(2)} = \langle\Psi_0^{(1)}|c_k^\dagger c_l|\Psi_0^{(1)}\rangle + \langle\Psi_0^{(2)}|c_k^\dagger c_l|\Phi_0\rangle + \langle\Phi_0|c_k^\dagger c_l|\Psi_0^{(2)}\rangle \tag{11.43}$$

As a consequence of the intermediate normalization supposed for $|\Psi_0\rangle$, the latter two terms vanish, and the evaluation of the remaining first term yields (see Exercise 11.1)

$$s_{kl}^{(2)} = \langle\Psi_0^{(1)}|c_k^\dagger c_l|\Psi_0^{(1)}\rangle = \delta_{kl} I_0^{(2)} - \sum_{a<b,j} v_{abkj} v_{ablj}^* \tag{11.44}$$

where $I_0^{(2)}$ is given by Eq. (11.34). Using Eqs. (11.40) and (11.42), Eq. (11.39) can be written as

$$\begin{aligned} M_{kl} =& \sum_{k',l'} (\boldsymbol{s}^{-1/2})_{kk'} \langle\Psi_0|c_{k'}^\dagger(\hat{H} - E_0)c_{l'}|\Psi_0\rangle (\boldsymbol{s}^{-1/2})_{l'l} \\ =& \langle\Psi_0|c_k^\dagger(\hat{H} - E_0)c_l|\Psi_0\rangle - \tfrac{1}{2}\sum_{l'} \langle\Psi_0|c_k^\dagger(\hat{H} - E_0)c_{l'}|\Psi_0\rangle s_{l'l}^{(2)} \\ & - \tfrac{1}{2}\sum_{k'} \langle\Psi_0|c_{k'}^\dagger(\hat{H} - E_0)c_l|\Psi_0\rangle s_{kk'}^{(2)} + O(3) \end{aligned} \tag{11.45}$$

It remains to collect the second-order contributions $M_{kl}^{(2)}$ arising on the right-hand side of the latter equation. For the last two terms, this is easily accomplished as here the second-order matrix elements $s_{ij}^{(2)}$ need to combine with zeroth-order contributions of the ground-state expectation values. The resulting second-order contribution to M_{kl}, termed $(C1)$, reads

$$(C1) \equiv \tfrac{1}{2}(\epsilon_k + \epsilon_l)s_{kl}^{(2)} \tag{11.46}$$

with $s_{kl}^{(2)}$ given by Eq. (11.44). There are three more second-order contributions stemming from the first term on the right-hand side of Eq. (11.45):

$$\begin{aligned}
(C2) &\equiv \langle \Psi_0^{(1)} | c_k^\dagger (\hat{H}_0 - E_0^{(0)}) c_l | \Psi_0^{(1)} \rangle \\
(C3) &\equiv \langle \Psi_0^{(1)} | c_k^\dagger \hat{H}_I c_l | \Phi_0 \rangle + \langle \Phi_0 | c_k^\dagger \hat{H}_I c_l | \Psi_0^{(1)} \rangle \\
(C4) &\equiv - E_0^{(2)} \delta_{kl}
\end{aligned}$$

Note that contributions involving $|\Psi_0^{(2)}\rangle$, such as $\langle \Psi_0^{(2)} | c_k^\dagger (\hat{H}_0 - E_0^{(0)}) c_l | \Phi_0 \rangle$, vanish (supposing intermediate normalization). Moreover, there are no contributions involving $E_0^{(1)}$, since here the respective expectation value factor vanishes, as for example, $\langle \Psi_0^{(1)} | c_k^\dagger c_l | \Phi_0 \rangle = 0$.

In evaluating $(C2)$ and $(C3)$, it is advantageous to treat the cases $k = l$ and $k \neq l$ separately. Supposing $k < l$, one obtains the following expressions

$$\begin{aligned}
(C2) &= - \sum_{a<b,j} v_{abkj} v_{ablj}^* (\epsilon_a + \epsilon_b - \epsilon_j - \epsilon_k - \epsilon_l) \\
(C3) &= + \sum_{a<b,j} V_{ab[kj]} v_{ablj}^* + \sum_{a<b,j} V_{ab[lj]}^* v_{abkj}
\end{aligned}$$

Now, the three non-vanishing second-order contributions have to be combined. Since $(C1)$, $(C2)$, and $(C3)$ differ only in orbital energy factors, they can easily be added to give

$$M_{kl}^{(2)} = \sum_{a<b,j} v_{abkj} v_{ablj}^* (\epsilon_a + \epsilon_b - \epsilon_j - \tfrac{1}{2}\epsilon_k - \tfrac{1}{2}\epsilon_l) \tag{11.47}$$

As can readily be established, this expression is no longer restricted to $k < l$ and comprises, in particular, the diagonal matrix elements, $k = l$, incorporating here also the $(C4)$ contribution. Thus, the final result for the ISR(2) secular matrix elements reads

$$M_{kl} = -\epsilon_k \delta_{kl} + M_{kl}^{(2)} \tag{11.48}$$

with $M_{kl}^{(2)}$ given by Eq. (11.47). Together with Eqs. (11.37) and (11.30), this constitutes the ISR(2) secular matrix at the second-order level.

The comparison with the ADC equations (10.33) shows that indeed the identity (11.26) between the ISR and the (negative) ADC secular matrices is valid through the second-order level.

In a similar way, the explicit ISR(2) expressions for

$$\begin{aligned}
\tilde{f}_{kp} &= \langle \tilde{\Psi}_k | c_p | \Psi_0 \rangle \\
\tilde{f}_{akl,p} &= \langle \tilde{\Psi}_{akl} | c_p | \Psi_0 \rangle
\end{aligned} \tag{11.49}$$

can be evaluated. The result confirms the identity (11.27) of the ADC and ISR transition amplitudes.

Of course, the ECO-IS basis can be used to represent $\hat{H}$ (instead of $\hat{H} - E_0$),

$$\tilde{H}_{IJ} = \langle \tilde{\Psi}_I | \hat{H} | \tilde{\Psi}_J \rangle = M_{IJ} + \delta_{IJ} E_0 \tag{11.50}$$

The second-order ISR(2) version of $\hat{H}$ is obtained from $\boldsymbol{M}$ by adding ground-state energy expansions to the diagonal elements in a consistent way, that is, $E_0^{(0)} + E_0^{(1)} + E_0^{(2)}$ in the $\boldsymbol{M}_{11}$ block, and just $E_0^{(0)}$ in the $\boldsymbol{M}_{22}$ block.

In view of the ISR(2) derivation just presented, it may be expected that matters will become rather tedious at the next higher, i.e., third-order, level. So while the ECO-ISR is the simpler concept, the ADC approach, based on the diagrammatic PT expansion of the electron propagator, is clearly preferable when it comes to the practical implementation.

11.3 Intermediate-State Representation of General Operators

Obviously, the intermediate-state representation introduced in the preceding two sections can be applied to operators other than the hamiltonian. This equips the ISR-ADC approach with the full flexibility of wave-function-based methods, allowing for applications not possible within the original propagator concept.

As an example for the additional opportunities, let us consider the treatment of ionic-state properties. Let $\hat{D}$ be a hermitian operator associated with the physical property of interest, e.g., the dipole moment along a particular axis. The ISR of $\hat{D}$ is given by the matrix elements of $\hat{D}$ with respect to the intermediate states $|\tilde{\Psi}_I\rangle$:

$$\tilde{D}_{IJ} = \langle \tilde{\Psi}_I | \hat{D} | \tilde{\Psi}_J \rangle \tag{11.51}$$

The corresponding matrix will be denoted by $\tilde{\boldsymbol{D}}$. For a particular ionic energy eigenstate $|\Psi_n^{N-1}\rangle$, the desired property is obtained as the expectation value

$$D_n = \langle \Psi_n^{N-1} | \hat{D} | \Psi_n^{N-1} \rangle = \boldsymbol{X}_n^\dagger \tilde{\boldsymbol{D}} \boldsymbol{X}_n$$

where $\boldsymbol{X}_n$ denotes the eigenvector associated with the ISR expansion

$$|\Psi_n^{N-1}\rangle = \sum_I X_{In} |\tilde{\Psi}_I\rangle \tag{11.52}$$

of $|\Psi_n^{N-1}\rangle$. In s similar way, transition moments involving two ionic states can be derived according to

$$D_{nm} = \langle\Psi_n^{N-1}|\hat{D}|\Psi_m^{N-1}\rangle = \boldsymbol{X}_n^\dagger \tilde{\boldsymbol{D}} \boldsymbol{X}_m \tag{11.53}$$

As another useful option enabled by the generalization of the ISR, the original hamiltonian $\hat{H}$ can be easily augmented with an additional (external) operator $\hat{U}$,

$$\hat{H} \rightarrow \hat{H}^x = \hat{H} + \hat{U} \tag{11.54}$$

The operator $\hat{U}$ may represent any perturbation of the system, such as an external field. The important point here is that $\hat{U}$ acts on the $(N-1)$-particle system, but does not affect the original N-electron ground state. The extended secular equations read

$$(\boldsymbol{M} + \tilde{\boldsymbol{U}})\boldsymbol{X} = \boldsymbol{X}\boldsymbol{\Omega}^x, \ \ \boldsymbol{X}^\dagger\boldsymbol{X} = \boldsymbol{1} \tag{11.55}$$

where $\boldsymbol{U}$ is the ISR of the external operator $\hat{U}$.

In the case of a time-dependent external potential, $\hat{U}(t)$, the time-dependent Schrödinger equation

$$i\frac{\partial}{\partial t}|\Psi(t)\rangle = (\hat{H} + \hat{U}(t))|\Psi(t)\rangle \tag{11.56}$$

becomes amenable to propagation schemes, such as,

$$\boldsymbol{x}(t+dt) = \boldsymbol{x}(t) - i(\boldsymbol{M} + \tilde{\boldsymbol{U}}(t))\boldsymbol{x}(t)dt \tag{11.57}$$

based on the IS representation of $\hat{U}(t)$; here, $\boldsymbol{x}(t)$ is the vector of the expansion coefficients $x_I(t) = \langle\tilde{\Psi}_I|\Psi(t)\rangle$.

Let us now turn to the explicit construction of the ECO-IS representation of an operator $\hat{D}$. The procedure is largely analogous to that of the ISR secular matrix discussed in the preceding section. Again, the ECO-IS construction establishes a PT expansion of $\tilde{\boldsymbol{D}}$:

$$\tilde{\boldsymbol{D}} = \tilde{\boldsymbol{D}}^{(0)} + \tilde{\boldsymbol{D}}^{(1)} + \tilde{\boldsymbol{D}}^{(2)} + \ldots \tag{11.58}$$

Consistent approximation schemes are obtained by truncating in a systematic way both the IS classes and the respective PT expansions. Being the counterpart to the ISR-ADC(2) secular equations, the second-order ISR(2) scheme comprises the excitation classes 1 and 2, and the PT expansions in the corresponding sub-blocks of $\tilde{\boldsymbol{D}}$ are as follows:

$$\begin{aligned}
\tilde{\boldsymbol{D}}_{11} &= \tilde{\boldsymbol{D}}_{11}^{(0)} + \tilde{\boldsymbol{D}}_{11}^{(1)} + \tilde{\boldsymbol{D}}_{11}^{(2)} \\
\tilde{\boldsymbol{D}}_{12} &= \tilde{\boldsymbol{D}}_{12}^{(0)} + \tilde{\boldsymbol{D}}_{12}^{(1)} \\
\tilde{\boldsymbol{D}}_{22} &= \tilde{\boldsymbol{D}}_{22}^{(0)}
\end{aligned} \tag{11.59}$$

The derivation of the explicit ISR(2) expressions is straightforward, though somewhat more tedious than in the case of the secular matrix. We dispense with a full description and limit ourselves to some guiding remarks. For a more detailed account of the procedure, the reader is referred to Ref. [5] where an analogous derivation for N-electron excitations has been presented. For simplicity, we shall confine ourselves to one-body operators, being of the form

$$\hat{D} = \sum d_{rs} c_r^\dagger c_s$$

The ISR construction can be applied to two-body (and higher-rank) operators as well, but obviously this is more cumbersome.

1. In deriving the ISR(2) expressions of a general one-body operator, the second-order term in the ground-state PT expansion,

$$|\Psi_0^{(2)}\rangle = |\Psi_{1p-1h}^{(2)}\rangle + |\Psi_{2p-2h}^{(2)}\rangle + |\Psi_{3p-3h}^{(2)}\rangle + |\Psi_{4p-4h}^{(2)}\rangle \tag{11.60}$$

   comes into play, though only with the $1p$-$1h$ part:

$$|\Psi_{1p-1h}^{(2)}\rangle = \sum_{ak} x_{ak}^{(2)} c_a^\dagger c_k |\Phi_0\rangle. \tag{11.61}$$

   Here, $x_{ak}^{(2)}$ denote the second-order expansion coefficients of the $1p$-$1h$ admixtures to the ground state,

$$x_{ak}^{(2)} = \langle \Phi_0 | c_k^\dagger c_a | \Psi_0^{(2)} \rangle \tag{11.62}$$

   Note that the $x_{ak}^{(2)}$ coefficients through second order can be identified with the p-h components of the one-particle density matrix,

$$\rho_{ak} = \langle \Psi_0 | c_k^\dagger c_a | \Psi_0 \rangle = x_{ak}^{(2)} + O(3) \tag{11.63}$$

   This relation can be used to replace the $1p$-$1h$ PT coefficient in the final ISR(2) expressions. As should be recalled, the density matrix elements ρ_{ak} can be derived from the electron propagator part $\boldsymbol{G}^-(\omega)$, for example, according to Eqs. (3.34) or (10.37).
2. The IS matrix elements turn out to be of the form

$$\tilde{D}_{IJ} = D_0 \delta_{IJ} + \tilde{D}'_{IJ} \tag{11.64}$$

   where

$$D_0 = I_0^{-1} \langle \Psi_0 | \hat{D} | \Psi_0 \rangle \tag{11.65}$$

   is the expectation value of $\hat{D}$ with respect to the (normalized) N-electron ground state, and the matrix $\tilde{\boldsymbol{D}}'$ is a representation of the subtracted operator $\hat{D}' = \hat{D} - D_0$. In the diagonal ISR(2) matrix elements of $\tilde{\boldsymbol{D}}_{11}$, the PT expansion of D_0

$$D_0(2) = D_0^{(0)} + D_0^{(2)} + O(3) \tag{11.66}$$

extends through second order, whereas in $\tilde{\boldsymbol{D}}_{22}$ only the zeroth-order term

$$D_0^{(0)} = \langle \Phi_0 | \hat{D} | \Phi_0 \rangle = \sum_j d_{jj} \tag{11.67}$$

is needed. Note that $D_0^{(1)} = 0$.

3. The zeroth-order contributions to $\tilde{\boldsymbol{D}}_{11}$, $\tilde{\boldsymbol{D}}_{12}$, and $\tilde{\boldsymbol{D}}_{22}$ can easily be evaluated:

$$\tilde{D}_{kl}^{(0)} = \langle \Phi_0 | c_k^\dagger \hat{D} c_l | \Phi_0 \rangle = \delta_{kl} D_0^{(0)} - d_{lk} \tag{11.68}$$

$$\begin{aligned} \tilde{D}_{akl,a'k'l'}^{(0)} &= \delta_{aa'} \delta_{kk'} \delta_{ll'} D_0^{(0)} + \delta_{kk'} \delta_{ll'} d_{aa'} \\ &\quad - (\delta_{aa'} \delta_{kk'} d_{ll'} + \delta_{aa'} \delta_{ll'} d_{kk'}) + (k' \leftrightarrow l') \end{aligned} \tag{11.69}$$

$$\tilde{D}_{k,ak'l'}^{(0)} = \delta_{kl'} d_{k'a} - \delta_{kk'} d_{l'a} \tag{11.70}$$

4. In first order, there are no contributions to $\tilde{\boldsymbol{D}}_{11}$,

$$\tilde{D}_{kl}^{(1)} = \langle \Psi_0^{(1)} | c_k^\dagger \hat{D} c_l | \Phi_0 \rangle + \langle \Phi_0 | c_k^\dagger \hat{D} c_l | \Psi_0^{(1)} \rangle = 0 \tag{11.71}$$

since the $1h/3h$-$2p$ coupling matrix elements of a one-particle operator vanish; here, the $3h$-$2p$ configurations arise by applying c_k to $|\Psi_0^{(1)}\rangle$.
As to the first-order contributions to $\tilde{\boldsymbol{D}}_{12}$, one has to evaluate the matrix elements

$$\tilde{D}_{k,ak'l'}^{(1)} = \langle \Psi_0^{(1)} | c_k^\dagger \hat{D} c_a^\dagger c_{k'} c_{l'} | \Phi_0 \rangle \tag{11.72}$$

combining a $2h$-$1p$ configuration $|\Phi_{ak'l'}\rangle$ and $3h$-$2p$ configurations $c_k|\Phi_{cdij}\rangle$ arising in $c_k|\Psi_0^{(1)}\rangle$. Note that terms of the type $\langle \Phi_0 | c_k^\dagger \hat{D} c_a^\dagger c_{k'} c_{l'} | \Psi_0^{(1)} \rangle$ vanish. The resulting expressions read

$$\tilde{D}_{k,ak'l'}^{(1)} = -\delta_{kk'} \sum_{b,j} v_{abjl'}^* d_{bj} + \delta_{kl'} \sum_{b,j} v_{abjk'}^* d_{bj} + \sum_b v_{abk'l'}^* d_{bk} \tag{11.73}$$

5. To determine the second-order matrix elements $\tilde{D}_{kl}^{(2)}$, we may set out from Eq. (11.45), replacing here $\hat{H} - E_0$ with $\hat{D}$ and using the zeroth-order ground-state wave function $|\Phi_0\rangle$ rather than $|\Psi_0\rangle$ in the second and third terms on the right-hand side:

$$\tilde{D}_{kl} = \langle \Psi_0 | c_k^\dagger \hat{D} c_l | \Psi_0 \rangle - \tfrac{1}{2} \sum_{l'} D_{kl'}^{(0)} s_{l'l}^{(2)} - \tfrac{1}{2} \sum_{k'} D_{k'l}^{(0)} s_{kk'}^{(2)} + O(3) \tag{11.74}$$

It remains to deal with the three second-order contributions to the first term on the right-hand side:

$$(1) \equiv \langle \Psi_0^{(1)} | c_k^\dagger \hat{D} c_l | \Psi_0^{(1)} \rangle \tag{11.75}$$

$$(2,3) \equiv \langle \Psi_0^{(2)} | c_k^\dagger \hat{D} c_l | \Phi_0 \rangle + \text{h.c.} \tag{11.76}$$

While the evaluation of (2) and (3) is rather simple, that of (1) is more demanding. In both cases, it is advisable to make use of the commutators,

$$[c_k^\dagger, \hat{D}] = -\sum_r d_{rk} c_r^\dagger, \; [\hat{D}, c_l] = -\sum_s d_{ls} c_s \tag{11.77}$$

We skip the somewhat lengthy derivation and jump directly to the final expressions [6]:

$$\tilde{D}_{kl} = D_0(2)\delta_{kl} - d_{lk} - \sum_a (d_{ak}\rho_{la}^{(2)} + d_{la}\rho_{ak}^{(2)}) + \tilde{D}_{kl}^{(2,1)} + \tilde{D}_{kl}^{(2,2)} + \tilde{D}_{kl}^{(2,3)}$$

where

$$\begin{aligned}
\tilde{D}_{kl}^{(2,1)} &= -\sum_{\substack{b,c,d \\ j}} v_{bdlj}^* v_{cdkj} d_{bc} \\
\tilde{D}_{kl}^{(2,2)} &= \tfrac{1}{2} \sum_{\substack{c,d \\ i,j}} v_{dclj}^* v_{dcki} d_{ij} \\
\tilde{D}_{kl}^{(2,3)} &= -\frac{1}{4} \sum_{\substack{c,d \\ i,j}} v_{dcij}^* v_{dcki} d_{lj} + \text{h.c.}
\end{aligned} \tag{11.78}$$

Here, $D_0(2)$ is given by Eq. (11.66), and $\rho_{rs}^{(2)}$ denotes the second-order contributions to the one-particle density matrix (Eq. 11.63).

The order relations and separability structure applying to $\tilde{\boldsymbol{D}}$ will be addressed in Sects. 12.1 and 12.2.

Exercises

11.1 Use ground-state PT to expand the overlap matrix for the $1h$ CE states (Eq. 11.7) through second order.

11.2 Derive the ISR(2) expressions for the transition amplitudes (11.49).

11.3 Use the direct ADC(2) secular equation to derive the ionization energy for a single-hole main state through second order, and compare the result to Eq. (8.68).

References

1. Schirmer J (1991) Phys Rev A 43:4647
2. Mertins F, Schirmer J (1996) Phys Rev A 53:2140
3. Manne R (1977) Chem Phys Lett 45:470
4. Dalgaard E (1979) Int J Quantum Chem 15:169
5. Schirmer J, Trofimov AB (2004) J Chem Phys 120:11449
6. Trofimov AB, Schirmer J (2005) J Chem Phys 123:144115

Chapter 12
Order Relations and Separability

The equivalence of the direct ADC approach and the ECO-ISR formulation rests on two common key features. The first is the so-called canonical PT order structure of the secular matrix [1], establishing the compactness of the nth-order ISR-ADC approximation schemes. The other is the separability [2] of the secular matrix with respect to two (or more) non-interacting sub-systems, which warrants size-consistency. While in the ADC context these features could have been substantiated using diagrammatic arguments, the ECO-ISR concept established in the preceding chapter allows for a stringent formulation and rigorous proofs. This is the topic of the present chapter. Section 12.1 (together with Appendix A.6) discusses the order structure of the secular matrix and the characteristic truncation errors thus entailed. The separability property will be treated in Sect. 12.2. Finally, in Sect. 12.3, we take a comparative look at the standard CI method, where the secular structure is simple but neither canonical nor separable.

12.1 Canonical Order Relations

In the ISR-ADC approach, the secular matrix elements are given in the form of PT expansions (see Eqs. 10.15, 11.22). Here, the remarkable finding is that these expansions do not necessarily begin at zeroth order. Rather, the lowest non-vanishing order in a matrix element M_{IJ} depends on the "distance" $[I] - [J]$ of the excitation classes, $[I]$, $[J]$, to which the configurations I and J belong. The rules fulfilled by the ISR-ADC secular matrix elements are as follows:

$$M_{IJ} \sim O(|[I] - [J]|) \tag{12.1}$$

J. Schirmer, *Many-Body Methods for Atoms, Molecules and Clusters*, Lecture Notes in Chemistry 94, https://doi.org/10.1007/978-3-319-93602-4_12

Fig. 12.1 Order structure of the ISR-ADC secular matrix $\boldsymbol{M}$ for $(N-1)$-electron excitations

	$1h$	$2h$-$1p$	$3h$-$2p$	$4h$-$3p$	$5h$-$4p$	...
$1h$	0	1	2	3	4	...
$2h$-$1p$	1	0	1	2	3	...
$3h$-$2p$	2	1	0	1	2	...
$4h$-$3p$	3	2	1	0	1	...
$5h$-$4p$	4	3	2	1	0	...
⋮	⋮	⋮	⋮	⋮	⋮	

This means that in the PT expansion of M_{IJ} the lowest non-vanishing contribution is of the order $|[I] - [J]|$. For example, the coupling matrix element $M_{i,abjkl}$ combining a $1h$ (class 1) and a $3h$-$2p$ (class 3) configuration is of second order:

$$M_{i,abjkl} \sim O(|[i] - [abjkl]|) = O(2) \tag{12.2}$$

The rules (12.1) are referred to as **canonical order relations** (COR). Fig. 12.1 depicts the COR structure of $\boldsymbol{M}$; here, the respective lowest non-vanishing PT order is assigned to the $\boldsymbol{M}_{\mu\nu}$ sub-blocks in the partitioning of $\boldsymbol{M}$ according to excitation classes, $\mu, \nu = 1, 2, \ldots$. A proof of these rules is given in Appendix A.6.

To better understand the essence of the COR, let us come back to the matrix element $M_{i,abjkl}$, where the excitation classes of the first and second entry differ by 2. In the CI secular matrix (see Sect. 12.3), the corresponding coupling is of first order (i.e., linear in the Coulomb repulsion integrals),

$$H_{i,abjkl} = \langle \Phi_i | \hat{H}_I | \Phi_{abjkl} \rangle = -\delta_{ij} V_{kl[ab]} + \delta_{ik} V_{jl[ab]} - \delta_{il} V_{jk[ab]} \tag{12.3}$$

By contrast, the first-order contribution to $M_{i,abjkl}$ vanishes as the result of a non-trivial cancelation, where the Gram–Schmidt orthogonalization adopted in $|\tilde{\Psi}_{abjkl}\rangle$ is crucial. According to

$$\begin{aligned} M^{(1)}_{i,abjkl} =& \langle \tilde{\Psi}_i | \hat{H} - E_0 | \tilde{\Psi}_{abjkl} \rangle^{(1)} \\ =& H_{i,abjkl} + \langle \tilde{\Psi}^{(1)}_i | \hat{H}_0 - E^{(0)}_0 | \Phi_{abjkl} \rangle + \langle \Phi_i | \hat{H}_0 - E^{(0)}_0 | \tilde{\Psi}^{(1)}_{abjkl} \rangle \end{aligned} \tag{12.4}$$

the first-order matrix element comprises in addition to the CI-type contribution (12.3) two terms involving the first-order wave functions $|\tilde{\Psi}^{(1)}_i\rangle$ and $|\tilde{\Psi}^{(1)}_{abjkl}\rangle$. The former simply reads

$$|\tilde{\Psi}_i^{(1)}\rangle = c_i|\Psi_0^{(1)}\rangle \tag{12.5}$$

where the first-order ground state, $|\Psi_0^{(1)}\rangle$, is given by Eq. (11.32). In first order, the Gram-Schmidt orthogonalization of the intermediate $3h$-$2p$ state with regard to the $1h$ states comes into play, giving rise to following three terms,

$$\begin{aligned}|\tilde{\Psi}_{abjkl}^{(1)}\rangle =& c_a^\dagger c_b^\dagger c_j c_k c_l|\Psi_0^{(1)}\rangle - |\Phi_j\rangle\langle\tilde{\Psi}_j^{(1)}|\Phi_{abjkl}\rangle - |\Phi_k\rangle\langle\tilde{\Psi}_k^{(1)}|\Phi_{abjkl}\rangle - |\Phi_l\rangle\langle\tilde{\Psi}_l^{(1)}|\Phi_{abjkl}\rangle \\ =& c_a^\dagger c_b^\dagger c_j c_k c_l|\Psi_0^{(1)}\rangle - |\Phi_j\rangle v_{abkl}^* + |\Phi_k\rangle v_{abjl}^* - |\Phi_l\rangle v_{abjk}^*\end{aligned} \tag{12.6}$$

Note that there are no first-order contributions from the orthogonalization with respect to $2h$-$1p$ states. Accordingly, the third term in Eq. (12.4) becomes

$$\langle\Phi_i|\hat{H}_0 - E_0^{(0)}|\tilde{\Psi}_{abjkl}^{(1)}\rangle = -\epsilon_i\,\langle\Phi_i|\tilde{\Psi}_{abjkl}^{(1)}\rangle = \epsilon_i\,(\delta_{ij}\,v_{abkl}^* - \delta_{ik}\,v_{abjl}^* + \delta_{il}\,v_{abjk}^*) \tag{12.7}$$

A related expression results from the second term,

$$\langle\tilde{\Psi}_i^{(1)}|\hat{H}_0 - E_0^{(0)}|\Phi_{abjkl}\rangle = \left(\delta_{ij}v_{abkl}^* - \delta_{ik}v_{abjl}^* + \delta_{il}v_{abjk}^*\right)(\epsilon_a + \epsilon_b - \epsilon_j - \epsilon_k - \epsilon_l) \tag{12.8}$$

Obviously, the latter two expressions can be combined to give $(-1)H_{i,abjkl}$ which cancels the CI term in Eq. (12.4), thus confirming the proposition

$$M_{i,abjkl}^{(1)} = 0 \tag{12.9}$$

In this derivation, the crucial role of Gram–Schmidt orthogonalization in effecting the COR could be seen explicitly. If symmetric orthonormalization (11.5) of the CE states is used rather than Gram–Schmidt, the $1h/3h$-$2p$ coupling matrix elements do not vanish in first order (see Exercise 12.1):

$$\langle\overline{\Psi}_i|\hat{H} - E_0|\overline{\Psi}_{abjkl}\rangle = \tfrac{1}{2}H_{i,abjkl} + O(2) \tag{12.10}$$

This example shows that a representation of $\hat{H} - E_0$ based on symmetrically orthonormalized CE states does not have the COR structure, nor the separability structure as will be discussed in Sect. 12.3.

The order relations determine the PT order of the error caused by truncating the (explicit) configuration space, supposing here a systematic truncation after a specified class, say μ. For example, the error in the $1h$ ionization energies caused by neglecting the $3h$-$2p$ (and higher) configurations is of fourth order. Here, the coupling of the $1h$ and $3h$-$2p$ configurations is at least of second order, so that the corresponding energy contribution (being quadratic in the coupling matrix element) is of fourth (or higher) order. Truncation after the $3h$-$2p$ configurations (class 3) leads to an error of the order 6, which reflects the COR value of 3 for the coupling between configurations of class 1 and 4.

The general formula for the truncation error order (TEO) in the $1h$-state energies is (see Appendix A.6)

$$O_{TE}(\mu) = 2\mu \tag{12.11}$$

where μ denotes the highest class included in the (explicit) configuration space.

This formula can be generalized to states other than the $1h$ (main) states. Here, we suppose that the final ionic states, $|\Psi_n^{N-1}\rangle$, can still be characterized according to their PT descent, that is, as originating from a given excitation class. To denote the respective PT descent, we will use the notation $[n]$, that is, $[n] = \nu$ if $|\Psi_n^{N-1}\rangle$ derives from a CI state of class ν. The generalized TEO formula then assumes the form

$$O_{TE}^{[n]}(\mu) = 2(\mu - [n] + 1), \;\; \mu \geq [n] \tag{12.12}$$

Obviously, this equation reduces to Eq. (12.11) in the case $[n] = 1$.

In a similar way, we may analyze the transition moments, more specifically, the spectroscopic factors

$$x_p^{(n)} = \langle \Psi_n^{N-1} | c_p | \Psi_0 \rangle = \boldsymbol{X}_n^{\dagger} \boldsymbol{f}_p \tag{12.13}$$

with respect to truncation errors. Order relations apply not only to the secular matrix but also to the ISR-ADC transition amplitudes (see Appendix A.6):

$$f_{Ip} \sim O([I] - 1) \tag{12.14}$$

As a direct consequence of the order structure (12.1) of the secular matrix, the COR can be established for the eigenvector matrix $\boldsymbol{X}$ as well:

$$X_{Jn} = \langle \tilde{\Psi}_J | \Psi_n^{N-1} \rangle \sim O(|[J] - [n]|) \tag{12.15}$$

A proof of the order structure of $\boldsymbol{X}$ is given in Appendix A.6. Using the latter order relations together with those for $\boldsymbol{f}$ in Eq. (12.13) results in the following general TEO formula for the spectroscopic factors:

$$O_{TE}^{[n]}(\mu) = 2\mu - [n] + 1, \;\; \mu \geq [n] \tag{12.16}$$

Note that for the $1h$ states, $[n] = 1$, the same TEO result applies both to the energies and spectroscopic factors.

In Sect. 11.3, we have discussed the ISR of a general one-particle operator $\hat{D}$, and it may be of interest to inspect the order structure in this case. Here, the order relations take on the form

$$\widetilde{D}_{IJ} \sim \begin{cases} O(|[I] - [J]| - 1), \;\; [I] \neq [J] \\ O(0), \;\; [I] = [J] \end{cases} \tag{12.17}$$

Fig. 12.2 Order structure of the intermediate-state representation $\widetilde{\boldsymbol{D}}$ of a one-particle operator $\hat{D}$

	$1h$	$2h$-$1p$	$3h$-$2p$	$4h$-$3p$	$5h$-$4p$	...
$1h$	0	0	1	2	3	...
$2h$-$1p$	0	0	0	1	2	...
$3h$-$2p$	1	0	0	0	1	...
$4h$-$3p$	2	1	0	0	0	...
$5h$-$4p$	3	2	1	0	0	...
⋮	⋮	⋮	⋮	⋮	⋮	

The corresponding structure is depicted in Fig. 12.2. The zeroth-order coupling between states I and J belonging to adjacent classes, $[I] = [J] \pm 1$, reflects the fact that the corresponding zeroth order (CI) coupling matrix elements $\langle \Phi_I | \hat{D} | \Phi_J \rangle$ need not vanish.

What is the effect of the somewhat weaker order relations (12.17) on the TEOs in the properties of an ionic state? Considering an expectation value

$$D_n = \langle \Psi_n^{N-1} | \hat{D} | \Psi_n^{N-1} \rangle = \boldsymbol{X}_n^\dagger \widetilde{\boldsymbol{D}} \boldsymbol{X}_n \tag{12.18}$$

the TEO can easily be determined by combining the order structure of $\widetilde{\boldsymbol{D}}$ with those of the respective eigenvector class. For $1h$ states ($[n] = 1$), the truncation error (truncation after class μ) is seen to be

$$O_{TE}(\mu) = 2\mu - 1 \tag{12.19}$$

This means that at the ISR-ADC(3) level, where the explicit configuration space comprises the classes 1 and 2, (one-particle) properties of the ionic $1h$ (main) states are treated consistently through second order only.

12.2 Separability of the ISR-ADC Secular Matrix

The size-consistency properties of a computational method can be analyzed in a stringent way by resorting to the **separate fragment model**, that is, a hypothetical system S consisting of two strictly non-interacting sub-systems or fragments, A and B. A method for treating electronic excitation, or ionization, electron attachment, etc., is size-consistent (here, more specifically, size-intensive), if for local excitations, say on fragment A, the computed excitation energies and transition moments do not depend on whether the method is applied to the fragment or the composite system. Obviously, the outcome in the artificial separate fragment model is indicative for the performance of the method in realistic systems, e.g., formed by interacting subsystems, or just extended systems beyond a certain size.

Let us begin with a few specifications of the separate fragment model. The total hamiltonian is the sum

$$\hat{H} = \hat{H}_A + \hat{H}_B \tag{12.20}$$

of the two fragment hamiltonians, $\hat{H}_A$ and $\hat{H}_B$, so that the ground state of the composite is given as the product

$$|\Psi_0\rangle = |\Psi_0^A\rangle|\Psi_0^B\rangle \tag{12.21}$$

of the fragment ground states, $|\Psi_0^A\rangle$ and $|\Psi_0^B\rangle$. It should be noted that in this and other product states inter-fragment antisymmetrization is irrelevant and can be waived.

The one-particle states (HF orbitals) of S are assumed to be local, that is, a given orbital either belongs to fragment A or B. As a consequence, the electron configurations in the set (11.2) can be partitioned into three different subsets, namely local excitations I_A on fragment A, local excitations I_B on fragment B, and mixed (or non-local) excitations I_{AB} involving both fragments A and B. A mixed excitation, for example, might consist of an ionization on A, accompanied by a neutral excitation on B. In analogy to the physical operators (11.2), we introduce the operator set

$$\{\hat{C}_J\} = \left\{c_a^\dagger c_k;\ c_a^\dagger c_b^\dagger c_k c_l, a < b, k < l;\ \ldots\right\} \tag{12.22}$$

associated with the neutral $1p$-$2h$, $2p$-$2h$, ..., excitations.

It should be noted that there are non-local excitations that do not conserve the charge of the fragments, such as a double ionization on A accompanied by electron attachment on B, resulting in an $A^{++}B^-$ structure. However, such charge-transfer excitations need not be regarded. In the separate fragment model, they are strictly decoupled from the fragment-charge conserving excitations to be considered in the following.

How is the structure of separated fragments reflected in the intermediate states $|\tilde{\Psi}_I\rangle$? For the CE states $|\Psi_J^0\rangle$, forming the starting point of the ECO- IS construction, the answer is trivial. Since the ground state is the product of the fragment ground states (Eq. 12.21) and the physical excitation operators $\hat{C}_J$ are operator products of local fermion operators, the CE states can be written as products of each two fragment states. A local state simply reads

$$|\Psi_{J_A}^0\rangle = \hat{C}_{J_A}|\Psi_0\rangle = \hat{C}_{J_A}|\Psi_0^A\rangle\,|\Psi_0^B\rangle \tag{12.23}$$

while for non-local states the factorization takes on the form

$$|\Psi^0_{J_{AB}}\rangle = \hat{C}_{J_{AB}}|\Psi_0\rangle = \hat{C}_{J_A}|\Psi^A_0\rangle\,\hat{C}_{J_B}|\Psi^B_0\rangle \tag{12.24}$$

But what about the final intermediate states? They result by applying an involved orthonormalization procedure to the CE states, and it is not obvious whether the outcome can be written as products of fragment states. The answer is given by the following **factorization theorem** stating that non-local ECO intermediate states can be written according to

$$|\tilde{\Psi}_{J_{AB}}\rangle = |\tilde{\Psi}^A_{J_A}\rangle\,|\tilde{\Psi}^B_{J_B}\rangle \tag{12.25}$$

as products of fragment ECO intermediate states. The factorization of local ECO intermediate states,

$$|\tilde{\Psi}_{J_A}\rangle = |\tilde{\Psi}^A_{J_A}\rangle\,|\Psi^B_0\rangle \tag{12.26}$$

can be seen as a special case of the general case (12.25).

A general proof of the factorization theorem has been given in Ref. [2] where the interested reader is referred to. To see how this factorization comes about, it is instructive to inspect the simple case of a non-local $2h$-$1p$ configuration. Let the configuration be $J \equiv a'j'k$, where k denotes an orbital associated with fragment A, and the primed indices refer to orbitals of fragment B. This is, J stands for a $1p$-$1h$ excitation on fragment B accompanying an electron vacancy (in orbital k) on fragment A. The corresponding CE state reads

$$|\Psi^0_{a'j'k}\rangle = c_k|\Psi^A_0\rangle\,c^\dagger_{a'}c_{j'}|\Psi^B_0\rangle \tag{12.27}$$

Gram–Schmidt orthogonalization with respect to the intermediate states of class 1 yields the precursor state

$$|\Psi^\#_{a'j'k}\rangle = |\Psi^0_{a'j'k}\rangle - \sum_l |\tilde{\Psi}_l\rangle\langle\tilde{\Psi}_l|\Psi^0_{a'j'k}\rangle \tag{12.28}$$

where the $1h$ intermediate states can be restricted to those being local on fragment A,

$$|\tilde{\Psi}_l\rangle = |\tilde{\Psi}^A_l\rangle|\Psi^B_0\rangle \tag{12.29}$$

Using the latter product form together with that of the CE state (12.27) in Eq. (12.28) gives

$$\begin{aligned}|\Psi^\#_{a'j'k}\rangle =& c_k|\Psi^A_0\rangle\,c^\dagger_{a'}c_{j'}|\Psi^B_0\rangle - \sum_l |\tilde{\Psi}^A_l\rangle\langle\tilde{\Psi}^A_l|c_k|\Psi^A_0\rangle\,|\Psi^B_0\rangle\langle\Psi^B_0|c^\dagger_{a'}c_{j'}|\Psi^B_0\rangle\\ =& c_k|\Psi^A_0\rangle\left(c^\dagger_{a'}c_{j'}|\Psi^B_0\rangle - |\Psi^B_0\rangle\langle\Psi^B_0|c^\dagger_{a'}c_{j'}|\Psi^B_0\rangle\right)\end{aligned} \tag{12.30}$$

To arrive at the second equation, we have used the identity

$$c_k|\Psi_0^A\rangle = \sum_l |\tilde{\Psi}_l^A\rangle\langle\tilde{\Psi}_l^A|c_k|\Psi_0^A\rangle \tag{12.31}$$

that is, the decomposition of the CE state $c_k|\Psi_0^A\rangle$ of fragment A with respect to the fragment A intermediate states of class 1. The term in brackets in the second line of Eq. (12.30) can be identified as the precursor state

$$|\Psi_{a'j'}^{B\#}\rangle = c_{a'}^\dagger c_{j'}|\Psi_0^B\rangle - |\Psi_0^B\rangle\langle\Psi_0^B|c_{a'}^\dagger c_{j'}|\Psi_0^B\rangle \tag{12.32}$$

of the $1p$-$1h$ excitation on fragment B which, being a neutral excitation, involves orthogonalization to the ground state of fragment B (see Sect. 14.1). This means that the $2h$-$1p$ precursor state (12.30) is the product

$$|\Psi_{a'j'k}^{\#}\rangle = |\Psi_k^{A\#}\rangle\,|\Psi_{a'j'}^{B\#}\rangle \tag{12.33}$$

of the respective fragment precursor states. Obviously, the ensuing symmetric orthonormalization of the (non-local) $2h$-$1p$ precursor states is obtained by independent orthonormalization of the fragment precursor states, establishing the product form

$$|\tilde{\Psi}_{a'j'k}\rangle = |\tilde{\Psi}_k^A\rangle\,|\tilde{\Psi}_{a'j'}^B\rangle \tag{12.34}$$

for the final intermediate states.

Equipped with the factorization theorem, we may now establish the **separability** of the secular matrix. The classification of the intermediate basis states according to their localization type, that is, as local excitations on fragment A, local excitations on fragment B, and non-local excitations involving both A and B, effects a corresponding partitioning of the secular matrix $\boldsymbol{M}$ into sub-blocks, $\boldsymbol{M}_{ZZ'}$ with $Z, Z' = A, B, AB$. The essential property of the ECO-ISR secular matrix is that all non-diagonal matrix blocks vanish. This structure, referred to as separability, is shown in Fig. 12.3.

Obviously, there is no coupling between states local on A and states local on B:

$\boldsymbol{M}_{AA}$	0	0
0	$\boldsymbol{M}_{BB}$	0
0	0	$\boldsymbol{M}_{AB,AB}$

Fig. 12.3 Separable block structure of the ISR-ADC secular matrix $\boldsymbol{M}$ with respect to the separate fragment model

$$M_{I_A,J_B} = \langle \tilde{\Psi}_{I_A} | \hat{H} | \tilde{\Psi}_{J_B} \rangle = \langle \tilde{\Psi}^A_{I_A} | \langle \Psi^B_0 | (\hat{H}_A + \hat{H}_B) | \tilde{\Psi}^B_{J_B} \rangle | \Psi^A_0 \rangle = 0 \tag{12.35}$$

as, for example, the states of fragment A, $|\tilde{\Psi}_{I_A}\rangle$ and $|\Psi^A_0\rangle$, are orthogonal (even relate to different electron numbers). The case of the coupling block $\boldsymbol{M}_{A,AB}$ is less trivial:

$$\begin{aligned} M_{I_A,J_{AB}} =& \langle \tilde{\Psi}_{I_A} | \hat{H} | \tilde{\Psi}_{J_{AB}} \rangle \\ =& \langle \tilde{\Psi}^A_{I_A} | \langle \Psi^B_0 | (\hat{H}_A + \hat{H}_B) | \tilde{\Psi}^A_{J_A} \rangle | \tilde{\Psi}^B_{J_B} \rangle \\ =& \langle \tilde{\Psi}^A_{I_A} | \hat{H}_A | \tilde{\Psi}^A_{J_A} \rangle \, \langle \Psi^B_0 | \tilde{\Psi}^B_{J_B} \rangle + \langle \tilde{\Psi}^A_{I_A} | \tilde{\Psi}^A_{J_A} \rangle \, \langle \Psi^B_0 | \hat{H}_B | \tilde{\Psi}^B_{J_B} \rangle = 0 \end{aligned} \tag{12.36}$$

In the last line, the first term vanishes because, by construction, $|\tilde{\Psi}^B_{J_B}\rangle$ is orthogonal to $|\Psi^B_0\rangle$, that is, $\langle \Psi^B_0 | \tilde{\Psi}^B_{J_B} \rangle = 0$. The second term vanishes according to

$$\langle \Psi^B_0 | \hat{H}_B | \tilde{\Psi}^B_{J_B} \rangle = E^B_0 \langle \Psi^B_0 | \tilde{\Psi}^B_{J_B} \rangle = 0 \tag{12.37}$$

as $|\Psi^B_0\rangle$ is the ground state of $\hat{H}_B$, and, again, $\langle \Psi^B_0 | \tilde{\Psi}^B_{J_B} \rangle = 0$.

It is important to note that Eq. (12.36) holds through all orders of PT. As a consequence, the separability property not only applies to the exact ISR but also to the systematic ISR(n) approximation schemes.

It remains to show that the diagonal block $\boldsymbol{M}_{AA}$ of $\boldsymbol{M}$ is identical with the secular matrix $\boldsymbol{M}^A$ of fragment A. This is easily seen as follows:

$$\begin{aligned} M_{I_A,J_A} =& \langle \tilde{\Psi}_{I_A} | \hat{H}_A + \hat{H}_B | \tilde{\Psi}_{J_A} \rangle - \delta_{I_A,J_A} E_0 \\ =& \langle \tilde{\Psi}^A_{I_A} | \hat{H}_A | \tilde{\Psi}^A_{J_A} \rangle + \delta_{I_A,J_A} E^B_0 - \delta_{I_A,J_A} E_0 \\ =& \langle \tilde{\Psi}^A_{I_A} | \hat{H}_A - E^A_0 | \tilde{\Psi}^A_{J_A} \rangle = M^A_{I_A,J_A} \end{aligned} \tag{12.38}$$

The full separability structure of the ECO-ISR secular matrix guarantees size-consistent results: For local excitations (ionizations) on one of the fragments, say A, the eigenvalues (ionization energies) of the secular matrix $\boldsymbol{M}$ of the composite system are identical with those of the fragment secular matrix $\boldsymbol{M}^A$.

Of course, this finding applies also to the spectroscopic factors. According to the separability structure of $\boldsymbol{M}$, the eigenvector of a local excitation, say on A, has the form

$$\boldsymbol{X}_n = \begin{pmatrix} \boldsymbol{X}^A_n \\ \boldsymbol{0} \\ \boldsymbol{0} \end{pmatrix} \tag{12.39}$$

where $\boldsymbol{X}^A_n$ denotes the corresponding eigenvector of $\boldsymbol{M}^A$. For an orbital p of fragment A, the vector $\boldsymbol{f}_p$ of transition amplitudes f_{Ip} takes on the same form,

$$\boldsymbol{f}_p = \begin{pmatrix} \boldsymbol{f}^A_p \\ \boldsymbol{0} \\ \boldsymbol{0} \end{pmatrix} \tag{12.40}$$

$\boldsymbol{D}_{AA}$	0	$\boldsymbol{D}_{A,AB}$
0	$\boldsymbol{D}_{BB}$	$\boldsymbol{D}_{B,AB}$
$\boldsymbol{D}_{AB,A}$	$\boldsymbol{D}_{AB,B}$	$\boldsymbol{D}_{AB,AB}$

Fig. 12.4 Block structure of the intermediate-state representation $\boldsymbol{D}$ of general operator $\hat{D}$ with respect to the separate fragment model

as, for example, $f_{I_B p} = \langle\Psi_0^A|c_p|\Psi_0^A\rangle\langle\tilde{\Psi}_{I_B}^B|\Psi_0^B\rangle = 0$. In Eq. (12.40), $\boldsymbol{f}_p^A$ denotes the vector of fragment A transition amplitudes, $f_{I_A p}^A = \langle\tilde{\Psi}_{I_A}^A|c_p|\Psi_0^A\rangle$. According to

$$x_p^{(n)} = \boldsymbol{X}_n^\dagger \boldsymbol{f}_p = \boldsymbol{X}_n^{A\dagger} \boldsymbol{f}_p^A = x_p^{A(n)} \tag{12.41}$$

the composite and fragment versions of the spectroscopic factor are identical.

In a similar way, one may analyze the representation of a general (not necessarily one-particle) operator $\hat{D} = \hat{D}_A + \hat{D}_B$. As seen in Fig. 12.4, the block structure resulting for the separate fragment partitioning is not separable. For example, the coupling matrix elements for local and non-local configurations

$$\langle\tilde{\Psi}_{I_A}|\hat{D}|\tilde{\Psi}_{J_{AB}}\rangle = \delta_{I_A,J_A}\langle\tilde{\Psi}_0^B|\hat{D}_B|\tilde{\Psi}_{J_B}^B\rangle \tag{12.42}$$

need not vanish. What does this mean for the computation of properties, which necessarily relate to the composite system? Let us consider a local excitation (ionization) $|\Psi_n^{N-1}\rangle = |\Psi_n^A\rangle|\Psi_0^B\rangle$, where the eigenvector is of the form given by Eq. (12.39). The corresponding expectation value of $\hat{D}$ becomes

$$D_n = \langle\Psi_n^{N-1}|\hat{D}_A + \hat{D}_B|\Psi_n^{N-1}\rangle = \boldsymbol{X}_n^{A\dagger}\boldsymbol{D}_{AA}\boldsymbol{X}_n^A = D_n^A + D_0^B \tag{12.43}$$

where $D_0^B = \langle\Psi_0^B|\hat{D}_B|\Psi_0^B\rangle$ is the ground-state expectation value of fragment B. The last equation reflects the fact that the $\boldsymbol{D}_{AA}$ block is given by

$$\boldsymbol{D}_{AA} = \boldsymbol{D}^A + D_0^B\mathbf{1} \tag{12.44}$$

where $\boldsymbol{D}^A$ denotes the ISR secular matrix of fragment A. According to Eq. (12.43), the ISR-ADC approach reproduces correctly a property pertaining to the composite as the sum of the fragment contributions, being here the property of the excited (ionized) fragment A in the excited (ionized) state and the property of fragment B in the ground state.

12.3 A Look at the CI Method

The benefit of the ISR-ADC approach presented in the preceding chapters has to be seen in comparison to conventional quantum-chemical methods, most notably the standard configuration–interaction (CI) treatment. Therefore, the performance of the CI method with regard to truncation errors and size-consistency shall briefly be inspected in the following.

In the CI treatment, here of an $(N-1)$-electron system, the Schrödinger equation

$$\hat{H}|\Psi_n^{N-1}\rangle = E_n^{N-1}|\Psi_n^{N-1}\rangle \tag{12.45}$$

is recast into an algebraic eigenvalue problem by expanding the ionic states according to

$$|\Psi_n^{N-1}\rangle = \sum_J X_{Jn}|\Phi_J\rangle \tag{12.46}$$

in terms of the CI states

$$|\Phi_J\rangle = \hat{C}_J|\Phi_0\rangle \tag{12.47}$$

as introduced by Eqs. (11.2), (11.3). In matrix notation, the algebraic eigenvalue equations read

$$\boldsymbol{HX} = \boldsymbol{X\Omega}, \quad \boldsymbol{X}^\dagger\boldsymbol{X} = \mathbf{1} \tag{12.48}$$

Here, $\boldsymbol{H}$ denotes the CI secular matrix,

$$H_{IJ} = \langle\Phi_I|\hat{H}|\Phi_J\rangle \tag{12.49}$$

$\boldsymbol{\Omega}$ is the diagonal matrix of ionic-state energies E_n^{N-1}, and $\boldsymbol{X}$ is the matrix of (column) eigenvectors $\boldsymbol{X}_n$. To form ionization energies, $I_n = E_n^{N-1} - E_0$, one has to obtain E_0 from a separate CI computation for the N-electron system.

Structure of the CI Matrix

Approximate CI treatments are obtained by limited CI expansions as opposed to full (FCI) expansions. In the following, we will be concerned with systematic truncations of the CI expansions, that is, expansions being complete through a given excitation class μ. These systematic truncation schemes can be examined with respect to the PT order of the induced error in the CI results. For this purpose, one has to inspect the structure of the CI secular matrix $\boldsymbol{H}$.

Being a fully variational method, the CI secular matrix $\boldsymbol{H}$ has a rather basic "order structure," as shown in Fig. 12.5. Owing to the $\hat{H}_0$ part of the hamiltonian, the diagonal matrix elements are of zeroth order, as indicated by the zeros in the diagonal blocks. Each excitation class μ is coupled through terms linear in the two-electron Coulomb integrals (first order) to the excitation classes $\mu \pm 1$ and $\mu \pm 2$, which is indicated by the entries 1 in the respective matrix blocks. There is no coupling between states differing by more than two excitation classes (entry "-").

Fig. 12.5 Order structure of the CI secular matrix $\boldsymbol{H}$

	$1h$	$2h$-$1p$	$3h$-$2p$	$4h$-$3p$	$5h$-$4p$	...
$1h$	0	1	1	-	-	...
$2h$-$1p$	1	0	1	1	-	...
$3h$-$2p$	1	1	0	1	1	...
$4h$-$3p$	-	1	1	0	1	...
$5h$-$4p$	-	-	1	1	0	...
⋮	⋮	⋮	⋮	⋮	⋮	

The characteristic structure of the CI secular matrix gives rise to comparatively large truncation errors in the final state energies. As the most important case, let us consider $1h$ (main) states. Due to the linear (first order) coupling to $3h$-$2p$ states, there is a second-order contribution to the $1h$ ionic energies arising from the admixture of $3h$-$2p$ excitations. This means that there is a second-order truncation error if the CI configuration space is confined to the $1h$ and $2h$-$1p$ configurations. The general formula for the TEO in the $1h$ state energies is given by

$$O_{TE}(\mu) = \begin{cases} \mu, & \mu \text{ even} \\ \mu + 1, & \mu \text{ odd} \end{cases} \tag{12.50}$$

where as before μ denotes the highest excitation class included in the CI expansion manifold.

The CI truncation errors are relatively large, which, in turn, implies that large CI expansions are required to meet specific accuracy levels. For example, in order to treat $1h$ states consistently through second order of PT, the CI configuration space must comprise the $3h$-$2p$ excitations ($\mu = 3$). By contrast, in the ISR-ADC method, a much smaller explicit configuration space, consisting of $1h$ and $2h$-$1p$ excitations, affords the same level of accuracy.

Non-separability

With regard to the separate fragment model and the corresponding partitioning, the CI secular matrix is not separable, as shown in Fig. 12.6. This is notwithstanding the obvious fact that all CI configurations $|\Phi_I\rangle$ assume the form of products of fragment states, e.g.,

$$
\begin{aligned}
|\Phi_0\rangle &= |\Phi_0^A\rangle|\Phi_0^B\rangle \\
|\Phi_{I_A}\rangle &= |\Phi_{I_A}^A\rangle|\Phi_0^B\rangle \\
|\Phi_{I_{AB}}\rangle &= |\Phi_{I_A}^A\rangle|\Phi_{I_B}^B\rangle
\end{aligned}
\tag{12.51}
$$

Unlike in the ECO-ISR case, there is no equivalent to Eq. (12.37) here. While the A and B states are strictly decoupled, $\boldsymbol{H}_{AB} = \mathbf{0}$, there can be a non-vanishing coupling between a local and a mixed excitation, e.g.,

$$
H_{I_A, J_{AB}} = \delta_{I_A J_A} \langle \Phi_0^B | \hat{H}_B | \Phi_{J_B}^B \rangle \tag{12.52}
$$

where $\langle \Phi_0^B | \hat{H}_B | \Phi_{J_B}^B \rangle$ is a righteous matrix element in the CI treatment of the ground state of fragment B.

Non-local coupling may arise explicitly in the CI matrix elements H_{IJ}, where the configurations I, J differ by two excitation classes, such as in the $1h$/$3h$-$2p$ matrix elements (12.3). Let l denote a $1h$ excitation on fragment A ($I_A \equiv l$) and $a'b'j'k'l \equiv J_{AB}$ a non-local $3h$-$2p$ excitations where the primed indices refer to one-particle states associated with fragment B; the coupling matrix element becomes

$$
H_{l,a'b'j'k'l} = V_{a'b'[j'k']} \tag{12.53}
$$

which obviously does not depend on the distance between the two fragments.

According to Eq. (12.10), the potentially non-local coupling between $1h$ and $3h$-$2p$ configurations arises also in the ISR version (11.5) based on symmetrically orthonormalized (SO) CE states. This shows that the SO procedure for the CE states does not result in a separable secular matrix either.

Given the structure shown in Fig. 12.6, there is no a priori decoupling of local excitations (say on fragment A) from non-local (or mixed) excitations. The CI treatment of the composite system S aims in an inextricable way at an optimal description of both fragments, that is, the ionic state of fragment A and the ground state of fragment B. In the exact (full) CI treatment of the composite system, the energy of a

$\boldsymbol{H}_{AA}$	0	$\boldsymbol{H}_{A,AB}$
0	$\boldsymbol{H}_{BB}$	$\boldsymbol{H}_{B,AB}$
$\boldsymbol{H}_{AB,A}$	$\boldsymbol{H}_{AB,B}$	$\boldsymbol{H}_{AB,AB}$

Fig. 12.6 Non-separable block structure of the CI secular matrix $\boldsymbol{H}$ with respect to the separate fragment model

local ionization, say on A, is of course given as the sum $E_n^{N-1} = E_n^A + E_0^B$ of the fragment state energies. Here, subtraction of the FCI result for the composite ground state, $E_0 = E_0^A + E_0^B$, would cancel E_0^B and lead to the correct ionization energy for the ionization of the local fragment A. In realistic applications, one has to resort to truncated CI expansions, and here, an energy computed for the composite system differs from the sum of the CI results for the respective fragment states. Due to this well-known size-consistency error, the CI method does not qualify as a genuine many-body method.

Exercises

12.1 (a) Consider the symmetrical orthonormalization (SO) of the CE states according to Eq. (11.5). Construct the $1h$ and $3h$-$2p$ states $|\overline{\Psi}_i\rangle$, $|\overline{\Psi}_{abjkl}\rangle$ through first order and verify Eq. (12.10). Hint: The overlap matrix $\boldsymbol{S}$ (Eq. 11.4) of the CE states can be expanded as $\boldsymbol{S} = \mathbf{1} + \boldsymbol{S}^{(1)} + \dots$ so that $\boldsymbol{S}^{-1/2} = \mathbf{1} - \frac{1}{2}\boldsymbol{S}^{(1)} + O(2)$.
(b) Sketch a proof for a factorization theorem analogous to the factorization of the ECO-ISR states according to Eq. (12.25). Discuss the separability properties of the SO-CE secular matrix.

12.2 Multiple non-interacting 2E-2O systems (see Exercise 2.4):
A well-known model (see for example Ref. [3]) to demonstrate the size-consistency problem in the (restricted) CI treatment is a system composed of M (> 1) mutually non-interacting 2E-2O sub-systems, such as, for example, an array of separate H_2 molecules in the minimal basis (two orbital) approximation.
(a) Apply the double excitation CI scheme (CID) to the ground state of the 2E-2O array. Establish that the configuration manifold comprises the reference state, $|\Phi_0^M\rangle = |\Phi_0^{(1)}\rangle|\Phi_0^{(2)}\rangle \dots |\Phi_0^{(M)}\rangle$, where $|\Phi_0^{(i)}\rangle$ denotes the HF ground state of the ith subsystem; and M doubly excited states of the form $|\Phi_i^M\rangle = \hat{C}_i|\Phi_0^M\rangle$, $i = 1, \dots, M$. Here, $\hat{C}_i = c^\dagger_{iu\alpha}c^\dagger_{iu\beta}c_{ig\beta}c_{ig\alpha}$ generates a double excitation in the subsystem i, and ig, iu denote the orbitals of subsystem i.
(b) Evaluate the $(M + 1)$-dimensional hamiltonian matrix, $\boldsymbol{H}$, and solve the corresponding eigenvalue equations for the ground state. Hint: the particular form of $\boldsymbol{H}$ allows for a simple analytical solution. Determine and characterize the remaining M solutions too.
(c) Compare the CID result, E_0^{CI}, with the exact ground-state energy of the model. How does the resulting correlation energy $E_c^{CI} = H_{00} - E_0^{CI}$ scale with M?
(d) Compare the exact ground state, $|\Psi_0^M\rangle$, with the CID result; consider here in particular the coefficient x_0 of the reference configuration (HF ground state), $|\Phi_0^M\rangle$, in the normalized exact ground state, $|\Psi_0^M\rangle$.

12.3 (a) Expand the CID result E_0^{CI} in a perturbation series in powers of V/Δ, (see Exercise 2.4c) and verify that the wrong scaling with M sets in with the quartic terms.

(b) Use RSPT (as in Exercise 4.2b) for the exact ground-state energy E_0 and analyze the M-dependence in the second- and fourth-order contributions.

12.4 Local ionization in the 2E-2O array:

(a) Use an analogous CI approach to determine the energy of the ionic state resulting upon removing one electron from subsystem 1. The corresponding one-hole ($1h$) configuration (spin symmetry $S_z = \frac{1}{2}$) reads $|\Phi_{1g}\rangle = c_{1g\beta}|\Phi_0^M\rangle$. In addition, there are $M-1$ $3h$-$2p$ configurations, $c_{1g\beta}\hat{C}_i|\Phi_0^M\rangle$, $i = 2, \ldots, M$, which differ from the 1h state by a double excitation on site i.

(b) Consider the ionization energy $\Delta E_{1g} = E_{1g}^{-} - E_0$ and compare the M-dependent CI result with the exact ionization energy (see Exercise 3.2). Examine the cases $M = 1$ and $M \gg 1$.

(c) Adapt the ADC(2) scheme (Eq. 10.33) to the 2E-2O array and verify the explicit size-consistency of the ionization energies.

References

1. Mertins F, Schirmer J (1996) Phys Rev A 53:2140
2. Schirmer J, Mertins F (1996a) Int J Quantum Chem 58:329
3. Meunier A, Levy B (1979) Int J Quantum Chem 16:955

Part IV
N-Electron Excitations

Chapter 13
Polarization Propagator

The electron propagator is the simplest many-body Green's functions, and accordingly, the formalism of diagrammatic perturbation theory was specifically presented for that paradigm in Chaps. 4–6. Also the ADC and ISR concepts in the design of practical approximation schemes (Chaps. 10–12) were explicitly formulated for the electron propagator and the underlying physics, being $(N-1)$- and $(N+1)$-electron excitations. In this and the two following Chaps. 14 and 15 we consider the polarization propagator designed for the treatment of N-electron excitations. By and large, the concepts developed for the electron propagator can be transferred to the case of the polarization propagator, but there are also important differences that need to be addressed. In this chapter we present the polarization propagator, discuss its physical significance, and outline the pertaining diagrammatic perturbation expansion.

13.1 Definition and Physical Significance

Reversing the order adopted in Sect. 3.1 introducing the electron propagator, we here begin with the spectral representation of the polarization propagator and from there go back to the original definition.

The polarization propagator is constituted by a matrix $\mathbf{\Pi}(\omega)$ of energy (or time)-dependent functions $\Pi_{rs,r's'}(\omega)$ which can be written in an explicit spectral representation form as follows:

$$\Pi_{rs,r's'}(\omega) = \sum_{m\neq 0} \frac{\langle\Psi_0|c_s^\dagger c_r|\Psi_m\rangle\langle\Psi_m|c_{r'}^\dagger c_{s'}|\Psi_0\rangle}{\omega - E_m + E_0 + i\eta} + \sum_{m\neq 0} \frac{\langle\Psi_0|c_{r'}^\dagger c_{s'}|\Psi_m\rangle\langle\Psi_m|c_s^\dagger c_r|\Psi_0\rangle}{-\omega - E_m + E_0 + i\eta} \tag{13.1}$$

In addition to notations already used in the analogous representation (3.17) of the electron propagator, E_m and $|\Psi_m\rangle$ denote the energies and energy eigenstates,

J. Schirmer, *Many-Body Methods for Atoms, Molecules and Clusters*, Lecture Notes in Chemistry 94, https://doi.org/10.1007/978-3-319-93602-4_13

respectively, of the N-electron system. It is important to note that the summations in Eq. (13.1) explicitly exclude the ground state, $m = 0$. According to Eq. (13.1), the polarization propagator functions consist of two parts,

$$\Pi_{rs,r's'}(\omega) = \Pi^+_{rs,r's'}(\omega) + \Pi^-_{rs,r's'}(\omega) \tag{13.2}$$

being analytic in the upper ($\boldsymbol{\Pi}^+$) and lower ($\boldsymbol{\Pi}^-$) halves of the complex ω-plane. In contrast to the electron propagator, here the two parts are interrelated,

$$\Pi^-_{rs,r's'}(\omega) = \Pi^+_{s'r',sr}(-\omega). \tag{13.3}$$

so that $\boldsymbol{\Pi}^-$ (and thus $\boldsymbol{\Pi}$) can be determined from $\boldsymbol{\Pi}^+$ and vice versa.

The physical information conveyed by $\boldsymbol{\Pi}^+$ (or $\boldsymbol{\Pi}^-$) comprises the (vertical) excitation energies,

$$\Delta E_m = E_m - E_0 \tag{13.4}$$

appearing as pole locations, and the transition amplitudes in the numerators,

$$x_{m,rs} = \langle \Psi_m | c_r^\dagger c_s | \Psi_0 \rangle \tag{13.5}$$

The transition amplitudes enter the computation of transition moments according to

$$T_m = \langle \Psi_m | \hat{D} | \Psi_0 \rangle = \sum_{r,s} x_{m,rs} d_{rs}. \tag{13.6}$$

where

$$\hat{D} = \sum_{r,s} d_{rs} c_r^\dagger c_s \tag{13.7}$$

is a pertinent transition operator.

Beyond the spectral information, the polarization propagator also provides an approach to time- or frequency-dependent *polarizabilities* and other *linear response properties*. A brief outline of linear response theory and the connection to the polarization propagator is given in Appendix A.7.

In analogy to Eq. (3.24), the components of $\boldsymbol{\Pi}^+(\omega)$ or $\boldsymbol{\Pi}^-(\omega)$ can be written as matrix elements of a resolvent operator,

$$\begin{aligned} \Pi^+_{rs,r's'}(\omega) =& \langle \Psi_0 | c_s^\dagger c_r (\omega - \hat{H} + E_0 + i\eta)^{-1} c_{r'}^\dagger c_{s'} | \Psi_0 \rangle \\ & - \langle \Psi_0 | c_s^\dagger c_r | \Psi_0 \rangle \langle \Psi_0 | c_{r'}^\dagger c_{s'} | \Psi_0 \rangle (\omega + i\eta)^{-1} \end{aligned} \tag{13.8}$$

where the second (subtraction) term cancels ground-state contributions ($m = 0$) comprised in the first term.

The time representation of the polarization propagator can be obtained by applying the inverse Fourier transform,

$$\Pi_{rs,r's'}(t,t') = \frac{1}{2\pi}\int_{-\infty}^{\infty} e^{-i\omega(t-t')}\Pi_{rs,r's'}(\omega)d\omega$$

to the spectral representation (13.1). Using Heisenberg operators as defined by Eq. (3.2), the result can then be brought into a compact form, which for $\Pi^{+}_{rs,r's'}(t,t')$ reads

$$\Pi^{+}_{rs,r's'}(t,t') = \theta(t-t')e^{-\eta(t-t')}\times \\ (-i)\left(\langle\Psi_0|c^{\dagger}_s[t]c_r[t]c^{\dagger}_{r'}[t']c_{s'}[t']|\Psi_0\rangle - \langle\Psi_0|c^{\dagger}_s c_r|\Psi_0\rangle\langle\Psi_0|c^{\dagger}_{r'}c_{s'}|\Psi_0\rangle\right)$$

The latter expressions can be combined with those for $\Pi^{-}_{rs,r's'}(t,t')$ to give the time representation of the polarization propagator in a form similar to the definition (3.3) of the one-particle Green's function:

$$\Pi_{rs,r's'}(t,t') = -i\langle\Psi_0|\hat{\mathcal{T}}\left(c^{\dagger}_s[t]c_r[t]c^{\dagger}_{r'}[t']c_{s'}[t']\right)|\Psi_0\rangle + i\langle\Psi_0|c^{\dagger}_s c_r|\Psi_0\rangle\langle\Psi_0|c^{\dagger}_{r'}c_{s'}|\Psi_0\rangle \tag{13.9}$$

Here, the $\hat{\mathcal{T}}$ time-ordering operator (3.5) incorporates the respective step functions $\theta(\tau)$, possibly together with the convergence factors $e^{\pm\eta(t-t')}$ as discussed in Sect. 3.1. The subtraction terms in $\Pi^{+}_{rs,r's'}(t,t')$ and $\Pi^{-}_{rs,r's'}(t,t')$ differ only in the respective step functions and convergence factors. Disregarding the latter, the step functions combine to unity, $\theta(t-t') + \theta(t'-t) = 1$. According to Eqs. (3.26), (3.27), the resulting substraction term on the right-hand side of Eq. (13.9) can be related to the product

$$\langle\Psi_0|c^{\dagger}_s c_r|\Psi_0\rangle\langle\Psi_0|c^{\dagger}_{r'}c_{s'}|\Psi_0\rangle = -G_{rs}(t,t^{+})G_{s'r'}(t',t'^{+}) \tag{13.10}$$

of one-particle Green's functions at equal times.

Relation to the Two-Particle Green's Function

The polarization propagator derives from the two-particle Green's function. The many-body Green's functions form a hierarchy in which the one-particle Green's function (3.3) is the lowest rank. The next higher member is the two-particle Green's function (2p-GF) defined according to (cf. Eq. 3.41)

$$G_{pq,uv}(t_1,t_2;t'_1,t'_2) = (-i)^2\langle\Psi_0|\hat{\mathcal{T}}\left(c_p[t_1]c_q[t_2]c^{\dagger}_v[t'_2]c^{\dagger}_u[t'_1]\right)|\Psi_0\rangle \tag{13.11}$$

Here, the notations are the ones introduced in Sect. 3.1. The 2p-GF allows one to introduce the entity

$$R_{pq,uv}(t_1,t_2;t'_1,t'_2) = G_{pq,uv}(t_1,t_2;t'_1,t'_2) - G_{pu}(t_1,t'_1)G_{qv}(t_2,t'_2) \tag{13.12}$$

referred to as particle–hole (p-h) response function [1]. Depending on four time arguments, the p-h response function obviously is a rather complex construct. A useful simplification is obtained by equating each two time arguments,

$$\Pi_{rs,r's'}(t,t') = i R_{rs',sr'}(t,t';t^+,t'^+) \tag{13.13}$$

and this expression, involving Eqs. (13.11), (13.12), constitutes the original definition of the polarization propagator. As the reader should check, the definition (13.13) does not depend on the time-ordering of the limits $t_1 \to t_1' = t$ and $t_2 \to t_2' = t'$; the particular ordering chosen on the right-hand side of Eq. (13.13) compares directly to the form of Eq. (13.9).

It is important to note that, unlike with the electron propagator, there is *no Dyson equation approach* to the polarization propagator. Of course, one might postulate a Dyson-type equation according to

$$\boldsymbol{\Pi}(\omega) = \boldsymbol{\Pi}^{(0)}(\omega) + \boldsymbol{\Pi}^{(0)}(\omega)\boldsymbol{P}(\omega)\boldsymbol{\Pi}(\omega) \tag{13.14}$$

thereby defining the quantity $\boldsymbol{P}(\omega)$ as an analogue to the self-energy in the Dyson equation part $\boldsymbol{\Sigma}(\omega)$. However, in contrast to $\boldsymbol{\Sigma}(\omega)$, there is no direct and diagram-based approach to determine $\boldsymbol{P}(\omega)$, which means that Eq. (13.14) is all but useless. Only in a rudimentary form, obtained by restricting $\boldsymbol{P}(\omega)$ to the constant first-order approximation, Eq. (13.14) acquires a meaning: Here, it represents the defining equation of the random-phase approximation (RPA) to the polarization propagator, which will be discussed in some detail in Chap. 15.

As an actual counterpart to the Dyson equation for the electron propagator one may see the *Bethe–Salpeter equation* [2] for the p-h response function (13.12), which in a shorthand notation can be written as

$$R_{12,1'2'} = G_{12'}G_{21'} - iG_{1\tau}G_{\sigma 1'}K^{ph}_{\tau\sigma',\sigma\tau'}R_{\tau'2,\sigma'2'} \tag{13.15}$$

Here, repeated indices imply summation over one-particle states and integration over time arguments; the quantity $\boldsymbol{K}^{ph}$ is referred to as the irreducible p-h vertex. Obviously, being an integral equation involving a fourfold time integration, the Bethe–Salpeter equation is by far more complex than the Dyson equation.

13.2 Diagrammatic Perturbation Theory for the Polarization Propagator

The diagrammatic perturbation theory formulated for the electron propagator in Chaps. 4–7 can be transferred with minor adjustments to the case of the polarization propagator. We recall the key steps as follows.

The starting point is the time representation of the polarization propagator

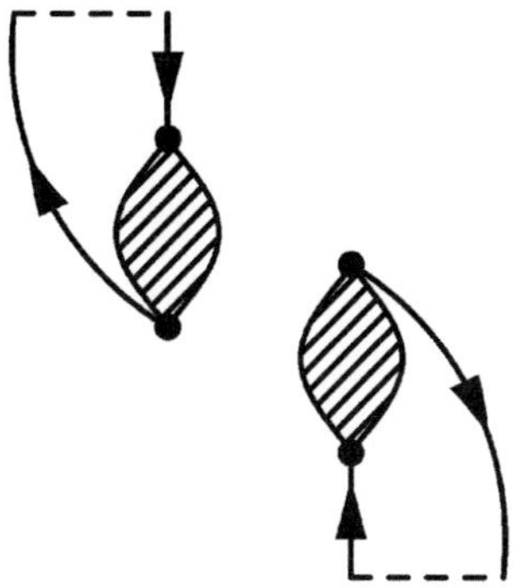

Fig. 13.1 Symbolic representation of disjoint diagrams arising in the subtraction term in Eq. (13.16)

$$\Pi_{rs,r's'}(t,t') = -i\langle\Psi_0|\hat{\mathcal{T}}\left(c_s^\dagger[t]c_r[t]c_{r'}^\dagger[t']c_{s'}[t']\right)|\Psi_0\rangle - iG_{rs}(t,t^+)G_{s'r'}(t',t'^+) \tag{13.16}$$

as given by Eqs. (13.9), (13.10). We will focus on the first term on the right-hand side, to be denoted hereafter as $\Pi'_{rs,r's'}$; the subtraction term, $-iG_{rs}(t,t^+)G_{s'r'}(t',t'^+)$, allows us to discard any diagrammatic contributions to $\Pi'_{rs,r's'}(t,t')$ which are of the 'disjoint' product structure, symbolically depicted in Fig. 13.1.

Repeating the development of Chap. 4 for $\Pi'_{rs,r's'}$ leads to a perturbation expansion of the *Gell-Mann and Low* type,

$$i\Pi'_{rs,r's'}(t,t') = \lim_{\epsilon\to 0}\sum_{n=0}^{\infty}\frac{(-i)^n}{n!}\int_{-\infty}^{\infty}\mathrm{d}t_1 \mathrm{e}^{-|t_1|}\ldots\int_{-\infty}^{\infty}\mathrm{d}t_n \mathrm{e}^{-|t_n|}$$
$$\frac{\langle\Phi_0|\hat{\mathcal{T}}\left[\hat{H}_I(t_1)\ldots\hat{H}_I(t_n)c_s^\dagger(t)c_r(t)c_{r'}^\dagger(t')c_{s'}(t')\right]|\Phi_0\rangle}{\langle\Phi_0|\hat{U}_\epsilon(\infty,-\infty)|\Phi_0\rangle} \tag{13.17}$$

which is the analogue of Eq. (4.58) for the electron propagator.

From here, we may leap forward to the *linked-cluster theorem* as presented in Sect. 5.3. The analysis given there can directly be transferred to the case of the polarization propagator. In analogy to Eq. (5.43), the perturbation expansion (13.17) assumes the much simpler form

$$i\Pi'_{rs,r's'}(t,t') = \sum_{n=0}^{\infty}\frac{(-i)^n}{n!}\int_{-\infty}^{\infty}\mathrm{d}t_1\ldots\int_{-\infty}^{\infty}\mathrm{d}t_n$$
$$\langle\Phi_0|\hat{\mathcal{T}}\left[\hat{H}_I(t_1)\ldots\hat{H}_I(t_n)c_s^\dagger(t)c_r(t)c_{r'}^\dagger(t')c_{s'}(t')\right]|\Phi_0\rangle_C \tag{13.18}$$

where $\langle\Phi_0|\ldots|\Phi_0\rangle_C$ means that only connected (linked) contributions are retained in the respective ground-state expectation values. Let us recall that in the linked contributions, the adiabatic limit $\epsilon\to 0$ can always safely be performed, which means

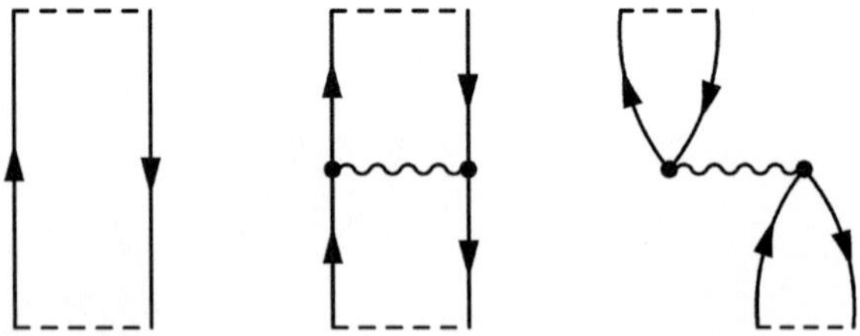

Fig. 13.2 Zeroth- and first-order Feynman diagrams for the polarization propagator

that the switching functions $e^{-\epsilon|t_k|}$ on the right-hand side of Eq. (13.17) are no longer needed.

The distinction of linked and unlinked contributions presupposes Wick's theorem, the second pillar in the formalism of diagrammatic perturbation theory. As discussed in Sects. 6.1 and 6.2, the ground-state expectation values arising, for example, in the PT expansion (13.18) of the polarization propagator can be evaluated in terms of contractions of operator pairs. This paves the way to the introduction of diagrams, as any possible contraction scheme can directly be assigned to a corresponding Feynman diagram.

To see the particular features of the Feynman diagrams for the polarization propagator, let us take a look at the zeroth-order term on the right-hand side of Eq. (13.18):

$$
\begin{aligned}
i\Pi'^{(0)}_{rs,r's'}(t,t') =&\langle\Phi_0|\hat{\mathcal{T}}\left[c_s^\dagger(t)c_r(t)c_{r'}^\dagger(t')c_{s'}(t')\right]|\Phi_0\rangle \\
=&(-1)c_r(t)^{\bullet}c_{r'}^\dagger(t')^{\bullet}\, c_{s'}(t')^{\bullet\bullet}c_s^\dagger(t)^{\bullet\bullet} + c_r(t)^{\bullet}c_s^\dagger(t)^{\bullet}\, c_{s'}(t')^{\bullet\bullet}c_{r'}^\dagger(t')^{\bullet\bullet}
\end{aligned}
$$

The ground-state expectation value gives rise to two contraction schemes, as specified in the second equation. Note the phase (-1) in the first term, which results from the need to reverse the order of the operators in the contraction $c_{s'}(t')^{\bullet}c_s^\dagger(t)^{\bullet}$. In the second term, the contractions relate to operators at equal time arguments t and t', respectively. Obviously, this term is canceled by zeroth-order contribution G^0G^0 to the subtraction term in the definition (13.16) of the polarization propagator. Accordingly, the zeroth-order polarization propagator can be written as

$$
\Pi^{(0)}_{rs,r's'}(t,t') = ic_r(t)^{\bullet}c_{r'}^\dagger(t')^{\bullet}\, c_{s'}(t')^{\bullet\bullet}c_s^\dagger(t)^{\bullet\bullet} = -iG^0_{rr'}(t,t')G^0_{s's}(t',t) \qquad (13.19)
$$

that is (up to a phase) the product of two free electron propagators corresponding to the zeroth-order Feynman diagram in Fig. 13.2.

To further familiarize ourselves with the diagrammatics in the case of the polarization propagator, let us consider also the first-order term in the expansion (13.18):

$$
\begin{aligned}
i\Pi'^{(1)}_{rs,r's'}(t,t') = (-i)\frac{1}{2}\sum_{u,v,w,z} V_{uvwz}\int_{-\infty}^{\infty} \mathrm{d}t_1 \\
\langle\Phi_0|\hat{\mathcal{T}}\left[c_u^\dagger(t_1)c_v^\dagger(t_1)c_z(t_1)c_w(t_1)c_s^\dagger(t)c_r(t)c_{r'}^\dagger(t')c_{s'}(t')\right]|\Phi_0\rangle_C \qquad (13.20)
\end{aligned}
$$

Here one has to deal with altogether 24 different contraction schemes (see Exercise 13.1). Of these schemes, 12 represent disjoint and/or unlinked contributions and can be discarded from the outset. Another 8 are of the form $G^0G^{(1)}$, which, as discussed in Sect. 6.2, can be disregarded if the PT is based on a HF representation and Møller–Plesset partitioning of the hamiltonian. The remaining four contraction schemes can be subsumed within the two first-order Feynman diagrams shown in Fig. 13.2. The corresponding analytical expressions read

$$\Pi^{(1)}_{rs,r's'}(t,t') = \int_{-\infty}^{\infty} dt_1\, (V_{rs'r's} - V_{rs'sr'})\, G^0_r(t,t_1)G^0_s(t_1,t)G^0_{r'}(t_1,t')G^0_{s'}(t',t_1) \tag{13.21}$$

where the free Green's functions are assumed to be diagonal (cf. Eq. 6.9).

The diagram rules (F1)–(F4) for the electron propagator stated in Sect. 6.1 can serve as a template for the case of the polarization propagator. Apart from some self-explanatory adjustments, there are two specific amendments concerning the rules (F1) and (F4). The first amendment, (F1'), reflects the subtraction term in the defining Eq. (13.16); the second, (F4'), is an additional phase rule applying to a continuous fermion line, which can be established analogously to the considerations before and after *Remark 5* in Sect. 6.1. The HF one-particle representation will be supposed in the following. As discussed in Sect. 6.2, this considerably simplifies the diagrammatic PT expansions. The findings of Sect. 6.2 can directly be transferred to the case of the polarization propagator.

Now, the Feynman diagram rules for the polarization propagator can be formulated as follows:

Feynman Diagram Rules for the Polarization Propagator

(F1) To generate the nth-order contribution to the polarization propagator, draw all topologically distinct connected diagrams with n wiggly interaction lines and $2n+2$ directed free fermion or G^0-lines, which start and end, respectively, at the external vertices $(r', s'; t')$ and $(r, s; t)$ with a pair of upwards and downwards directed free fermion lines (labeled (r', s') and (r, s)).

(F1') Skip all disjoint diagrams, i.e., diagrams of the structure shown in Fig. 13.1. Such diagrams contribute to the subtraction term $-iG_{rs}(t,t^+)G_{s'r'}(t',t'^+)$.

(F2) To evaluate a given diagram, assign one-particle indices and time arguments to the interaction lines (inner vertices), thereby defining the one-particle indices and time arguments of the free fermion lines. The arrows specify order of the one-particle indices and time arguments in the G^0-functions. Replace the graphical symbols by the corresponding analytical expressions, V_{ouvw} and $G^0_u(t_i,t_j)$ (supposing diagonal G^0 functions). In the case of a G^0-function with equal time arguments, the limit $G^0(t_i,t_i^+)$ applies according to *Remark 1* in Sect. 5.2.

(F3) Sum over indices and integrate over time arguments of the inner vertices.

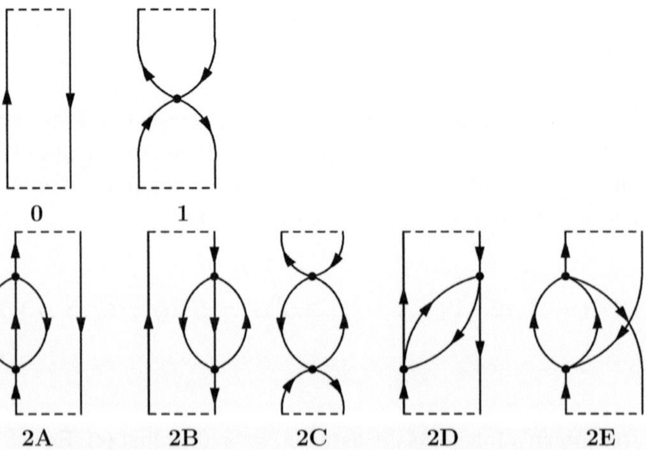

Fig. 13.3 Feynman diagrams (in Abrikosov form) for the polarization propagator through second order

(F4) Multiply by a sign factor $(-1)^L$, where L is the number of closed (fermion) loops; multiply by a factor $i^{n+1} = (-i)(-i)^n(+i)^{2n+2}$, where the three factors relate to the definition of the polarization propagator, the n interaction lines, and the $2n+2$ free fermion lines.

(F4') An additional factor (-1) applies if one continuous fermion line runs from the bottom (lower external vertex) to the top (upper external vertex) of the diagram; note that then another fermion line runs from the top to the bottom. (The opposite case is when one fermion lines run from top to top and another one from bottom to bottom.)

As a first application, the reader may use these rules to confirm the first-order expression (13.21).

Abrikosov Diagrams

In Sect. 6.3, we have discussed how the Abrikosov notation, replacing the wiggly interaction lines of the original Feynman diagrams by interaction dots, leads to a more compact diagrammatic presentation. Obviously, this procedure applies as well to the case of the polarization propagator.

Figure 13.3 shows the Abrikosov diagrams for the polarization propagator through second order of PT. In first order, for example, the two Feynman diagrams shown in Fig. 13.2 have merged into a single Abrikosov diagram. The Abrikosov diagrams are based on the original Feynman diagrams, and the rules for drawing and evaluating Abrikosov diagrams can be derived from those for the Feynman diagrams in a straightforward way. Therefore, we may dispense with a separate presentation of the Abrikosov rules for the polarization propagator. The reader may revisit the derivation in Sect. 6.3 of the Abrikosov rules (A1)–(A5) for the electron propagator.

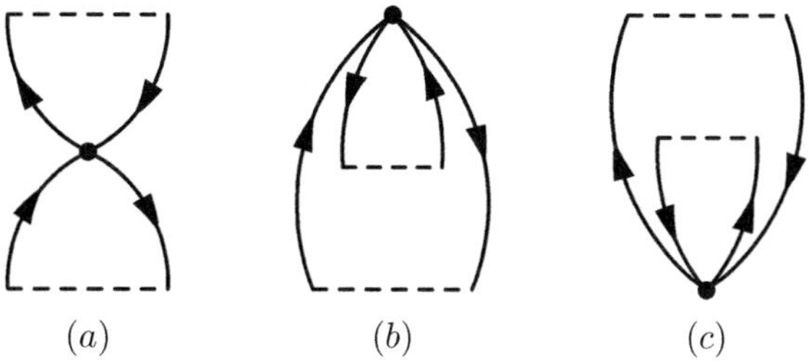

Fig. 13.4 First-order time-ordered (Goldstone) diagrams contributing to $\mathbf{\Pi}^+$

Time-Ordered or Goldstone Diagrams

The time-ordered or Goldstone diagrams considered in Chap. 7 allow one to write down in a direct diagrammatic way the final outcome of a Feynman diagram after performing the n internal time- or ω-integrations required at the nth-order level. The Goldstone diagram rules (G1)–(G4) stated in Sect. 7.2 apply as well to the case of the polarization propagator aside from a minor adjustment of the sign rule (G4) as follows:

(G4) Each hole line introduces the factor (-1). Thus, multiply by a sign factor $(-1)^{L+M}$, where L is the number of closed loops and M is the number of hole lines. Each (inner) vertex contributes a factor $(-i)$, each cut a factor $(+i)$, so that, together with the factor $(-i)$ from the definition of the polarization propagator, the overall phase factor simply becomes unity: $(-i)(-i)^n(+i)^{n+1} = 1$. According to the Feynman rule (F4'), an additional factor (-1) applies if one (or two) continuous fermion line(s) run between the external vertices.

For an illustration, we consider the third time-ordering (c) of the first-order Abrikosov diagram shown in Fig. 13.4. The analytic expression reads

$$D^{(1c)} \equiv \frac{-1}{\omega - \epsilon_r + \epsilon_s} \frac{V_{rs'[r's]}}{\epsilon_{r'} + \epsilon_s - \epsilon_r - \epsilon_{s'}} \bar{n}_r n_s \, n_{r'} \bar{n}_{s'} \tag{13.22}$$

As in the case of the electron propagator (see Sect. 7.2), the time-ordered diagrams for the polarization propagator can be divided into two classes, I and II, corresponding to the time-orderings, $t > t'$ and $t < t'$, respectively, of the time arguments of the external vertices. The diagrams of class I contribute exclusively to $\mathbf{\Pi}^+$, those of class II exclusively to $\mathbf{\Pi}^-$. This establishes distinct diagrammatic perturbation expansions for the two parts, allowing one to establish direct ADC formulations for $\mathbf{\Pi}^+$ or $\mathbf{\Pi}^-$. In view of the redundancy of the two parts, here one may confine oneself to the case of $\mathbf{\Pi}^+$. The corresponding ADC and ISR approaches to the polarization propagator are presented in the ensuing Chap. 14.

Exercises

13.1 Inspect all possible contraction schemes in the first-order term (13.20). Use a diagrammatic representation and identify contraction schemes that are either unlinked or disjoint (as in Fig. 13.1).

13.2 Draw and evaluate the Feynman diagrams comprised in the second-order Abrikosov diagram 2E in Fig. 13.3.
13.3 Evaluate the first-order Goldstone diagrams (*a*) and (*b*) in Fig. 13.4.
13.4 Adapt the matrix representation of diagrams presented in Sect. 6.3 to the Abrikosov diagrams for the polarization propagator. Use the matrix concept to establish the five second-order diagrams in Fig. 13.3.

References

1. Baym G (1962) Phys. Rev. 127:1391
2. Salpeter EE, Bethe HA (1951) Phys. Rev. 84:1232

Chapter 14
ADC and ISR Approaches to the Polarization Propagator

The direct ADC procedure presented in Chap. 10 for the $\boldsymbol{G}^{\pm}(\omega)$ parts of the electron propagator can essentially be transferred to the polarization propagator, where it suffices to deal with $\boldsymbol{\Pi}^{+}(\omega)$ due to the redundancy of the two parts. This will be demonstrated in the following. Moreover, in analogy to Chap. 11, the intermediate-state representation (ISR), here based on the correlated excited N-electron states, will be established and discussed.

14.1 General Framework

The spectral representation for the $\boldsymbol{\Pi}^{+}(\omega)$ part of the polarization propagator, given by the first term on the right-hand side of Eq. (13.1), can be written in matrix notation as

$$\boldsymbol{\Pi}^{+}(\omega) = \boldsymbol{x}^{\dagger}(\omega - \boldsymbol{\Omega})^{-1}\boldsymbol{x} \tag{14.1}$$

where $\boldsymbol{\Omega}$ is the diagonal matrix of excitation energies, $\omega_m = E_m - E_0$, and $\boldsymbol{x}$ denotes the matrix of transition amplitudes $x_{m,rs}$ specified by Eq. (13.5). The ADC formulation, by contrast, sets out from a non-diagonal representation

$$\boldsymbol{\Pi}^{+}(\omega) = \boldsymbol{f}^{\dagger}(\omega - \boldsymbol{M})^{-1}\boldsymbol{f} \tag{14.2}$$

deriving from a set of intermediate states $|\tilde{\Psi}_J\rangle$ to be further specified below. Here, the secular matrix $\boldsymbol{M}$ is defined according to

$$M_{IJ} = \langle\tilde{\Psi}_I|\hat{H} - E_0|\tilde{\Psi}_J\rangle \tag{14.3}$$

as the intermediate-state representation (ISR) of the (shifted) hamiltonian $\hat{H} - E_0$. The matrix $\boldsymbol{f}$ denotes a matrix of "effective" transition amplitudes

J. Schirmer, *Many-Body Methods for Atoms, Molecules and Clusters*, Lecture Notes in Chemistry 94, https://doi.org/10.1007/978-3-319-93602-4_14

$$f_{I,rs} = \langle \tilde{\Psi}_I | c_r^\dagger c_s | \Psi_0 \rangle \tag{14.4}$$

For a given, exact or approximate, secular matrix $\boldsymbol{M}$, the excitation energies ω_m can be computed by solving the (hermitian) eigenvalue problem

$$\boldsymbol{M}\boldsymbol{X} = \boldsymbol{X}\,\boldsymbol{\Omega}, \quad \boldsymbol{X}^\dagger \boldsymbol{X} = \mathbf{1} \tag{14.5}$$

Here, $\boldsymbol{\Omega}$ and $\boldsymbol{X}$ denote the diagonal matrix of eigenvalues and the eigenvector matrix, respectively. Using the expansion

$$|\Psi_m\rangle = \sum_J X_{Jm} |\tilde{\Psi}_J\rangle \tag{14.6}$$

the transition moments (Eq. 13.6) can be written as

$$T_m = \sum_{r,s} x_{m,rs} d_{rs}. \tag{14.7}$$

with the transition amplitudes given by

$$x_{m,rs} = \sum_J X_{Jm}^* f_{J,rs} \tag{14.8}$$

Now let us consider the construction of the intermediate states. As discussed in Sect. 11.1, one starts from the correlated excited (CE) states

$$|\Psi_J^0\rangle = \hat{C}_J |\Psi_0\rangle \tag{14.9}$$

Here, the operators $\hat{C}_J$ relate to neutral $1p$-$1h$, $2p$-$2h$, ... excitations:

$$\{\hat{C}_J\} = \left\{ c_a^\dagger c_k;\ c_a^\dagger c_b^\dagger c_k c_l,\ a < b, k < l;\ \ldots \right\} \tag{14.10}$$

Again, we may number the successive excitation classes, $\mu = 1, 2, 3, \ldots$, and use the notation $[J]$ to specify the excitation class to which the configuration J belongs; that is, $[J] = \mu$ if J belongs to the class of μh-μp excitations.

Unlike the ECO construction of $(N-1)$ states considered in Sect. 11.1, now the exact ground state $|\Psi_0\rangle$ has to be taken into account in order to ensure that the intermediate states are orthogonal to $|\Psi_0\rangle$,

$$\langle \Psi_0 | \tilde{\Psi}_J \rangle = 0 \tag{14.11}$$

To this end the rule (1), addressing the precursor states (cf. Eq. 11.11) has to be extended as follows:

(1) Assume that the intermediate states $|\tilde{\Psi}_K\rangle$ of the classes $1, \dots, \nu - 1$ have been constructed. Then, orthogonalize the CE states $|\Psi_J^0\rangle$ of class ν with respect to the intermediate states of class $1, \dots, \nu - 1$ and to $|\Psi_0\rangle$ (which may be seen as constituting a zeroth excitation class, $\nu = 0$) according to

$$|\Psi_J^\#\rangle = |\Psi_J^0\rangle - |\Psi_0\rangle\langle\Psi_0|\Psi_J^0\rangle - \sum_{[K]<\nu} |\tilde{\Psi}_K\rangle\langle\tilde{\Psi}_K|\Psi_J^0\rangle, \ [J] = \nu \tag{14.12}$$

Note that in the ground-state Gram-Schmidt term $|\Psi_0\rangle$ is supposed to be normalized to 1. The second rule can be literally transcribed from Sect. 11.1:

(2) The "precursor" states $|\Psi_J^\#\rangle$ of class ν may then be orthonormalized symmetrically, yielding

$$|\tilde{\Psi}_J\rangle = \sum_{[I]=\nu} |\Psi_I^\#\rangle(\boldsymbol{S}_\nu^{-1/2})_{IJ} \tag{14.13}$$

where $\boldsymbol{S}_\nu$ is the overlap matrix of the precursor states of class ν,

$$(\boldsymbol{S}_\nu)_{IJ} = \langle\Psi_I^\#|\Psi_J^\#\rangle, \ [I] = [J] = \nu \tag{14.14}$$

As discussed in Sect. 11.1, the ECO construction establishes PT expansions of the secular matrix elements,

$$\boldsymbol{M} = \boldsymbol{M}^{(0)} + \boldsymbol{M}^{(1)} + \boldsymbol{M}^{(2)} + \dots \tag{14.15}$$

and the effective transition amplitudes,

$$\boldsymbol{f} = \boldsymbol{f}^{(0)} + \boldsymbol{f}^{(1)} + \boldsymbol{f}^{(2)} + \dots \tag{14.16}$$

The explicit PT expressions in these expansions can be obtained by using the PT expansion of the ground state, $|\Psi_0\rangle$, and ground-state energy, E_0, in the CE states, $|\Psi_I^0\rangle$. This has been described in Sect. 11.2 for the $(N-1)$-electron states. Alternatively, one may resort to the ADC approach in which the PT expansion of the right-hand side of Eq. (14.2) is compared with the diagrammatic PT expansion for $\boldsymbol{\Pi}^+(\omega)$ through successively higher order.

A hierarchy of higher-order approximation (ADC(n)) schemes for N-electron excitations can then be devised by letting the explicit IS or ADC configuration space comprise ever higher excitation classes and truncating the PT expansions of the matrix elements in a coordinated and consistent way. In the next section, the construction of the first- and second-order ADC schemes for the polarization propagator will be demonstrated.

14.2 Explicit ADC Schemes for the Polarization Propagator

In the ADC formulation, the secular matrix is conveniently written as

$$\begin{aligned}\boldsymbol{M} &= \boldsymbol{K} + \boldsymbol{C} \\ &= \boldsymbol{K} + \boldsymbol{C}^{(1)} + \boldsymbol{C}^{(2)} + \dots\end{aligned} \tag{14.17}$$

where

$$\boldsymbol{K} = \boldsymbol{M}^{(0)} \tag{14.18}$$

denotes the zeroth-order part of $\boldsymbol{M}$. According to

$$M_{IJ}^{(0)} = \langle \Phi_I | \hat{H}_0 - E_0^{(0)} | \Phi_J \rangle = K_{IJ} \tag{14.19}$$

$\boldsymbol{K}$ is the diagonal matrix of HF excitation energies,

$$\begin{aligned}K_{ak,ak} &= \epsilon_a - \epsilon_k \\ K_{abkl,abkl} &= \epsilon_a + \epsilon_b - \epsilon_k - \epsilon_l \\ &\vdots\end{aligned} \tag{14.20}$$

In **zeroth order**, the ADC term

$$\boldsymbol{\Pi}^{(0)+}(\omega) = \boldsymbol{f}^{(0)\dagger}(\omega - \boldsymbol{K})^{-1}\boldsymbol{f}^{(0)} \tag{14.21}$$

is to be compared with the diagrammatic expression (Eq. 13.19),

$$\Pi_{ak,a'k'}^{(0)+}(\omega) = \delta_{aa'}\delta_{kk'}(\omega - \epsilon_a + \epsilon_k)^{-1} \tag{14.22}$$

from which one obtains

$$f_{ak,rs}^{(0)} = \delta_{ar}\delta_{ks} \tag{14.23}$$

Moreover, the comparison in zeroth order implies $\boldsymbol{f}_\mu^{(0)} = \boldsymbol{0}$ for $\mu > 1$; that is, the zeroth-order matrix elements $f_{I,rs}^{(0)}$ vanish unless I is a $1p$-$1h$ configuration.

First-Order ADC Scheme

Unlike in the $(N-1)$-electron case discussed in Chap. 10, there is a non-trivial first-order ADC(1) scheme here. The first-order ADC form reads

$$\begin{aligned}\boldsymbol{\Pi}^{(1)+}(\omega) = \boldsymbol{f}_1^{(1)\dagger}(\omega - \boldsymbol{K}_1)^{-1}\boldsymbol{f}_1^{(0)} + \boldsymbol{f}_1^{(0)\dagger}(\omega - \boldsymbol{K}_1)^{-1}\boldsymbol{f}_1^{(1)} \\ + \boldsymbol{f}_1^{(0)\dagger}(\omega - \boldsymbol{K}_1)^{-1}\boldsymbol{C}_{11}^{(1)}(\omega - \boldsymbol{K}_1)^{-1}\boldsymbol{f}_1^{(0)}\end{aligned} \tag{14.24}$$

where the respective $1p$-$1h$ matrix blocks are indicated by the subscript 1. The three terms on the right-hand side are matched by the first-order Goldstone diagrams (b), (c), and (a) shown in Fig. 13.4, which allows one to simply read off the quantities $\boldsymbol{C}_{11}^{(1)}$ and $\boldsymbol{f}_{1}^{(1)}$ from the diagrammatic expressions. In particular, the expressions for diagram (a) and diagram (b) (see Eq. 13.22) yield

$$C_{ak,a'k'}^{(1)} = -V_{ak'[a'k]} \tag{14.25}$$

and

$$f_{ak,rs}^{(1)} = \frac{V_{as[rk]}n_r\bar{n}_s}{\epsilon_a + \epsilon_s - \epsilon_k - \epsilon_r} \tag{14.26}$$

Note that $f_{ak,rs}^{(1)}$ vanishes unless the index pair rs is of h-p type.

Of course, these first-order results could have been obtained equally well using the ISR formulation. Through first order, the $1p$-$1h$ intermediate states simply read

$$|\tilde{\Psi}_{ak}\rangle = |\Phi_{ak}\rangle + c_a^\dagger c_k|\Psi_0^{(1)}\rangle + O(2) \tag{14.27}$$

Consequently, the secular matrix

$$M_{ak,a'k'} = \langle\tilde{\Psi}_{ak}|\hat{H} - E_0|\tilde{\Psi}_{a'k'}\rangle = K_{ak,a'k'} + \langle\Phi_{ak}|\hat{H}_I - E_0^{(1)}|\Phi_{a'k'}\rangle + O(2) \tag{14.28}$$

can easily be expanded through first order, reproducing the ADC result of Eq. (14.25). In a similar way, the expansion of the IS transition amplitudes

$$\begin{aligned} f_{ak,rs} &= \langle\tilde{\Psi}_{ak}|c_r^\dagger c_s|\Psi_0\rangle \\ &= \langle\Phi_{ak}|c_r^\dagger c_s|\Phi_0\rangle + \langle\Phi_{ak}|c_r^\dagger c_s|\Psi_0^{(1)}\rangle + O(2) \end{aligned} \tag{14.29}$$

recovers Eqs. (14.23) and (14.26).

The ADC(1) scheme, constituted by the secular matrix

$$M_{ak,a'k'} = (\epsilon_a - \epsilon_k)\delta_{aa'}\delta_{kk'} - V_{ak'[a'k]} \tag{14.30}$$

and the effective transition amplitudes given by Eqs. (14.23) and (14.26), can be compared to the CIS (single excitation configuration interaction) method (see Sect. 12.3), also referred to as *Tamm–Dancoff approximation* (TDA). Eq. (14.28) shows that the CIS secular matrix is identical with the ADC(1) matrix (up to the shift by $-E_0(1)$ in the diagonal). The quality of the ADC(1) or CIS computational schemes is rather modest, as the excitation energies of the $1p$-$1h$ states are treated consistently through first order only. With regard to the transition moments, the CIS description based on the expression

$$T_m = \sum_{a'k'} X_{a'k',m}^* d_{a'k'} \tag{14.31}$$

affords only zeroth-order consistency. The full first-order contribution (to a final state deriving from the HF excitation $|\Phi_{ak}\rangle$) given by

$$T_{ak}^{(1)} = \sum_{a'k'} X_{a'k',ak}^{(1)*} d_{a'k'} + \sum_{a'k'} f_{ak,k'a'}^{(1)} d_{k'a'} \tag{14.32}$$

comprises a second term deriving from the first-order contribution to the ground state, $|\Psi_0^{(1)}\rangle$. By construction, the ADC(1) result for the transition moment is correct through first order. A more detailed PT analysis of the physics encountered in single electron excitations is given in Sect. 15.1.

Second- and Third-Order ADC Approximations

At second order, the ADC procedure for the polarization propagator is more involved than in the case of the electron propagator treated in Sect. 10.2. Now we have to deal with five second-order diagrams (rather than one for $\boldsymbol{G}^{(2)}$), and, moreover, there are more terms in the ADC form as a consequence of non-vanishing first-order contributions to $\boldsymbol{C}_{11}$ and $\boldsymbol{f}_1$. Nevertheless, the procedure is rather straightforward, and we may confine ourselves to a brief outline.

In the second-order ADC form, there are eight distinct (non-vanishing) contributions, (A)–(H), listed in Table 14.1. Note that in a formal sense there are more terms, for example, $\boldsymbol{f}_3^{(1)\dagger}\boldsymbol{\omega}_3^{-1}\boldsymbol{f}_3^{(1)}$, that is, terms of the type (F) for $3p$-$3h$ and higher configurations. However, as there are no diagrammatic counterparts, those terms simply vanish, which means $\boldsymbol{f}_\mu^{(1)} = 0$ for $\mu \geq 3$.

The terms of Table 14.1 have to be compared to the second-order diagrams of Fig. 13.3. As indicated by the terms (F), (G), and (H), now the $2p$-$2h$ excitations ($\mu = 2$) come explicitly into play, requiring to determine $\boldsymbol{C}_{12}^{(1)}$ and $\boldsymbol{f}_2^{(1)}$. The other ADC quantities to be derived from the second-order diagrams are $\boldsymbol{C}_{11}^{(2)}$ and $\boldsymbol{f}_1^{(2)}$. Each of the five second-order diagrams $2A, \ldots, 2E$ gives rise to 12 Goldstone diagrams contributing to $\boldsymbol{\Pi}^+(\omega)$, so that altogether a manifold of 60 diagrams has to be inspected. As an example, we consider diagram 2C, which, by the way, belongs to the series of RPA diagrams (see Sect. 15.1). Its 12 Goldstone diagrams are shown in Fig. 14.1. Most of these diagrams can directly be identified with corresponding

Table 14.1 Non-trivial contributions to the ADC expansion for $\boldsymbol{\Pi}^+(\omega)$ at second-order

$\boldsymbol{f}_1^{(2)\dagger}\boldsymbol{\omega}_1^{-1}\boldsymbol{f}_1^{(0)}$ + h.c.	(A)
$\boldsymbol{f}_1^{(1)\dagger}\boldsymbol{\omega}_1^{-1}\boldsymbol{f}_1^{(1)}$	(B)
$\boldsymbol{f}_1^{(1)\dagger}\boldsymbol{\omega}_1{}^{-1}\boldsymbol{C}_{11}^{(1)}\boldsymbol{\omega}_1^{-1}\boldsymbol{f}_1^{(0)}$ + h.c.	(C)
$\boldsymbol{f}_1^{(0)\dagger}\boldsymbol{\omega}_1{}^{-1}\boldsymbol{C}_{11}^{(1)}\boldsymbol{\omega}_1^{-1}\boldsymbol{C}_{11}^{(1)}\boldsymbol{\omega}_1{}^{-1}\boldsymbol{f}_1^{(0)}$	(D)
$\boldsymbol{f}_1^{(0)\dagger}\boldsymbol{\omega}_1{}^{-1}\boldsymbol{C}_{11}^{(2)}\boldsymbol{\omega}_1^{-1}\boldsymbol{f}_1^{(0)}$	(E)
$\boldsymbol{f}_2^{(1)\dagger}\boldsymbol{\omega}_2{}^{-1}\boldsymbol{f}_2^{(1)}$	(F)
$\boldsymbol{f}_2^{(1)\dagger}\boldsymbol{\omega}_2{}^{-1}\boldsymbol{C}_{21}^{(1)}\boldsymbol{\omega}_1^{-1}\boldsymbol{f}_1^{(0)}$ + h.c.	(G)
$\boldsymbol{f}_1^{(0)\dagger}\boldsymbol{\omega}_1{}^{-1}\boldsymbol{C}_{12}^{(1)}\boldsymbol{\omega}_2^{-1}\boldsymbol{C}_{21}^{(1)}\boldsymbol{\omega}_1{}^{-1}\boldsymbol{f}_1^{(0)}$	(H)

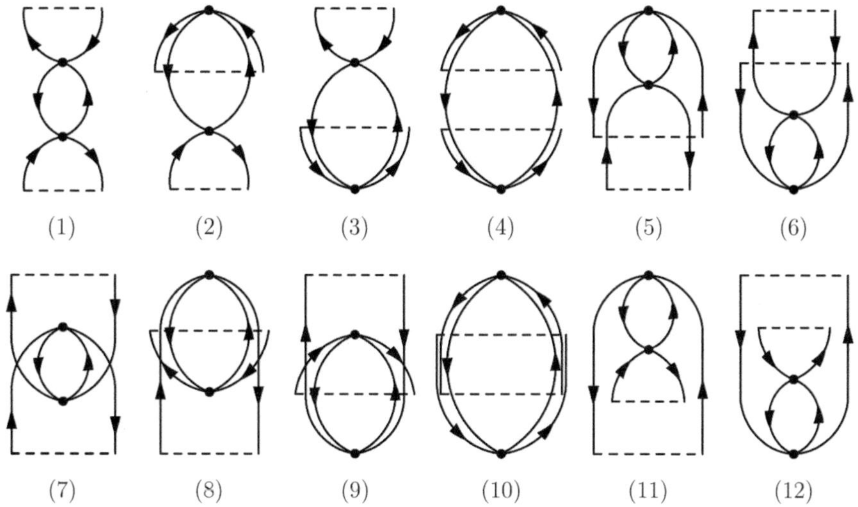

Fig. 14.1 Time-ordered (or Goldstone) diagrams of second-order diagram 2C contributing to $\Pi^+(\omega)$

ADC terms in Table 14.1. Obviously, diagram (1) corresponds to term (D), and is redundant for the ADC analysis as its constituents have already been determined at the ADC(1) level. Such repetitive diagrams can simply be skipped. This is the case for diagram (2) and its hermitian conjugate (3) matching term (C), as well as for diagram (4) to be associated with term (B). The diagrams (5) and (11) and their hermitian conjugated counterparts (6) and (12) are of the form (A), allowing for the specification of contributions to $\boldsymbol{f}_1^{(2)}$.

The remaining four diagrams (7)–(10) do not fit individually to the ADC terms and, moreover, introduce denominators of $3p$-$3h$ type. However, as explicitly demonstrated in Sect. 10.2 after Eq. (10.29), some simple algebra can be used to transform the sum of these four structurally similar diagrams into the forms (E) and (A) and derive contributions to $\boldsymbol{C}_{11}^{(2)}$ and $\boldsymbol{f}_1^{(2)}$, respectively:

$$(7) + (8) + (9) + (10)|_{ak,a'k'} \rightarrow \begin{cases} C_{ak,a'k'}^{(2)} \\ f_{ak,a'k'}^{(2)} \end{cases} \tag{14.33}$$

In a similar way, one may treat the time-orderings (7)–(10) of the diagrams $2A$ and $2B$.

The secular matrix of the ADC(2) scheme is given as follows:

$1p$-$1h$ block:

$$K_{ak,a'k'} = (\epsilon_a - \epsilon_k)\delta_{aa'}\delta_{kk'} \tag{14.34}$$

$$C_{ak,a'k'}^{(1)} = -V_{ak'[a'k]} \tag{14.35}$$

$$C_{ak,a'k'}^{(2)} = C_{ak,a'k'}^{(A)} + C_{ak,a'k'}^{(B)} + C_{ak,a'k'}^{(C)} \tag{14.36}$$

where

$$C^{(A)}_{ak,a'k'} = \tfrac{1}{2}\delta_{kk'} \sum_{c,i,j} v_{acij} v_{ija'c} \left(\epsilon_i + \epsilon_j - \epsilon_c - \tfrac{1}{2}(\epsilon_a + \epsilon_{a'})\right) \tag{14.37}$$

$$C^{(B)}_{ak,a'k'} = \tfrac{1}{2}\delta_{aa'} \sum_{c,d,i} v_{cdki} v_{k'icd} \left(\tfrac{1}{2}(\epsilon_k + \epsilon_{k'}) + \epsilon_i - \epsilon_c - \epsilon_d\right) \tag{14.38}$$

$$C^{(C)}_{ak,a'k'} = \sum_{c,i} v_{k'ia'c} v_{acik} \left(\tfrac{1}{2}(\epsilon_k + \epsilon_{k'} - \epsilon_a - \epsilon_{a'}) + \epsilon_i - \epsilon_c\right) \tag{14.39}$$

using again the abbreviations v_{rsuv} as defined in Eq. (10.31).

$1p$-$1h/2p$-$2h$ coupling block:

$$C^{(1)}_{ak,a'b'k'l'} = \delta_{aa'} V_{k'l'[kb']} - \delta_{ab'} V_{k'l'[ka']} - \delta_{kk'} V_{al'[a'b']} + \delta_{kl'} V_{ak'[a'b']} \tag{14.40}$$

$2p$-$2h$ block:

$$K_{abkl,a'b'k'l'} = (\epsilon_a + \epsilon_b - \epsilon_k - \epsilon_l)\delta_{aa'}\delta_{bb'}\delta_{kk'}\delta_{ll'} \tag{14.41}$$

The relative signs have been chosen such that the first-order contributions $C^{(1)}_{IJ}$ are consistent with $\langle\Phi_I|\hat{H}_I - E_0(1)|\Phi_J\rangle$. The ADC(2) expressions for the transition amplitudes $\boldsymbol{f}_1^{(0-2)}$ and $\boldsymbol{f}_2^{(1)}$ are listed in Appendix A.9.

Figure 14.2 illustrates the ADC(2) secular matrix and the matrix of effective transition amplitudes. Referring to the remarks in Sect. 10.3, the computational cost at the ADC(2) level scales as m^5. The energies of single ($1p$-$1h$) excitations are treated consistently through second order, which translates into typical errors in the range of ±0.5 eV. Double excitations are described only poorly, through zeroth order, that is. A consistent first-order treatment of the double excitations is achieved in the ADC(2)-x extension. Here, the zeroth-order $2p$-$2h$ diagonal secular matrix block, $\boldsymbol{K}_{22}$, is extended by the first-order contribution, $\boldsymbol{C}^{(1)}_{22}$, which can either be anticipated from the ADC(3) scheme or directly identified as the CI term

$$C^{(1)}_{abkl,a'b'k'l'} = \langle\Phi_{abkl}|\hat{H}_I - E_0^{(1)}|\Phi_{a'b'k'l'}\rangle$$

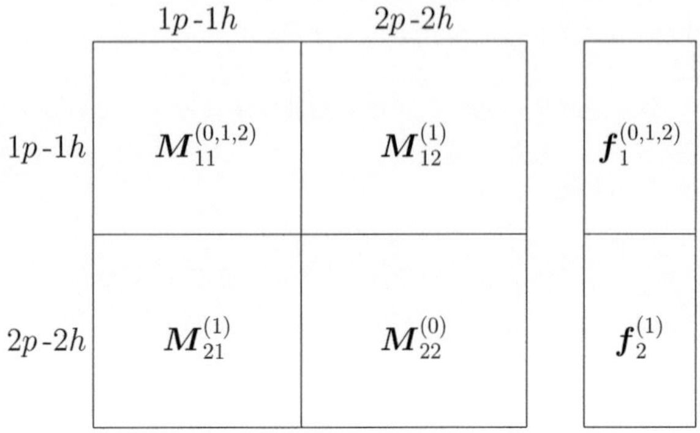

Fig. 14.2 Structure of the ADC(2) matrices $\boldsymbol{M}$ and $\boldsymbol{f}$

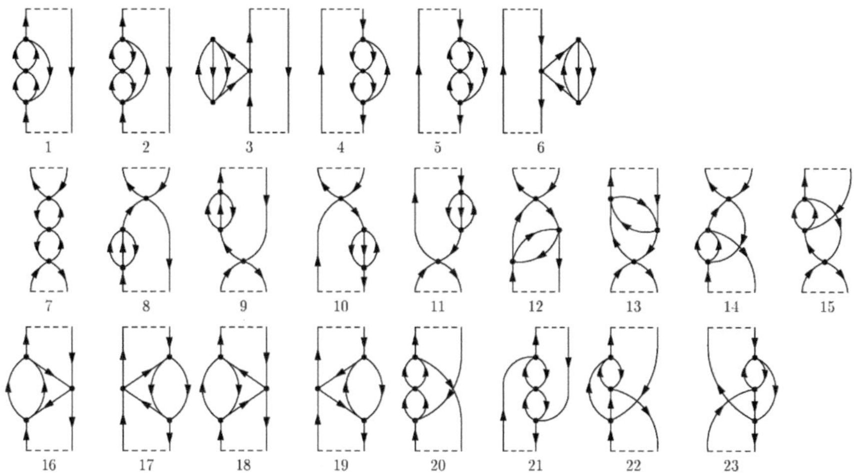

Fig. 14.3 Third-order Feynman diagrams (in Abrikosov form) for the polarization propagator

using the ISR construction. While distinctly improving the description of the double excitations and, thereby of single excitations with large admixtures of double excitations, the ADC(2)-x scheme does not afford a systematic improvement of single excitations in general. To this end, one has to resort to a consistent advancement such as offered by the next higher ADC(3) scheme.

While the derivation of the ADC(2) expressions could be done by anyone familiar with Feynman diagrams on a "rainy sunday afternoon", the corresponding task at the third-order level is nothing less than a small research project [1]. We may confine ourselves here to a few remarks and refer the reader to the original article. As depicted in Fig. 14.3, there are 23 third-order Feynman/Abrikosov diagrams, each contributing 60 Goldstone diagrams to $\mathbf{\Pi}^+(\omega)$, so that altogether one encounters a diagram manifold of 1380 Goldstone diagrams. While only a fraction of the diagrams really has to be considered in the ADC procedure, the number of key diagrams needed to determine the additional ADC(3) terms is still considerable.

The explicit ADC(3) configuration space is as in the ADC(2) case spanned by the $1p$-$1h$ excitations (class $\mu = 1$) and the $2p$-$2h$ excitations (class $\mu = 2$), and the PT expansions for the different blocks are as follows:

$$\begin{aligned} \boldsymbol{C}_{11} &= \boldsymbol{C}_{11}^{(1)} + \boldsymbol{C}_{11}^{(2)} + \boldsymbol{C}_{11}^{(3)} \\ \boldsymbol{C}_{12} &= \boldsymbol{C}_{12}^{(1)} + \boldsymbol{C}_{12}^{(2)} \\ \boldsymbol{C}_{22} &= \boldsymbol{C}_{22}^{(1)} \end{aligned} \tag{14.42}$$

The additional contributions to be determined at the third-order level are $\boldsymbol{C}_{11}^{(3)}$, $\boldsymbol{C}_{12}^{(2)}$, and $\boldsymbol{C}_{22}^{(1)}$. The corresponding expansions of the transition amplitudes read

$$f_1 = f_1^{(0)} + f_1^{(1)} + f_1^{(2)} + f_1^{(3)}$$
$$f_2 = f_2^{(1)} + f_2^{(2)}, \tag{14.43}$$

and the additional ADC(3) contributions are $f_1^{(3)}$ and $f_2^{(2)}$. One may note that the explicit third-order expressions are already somewhat lengthy. The $\boldsymbol{C}_{11}^{(3)}$ matrix elements, for example, consist of 29 individual terms.

The scaling of the ADC(3) computational cost is m^6, that is, only one order higher than in the ADC(2) scheme, where of course the applicable prefactors will differ substantially. The ADC(3) results are consistently more accurate than those of the ADC(2) level, the typical error being in the order of ±0.2 eV. For exemplary applications of the ADC(3) scheme, the reader is referred to Ref. [2] and a benchmark study by Harbach, Wormit, and Dreuw [3].

14.3 Properties of the ISR-ADC Schemes

In Chap. 12, we have addressed the particular features of the ISR-ADC schemes, that is, the *canonical order relations* for the secular matrix elements and the *separability* of the secular matrix with regard to local and non-local excitations as encountered in the separate fragments model. While the discussion in Chap. 12 was focussed on the case of $N-1$ electrons, the essential findings are entirely general and can directly be transferred to the N-electron excitations considered here.

Canonical Order Relations and Separability

The N-electron ISR-ADC matrix elements (14.3) fulfill the canonical order relations,

$$M_{IJ} \sim O(|[I] - [J]|) \tag{14.44}$$

being completely analogous to Eq. (12.1) for the $(N-1)$-electron case. Figure 14.4, depicting the order structure of the N-electron ISR-ADC matrix, is essentially a replicate of Fig. 12.1, differing only in the notations (though not the numbering) of the excitation classes.

As discussed in Sect. 12.1, the canonical order relations allow one to specify truncation errors, i.e., the errors arising from truncating the explicit ISR-ADC configuration spaces (see Eqs. 12.11, 12.12). For example, the error in the single ($1p$-$1h$) excitation energies for a truncation after class μ is of the order

$$O_{TE}(\mu) = 2\mu \tag{14.45}$$

In the ADC(2) and ADC(3) schemes, the configuration space comprises the excitation classes $\mu = 1, 2$ so that the truncation error here is of fourth order.

Fig. 14.4 Order structure of the ISR-ADC secular matrix ***M***

	$1p$-$1h$	$2p$-$2h$	$3p$-$3h$	$4p$-$4h$	$5p$-$5h$	...
$1p$-$1h$	0	1	2	3	4	...
$2p$-$2h$	1	0	1	2	3	...
$3p$-$3h$	2	1	0	1	2	...
$4p$-$4h$	3	2	1	0	1	...
$5p$-$5h$	4	3	2	1	0	...
⋮	⋮	⋮	⋮	⋮	⋮	

For the ISR-ADC transition amplitudes, the order relations are as given by Eq. (12.14). Together with Eq. (12.15) for the order relations in the eigenvector matrix $\boldsymbol{X}$, one recovers the TEO formula (12.16) for the truncation errors in the transition amplitudes (14.8) and the transition moments (13.6).

The discussion in Sect. 12.2 of the separability property of the ISR-ADC secular matrix applies *mutatis mutandis* to the secular matrix (14.3) for the N-electron excitations and needs not be repeated here. In particular, Eqs. (12.35), (12.36), and (12.38) can be transferred directly to the N-electron case. The corresponding block structure of $\boldsymbol{M}$ is as shown in Fig. 12.3.

ISR of General Operators

Let us recall that the intermediate states $|\tilde{\Psi}_I\rangle$ established by the ECO procedure according to Eqs. (14.12)–(14.14) represent (together with the ground state $|\Psi_0\rangle$) a well-defined basis for N-electron states . Accordingly, they can be used to represent, besides the hamiltonian, any other operator of physical interest. Let $\hat{D}$ denote an operator, then the matrix $\tilde{\boldsymbol{D}}$ of elements

$$\tilde{D}_{IJ} = \langle \tilde{\Psi}_I | \hat{D} | \tilde{\Psi}_J \rangle \tag{14.46}$$

is the IS representation of $\hat{D}$. In Sect. 11.3, the ISR of arbitrary operators in the case of $(N-1)$-electron states has been discussed. The essential features are quite general and apply to the N-electron case as well. The explicit construction of the consistent second-order (ISR(2)) approximation scheme for the ISR of a one-particle operator, while basically analogous to the discussion in Sect. 11.3, is more demanding in the N-electron case, and we refer the reader to Ref. [4] for details and the resulting expressions. Here, it may suffice to briefly inspect the structure of $\tilde{\boldsymbol{D}}$ in the ISR(2) approximation:

$$\begin{aligned}\tilde{\boldsymbol{D}}_{11} &= \tilde{\boldsymbol{D}}_{11}^{(0)} + \tilde{\boldsymbol{D}}_{11}^{(1)} + \tilde{\boldsymbol{D}}_{11}^{(2)} \\ \tilde{\boldsymbol{D}}_{12} &= \tilde{\boldsymbol{D}}_{12}^{(0)} + \tilde{\boldsymbol{D}}_{12}^{(1)} \\ \tilde{\boldsymbol{D}}_{22} &= \tilde{\boldsymbol{D}}_{22}^{(0)}\end{aligned} \tag{14.47}$$

As in the ADC(2) and ADC(3) secular equations, the configuration space comprises the $1p$-$1h$ and $2p$-$2h$ excitations (classes 1 and 2), and the PT expansions in the various sub-blocks essentially match those of the ADC(2) secular matrix.

The ISR of an operator $\hat{D}$ can be used to compute the associated property of an excited state $|\Psi_m\rangle$ as the expectation value

$$D_m = \langle \Psi_m | \hat{D} | \Psi_m \rangle = \boldsymbol{X}_m^\dagger \tilde{\boldsymbol{D}} \boldsymbol{X}_m \tag{14.48}$$

Here, $\boldsymbol{X}_m$ is the ADC eigenvector (see Eq. 14.5) pertaining to $|\Psi_m\rangle$. In a similar way, transition moments involving two states are obtained according to

$$D_{mn} = \langle \Psi_m | \hat{D} | \Psi_n \rangle = \boldsymbol{X}_m^\dagger \tilde{\boldsymbol{D}} \boldsymbol{X}_n \tag{14.49}$$

In such applications, usually one will ensure that the approximation schemes for the two ingredients are consistent, e.g., in combining the ISR(2) operator matrix with ADC(2) eigenvectors. Of course, one may as well combine the more accurate ADC(3) eigenvectors with the second-order ISR operator expressions..

Another benefit afforded by the ISR of general operators, already addressed in Sect. 11.3, is the option to augment the original Hamiltonian $\hat{H}$ with an additional operator $\hat{U}$, for example, associated with an external field:

$$\hat{H} \rightarrow \hat{H}^x = \hat{H} + \hat{U} \tag{14.50}$$

In the N-electron case considered here, the intermediate states can couple to the ground state via $\hat{U}$. Accordingly, the IS configurations must be enlarged by $|\Psi_0\rangle$, and the representation of $\hat{H}^x - E_0$ has an additional row (and column), the elements being

$$M_{00}^x = U_0, \ M_{I0}^x = F_I(U), \ I \neq 0 \tag{14.51}$$

Here, $U_0 = \langle \Psi_0 | \hat{U} | \Psi_0 \rangle$ is the ground-state expectation value of $\hat{U}$, and $F_I(U)$ denote the IS transition moments for the operator $\hat{U}$,

$$F_I(U) = \langle \tilde{\Psi}_I | \hat{U} | \Psi_0 \rangle = \sum f_{I,rs} u_{rs} \tag{14.52}$$

supposing here a one-particle operator of the form $\hat{U} = \sum u_{rs} c_r^\dagger c_s$. The matrix elements in the main block are given by

$$M_{IJ}^x = M_{IJ} + \tilde{U}_{IJ}, \ I, J \neq 0 \tag{14.53}$$

where $\tilde{U}_{IJ}$ are the ISR matrix elements of $\hat{U}$.

The order structure of $\tilde{\boldsymbol{D}}$ (in case of a one-particle operator) is as shown in Fig. 12.2 (with the respective N-electron excitation classes replacing the original ones). Figure 12.4 shows the (non-separable) block structure of $\tilde{\boldsymbol{D}}$ generated in the partitioning scheme of the separate fragment model. The conclusions with regard to the size-consistency of excited-state properties (14.48) and transition moments (14.49) can be inferred from the corresponding discussion in Sect. 12.2.

The Dipole Sum Rule and the Equivalence of Length and Velocity Forms of the Transition Moments

In the following, we dispense from supposing atomic units and display the equations in their more familiar explicit form.

The first spectral moment of the dipole operator, more specifically its z-component,

$$\hat{Z} = \sum_{i=1}^{N} \hat{z}^{(i)} = \sum_{r,s} z_{rs} c_r^\dagger c_s \tag{14.54}$$

is given by

$$S_1(Z) = \sum_n (E_n - E_0) \left| \langle \Psi_n | \hat{Z} | \Psi_0 \rangle \right|^2 \tag{14.55}$$

As is easily seen, $S_1(Z)$ can be expressed as the ground-state expectation value

$$S_1(Z) = \tfrac{1}{2} \langle \Psi_0 | [\hat{Z}, [\hat{H}, \hat{Z}]] | \Psi_0 \rangle. \tag{14.56}$$

of a double commutator, which can directly be evaluated to give a constant,

$$[\hat{Z}, [\hat{H}, \hat{Z}]] = [\hat{Z}, -i \hat{P}_z] \hbar / m_e = N \hbar^2 / m_e \tag{14.57}$$

Here, $\hat{P}_z$ is the z-component of the momentum operator, coming into play as the result of the commutator

$$[\hat{H}, \hat{Z}] = -i \hat{P}_z \hbar / m_e \tag{14.58}$$

Accordingly, the final result for $S_1(Z)$ reads

$$S_1(Z) = \tfrac{1}{2} N \hbar^2 / m_e \tag{14.59}$$

which is the well-known Thomas–Reiche–Kuhn (TRK) or **dipole sum rule**.

In the ADC formulation, the moment $S_1(Z)$ can be written in the following compact form:

$$S_1(Z) = \boldsymbol{F}(Z)^\dagger \boldsymbol{M} \boldsymbol{F}(Z) \tag{14.60}$$

Here, $\boldsymbol{F}(Z)$ denotes the vector of effective transition moments for the dipole operator $\hat{Z}$,

$$F_J(Z) = \sum_{r,s} f_{J,rs} z_{rs} \tag{14.61}$$

The equivalence of the ADC form (14.60) and Eq. (14.55) can be seen by writing the secular equation (14.5) as

$$\boldsymbol{M} = \boldsymbol{X}\boldsymbol{\Omega}\boldsymbol{X}^{\dagger} \tag{14.62}$$

and using Eq. (14.7) with the transition operator $\hat{Z}$. Let us note that the ADC form can readily be extended to other spectral moments, $S_k(D)$, $k = 0, 1, 2, \ldots$:

$$S_k(D) = \boldsymbol{F}(D)^{\dagger}\boldsymbol{M}^k\boldsymbol{F}(D) \tag{14.63}$$

Here D refers to a given one-particle operator $\hat{D}$.

The dipole sum rule (14.59) may serve as a test of the accuracy of the computational scheme. Deviations from the exact value can reflect shortcomings of the method as well as an incompleteness of the molecular basis set used to compute the excitation spectrum. Interestingly, in the random-phase approximation (RPA), to be considered in Chap. 15, the dipole sum rule is fulfilled exactly (apart from the basis set error) although the RPA excitation energies and transition moments (of singly excited states) are consistent only through first order. We will come back to this point in Sect. 15.1.

Another significant quality test is the **equivalence** of the **dipole length** (L) and **dipole velocity** (V) forms of the transition moments for exact states. This refers to the well-known identity

$$\langle\Psi_m|[\hat{H}, \hat{Z}]|\Psi_0\rangle = (E_m - E_0)\langle\Psi_m|\hat{Z}|\Psi_0\rangle = -i\langle\Psi_m|\hat{P}_z|\Psi_0\rangle\,\hbar/m_e \tag{14.64}$$

The second term is referred to as the L-form of the transition moment; the third term, obtained by evaluating the commutator according to Eq. (14.58), represents the V-form. In the ADC formulation, the L-form can be written as

$$(E_m - E_0)\langle\Psi_m|\hat{Z}|\Psi_0\rangle = \boldsymbol{X}_m^{\dagger}\boldsymbol{M}\boldsymbol{F}(Z) \tag{14.65}$$

while the V-form of the transition moment is given by

$$\langle\Psi_m|\hat{P}_z|\Psi_0\rangle = \boldsymbol{X}_m^{\dagger}\boldsymbol{F}(P_z) \tag{14.66}$$

Here, $\boldsymbol{F}(P_z)$ is the vector of effective transition moments for $\hat{P}_z$. By abstracting the eigenvector $\boldsymbol{X}_m^{\dagger}$ on the right-hand side of the latter two equations, one yields the general identity

$$\boldsymbol{M}\boldsymbol{F}(Z) = -i\boldsymbol{F}(P_z)\,\hbar/m_e \tag{14.67}$$

which is no longer restricted to a particular transition.

As in the case of the dipole sum rule, deviations from this global form of length-velocity equivalence encountered in actual computations are an indication of the accuracy afforded at the respective level of approximation and the quality of the one-particle basis set used. Interestingly, for a consistent nth-order treatment, obtained, for example, at the ADC(n) level, the deviation error is of order n, rather than $n+1$ as one might expect. For an explanation, we have to take a closer look at how the PT expansions conform to the exact spectral identities.

The difference Δ_m between the length and velocity transition moments written in the form

$$\Delta_m = \langle\Psi_m|[\hat{H}, \hat{Z}]|\Psi_0\rangle + i\langle\Psi_m|\hat{P}_z|\Psi_0\rangle\, \hbar/m_e = 0 \tag{14.68}$$

has a non-trivial PT expansion

$$\Delta_m = \Delta_m^{(0)} + \Delta_m^{(1)} + \Delta_m^{(2)} + \ldots \tag{14.69}$$

with non-vanishing terms $\Delta_m^{(\nu)} \neq 0$, in spite of the fact that Δ_m vanishes. There is a compensation of successive terms such that a finite expansion, say through order n, has a non-vanishing residue of order n,

$$\sum_{\nu=0}^{n} \Delta_m^{(\nu)} = O(n) \tag{14.70}$$

which is compensated by a corresponding contribution of $\Delta_m^{(n+1)}$. The reason for this behavior is that the commutator relation (14.58) holds for the total hamiltonian $\hat{H}$ but not individually for the parts $\hat{H}_0$ and $\hat{H}_I$ of the Møller–Plesset (MP) partitioning

$$\hat{H} = \hat{T} + \hat{V} = \hat{H}_0 + \hat{H}_I \tag{14.71}$$

underlying the PT expansions. Let us consider the HF operator

$$\hat{H}_0 = \sum_{i=1}^{N} \hat{f}^{(i)} \tag{14.72}$$

where the one-particle Fock operators are of the form (see Eq. 4.5)

$$\hat{f} = \hat{t} - \hat{w} \tag{14.73}$$

While the commutator with $\hat{t}$ (comprising the operator of kinetic energy and the electron-nuclei interaction) yields

$$[\hat{t}, \hat{z}] = -i\,\hat{p}_z\, \hbar/m_e \tag{14.74}$$

the second part

$$\hat{w} = -\sum_r (\hat{J}_r - \hat{K}_r) n_r \tag{14.75}$$

does not commute with $\hat{z}$ because of the presence of the non-local exchange operators $\hat{K}_r$:

$$[\hat{w}, \hat{z}] = \hat{q} \neq 0 \tag{14.76}$$

where $\hat{q}$ can readily be evaluated (see Exercise 14.3). Accordingly, the commutator for the corresponding N-particle operator $\hat{W} = \sum_i \hat{w}^{(i)}$ becomes

$$[\hat{W}, \hat{Z}] = \hat{Q} \tag{14.77}$$

where $\hat{Q} = \sum_i \hat{q}^{(i)}$. Noting that $\hat{H}_0 = \hat{T} - \hat{W}$ and $\hat{H}_I = \hat{V} + \hat{W}$, the individual commutators become

$$\begin{aligned}[\hat{H}_0, \hat{Z}] &= -\, i P_z \hbar/m_e - \hat{Q} \\ [\hat{H}_I, \hat{Z}] &= \hat{Q}\end{aligned} \tag{14.78}$$

Now we apply the MP partitioning to Eq. (14.68) which then reads

$$\Delta_m = \langle \Psi_m | [\hat{H}_0, \hat{Z}] + i \hat{P}_z \hbar/m_e | \Psi_0 \rangle + \langle \Psi_m | [\hat{H}_I, \hat{Z}] | \Psi_0 \rangle \tag{14.79}$$

Using the commutator relations (14.78), the first term becomes $-\langle \Psi_m | \hat{Q} | \Psi_0 \rangle$, and the same expression applies to the second term, albeit with the opposite sign so that the two terms cancel each other. In the PT expansion of Δ_m, however, this cancelation involves successive PT orders. For any nth-order contribution to the first term, there is an analogous $(n+1)$st contribution (of opposite sign) to the second term.

A similar analysis applies to the dipole sum rule (14.59). The MP partitioning of $\hat{H}$ splits the double commutator in Eq. (14.56) into two parts,

$$[\hat{Z}, [\hat{Z}, \hat{H}_0]] = \tfrac{1}{2} N \hbar^2/m_e + \hat{Q}' \tag{14.80}$$

$$[\hat{Z}, [\hat{Z}, \hat{H}_I]] = -\, \hat{Q}' \tag{14.81}$$

where $\hat{Q}' = [\hat{Q}, \hat{Z}]$. As a consequence, there is a non-trivial PT expansion of the first spectral moment,

$$S_1(Z) = S_1^{(0)}(Z) + S_1^{(1)}(Z) + S_1^{(2)}(Z) + \cdots = \tfrac{1}{2} N \hbar^2/m_e \tag{14.82}$$

As in Eq. (14.70), summing these terms through order n,

$$\sum_{\nu=0}^{n} S_1^{(\nu)}(Z) = O(n) \quad (\text{for } n > 1) \tag{14.83}$$

results in a non-vanishing contribution of the order n, which is canceled by a part of the next higher term, $S_1^{(n+1)}(Z)$. As should be noted, the expansion through first order yields the exact result

$$S_1^{(0)}(Z) + S_1^{(1)}(Z) = \tfrac{1}{2} N \hbar^2 / m_e \tag{14.84}$$

This is a consequence of the fact that the matrix elements

$$\langle \Phi_0 | \hat{Q}' | \Psi_0^{(1)} \rangle = \langle \Psi_0^{(1)} | \hat{Q}' | \Phi_0 \rangle = 0 \tag{14.85}$$

vanish since $\hat{Q}'$ is a one-particle operator and $|\Psi_0^{(1)}\rangle$ a sum of $2p$-$2h$ excitations.

Exercises

14.1 Evaluate the time-orderings (7)–(10) of the second-order diagram 2C shown in Fig. 14.1. Perform the ADC analysis and derive the corresponding contributions to $C_{ak,a'k'}^{(2)}$ and $f_{ak,a'k'}^{(2)}$.

14.2 Consider the IS transition moments $\langle \tilde{\Psi}_I | \hat{N} | \Psi_0 \rangle$ (see Eq. 14.52) for the particle number operator $\hat{N}$. Verify the resulting conditions on the diagonal amplitudes $f_{I,rr}$ in the ADC(2) expressions listed in Appendix A.9.

14.3 In spin-free (spatial) form, the operator $\hat{w}$ (Eq. 14.75) is given by $\hat{w} = -\sum(2\hat{J}_i - \hat{K}_i)n_i$, where $\hat{J}_i$ and $\hat{K}_i$ denote the Coulomb and exchange operators for the (occupied) spatial orbital i. Apply the commutator $\hat{q} = [\hat{w}, \hat{z}]$ to a given spatial orbital, $\phi(\boldsymbol{x})$. Evaluate the (spatial) matrix element $q_{ak} = \langle \phi_a | \hat{q} | \phi_k \rangle$ in the form $q_{ak} = \sum_{i,r}(V_{aiir} d_{rk} - d_{ar} V_{riik})$, where $d_{rs} = \langle \phi_r | \hat{z} | \phi_s \rangle$, by using the resolution of the identity for the basis of spatial orbitals $\phi_r(\boldsymbol{x})$.

References

1. Trofimov AB, Stelter G, Schirmer J (1999) J Chem Phys 111:9982
2. Trofimov AB, Stelter G, Schirmer J (2002) J Chem Phys 117:6402
3. Harbach PHP, Wormit M, Dreuw A (2014) J Chem Phys 141:064113
4. Schirmer J, Trofimov AB (2004) J Chem Phys 120:11449

Chapter 15
Random-Phase Approximation (RPA)

In this chapter, we will take a look at the famous random-phase approximation (RPA) to the polarization propagator. The computational benefit afforded by the RPA is rather modest, at least for atoms and molecules, as the resulting excitation energies and transition moments are only consistent through first order of perturbation theory. From a theoretical point of view, though, the RPA represents a highly interesting concept, to be seen as an integral part of general knowledge in many-body physics. Let us note some essential features:

- The RPA was originally devised in solid-state physics [1, 2] as a means to study collective excitations (*plasmons*) of the interacting electron gas.
- The RPA can be obtained by summing a distinct class of diagrams (RPA diagrams) through infinite order (see for example Thouless [3]), representing the paradigm of an infinite partial summation of diagrams in propagator perturbation theory.
- There are other independent derivations of the RPA, such as via the time-dependent Hartree–Fock (TDHF) approach (going back to Dirac [4]), or the equation-of-motion (EOM) method (see Sect. 16.3).
- The RPA introduces interesting mathematics, featuring a specific pseudo-eigenvalue secular problem.
- The RPA solutions fulfill the dipole sum rule and the equivalence of the length and velocity forms of the transition moments.

As a point of interest in the present context, the ADC procedure can be specifically applied to the series of RPA diagrams. The corresponding reformulation of the RPA and the derivation of ADC(n) schemes, approximating the full RPA solution, are discussed in Sect. 15.2.

J. Schirmer, *Many-Body Methods for Atoms, Molecules and Clusters*, Lecture Notes in Chemistry 94, https://doi.org/10.1007/978-3-319-93602-4_15

15.1 Derivation and Properties of the RPA Equations

The random-phase approximation (RPA) for the polarization propagator is established by a specific subset of diagrams shown in Fig. 15.1, referred to as RPA diagrams. The RPA diagrams are complete through first order, whereas in second and higher order they represent only a small fraction of the full set of diagrams. As seen in Fig. 13.3, the RPA diagram is one of altogether five second-order diagrams. The RPA diagrams can be summed completely through infinite order owing to their simple recursive construction principle, which is depicted in Fig. 15.2. In energy representation, this recursion relation takes on the simple algebraic form (see Exercise 15.1)

$$\mathbf{\Pi}^{RPA}(\omega) = \mathbf{\Pi}^{(0)}(\omega) + \mathbf{\Pi}^{(0)}(\omega)\boldsymbol{R}\,\mathbf{\Pi}^{RPA}(\omega) \tag{15.1}$$

Here, $\mathbf{\Pi}^{(0)}(\omega)$ is the free (or zeroth-order) polarization propagator (13.19), reading in energy representation

$$\Pi^{(0)}_{rs,r's'}(\omega) = \delta_{rr'}\delta_{ss'}\left(\frac{\bar{n}_r n_s}{\omega - \epsilon_r + \epsilon_s + i\eta} - \frac{n_r \bar{n}_s}{\omega - \epsilon_r + \epsilon_s - i\eta}\right) \tag{15.2}$$

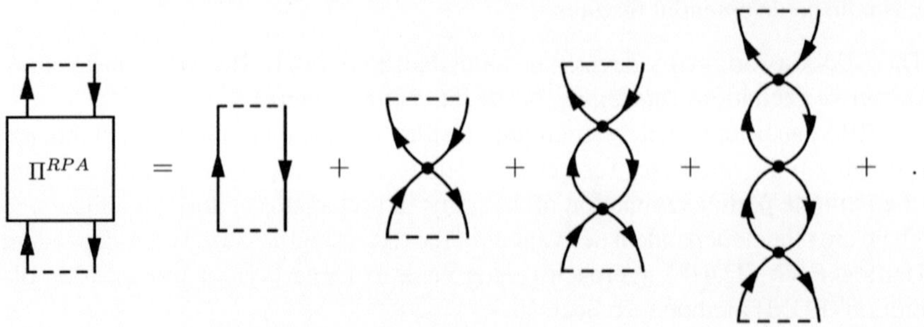

Fig. 15.1 Series of RPA diagrams through third order

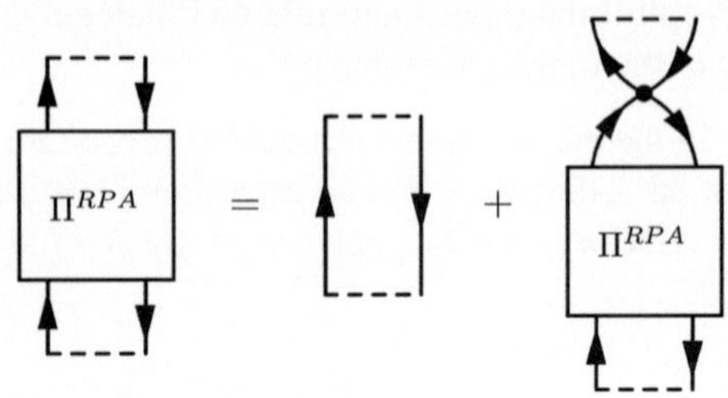

Fig. 15.2 Recursive construction of the RPA diagram series

and $\boldsymbol{R}$ is a constant matrix of Coulomb integrals,

$$R_{rs,r's'} = -V_{rs'[r's]} \tag{15.3}$$

The configuration space of $\boldsymbol{R}$ (as of $\boldsymbol{\Pi}^{(0)}$ and $\boldsymbol{\Pi}^{RPA}$) is spanned by the $1p$-$1h$ configurations ($\bar{n}_r n_s = 1$) and $1h$-$1p$ configurations ($n_r \bar{n}_s = 1$). Using Eqs. (15.2) and (15.3), the formal solution of Eq. (15.1) can be written as

$$\boldsymbol{\Pi}^{RPA}(\omega) = (\omega \boldsymbol{S} - \boldsymbol{\mathcal{M}})^{-1} \tag{15.4}$$

where $\boldsymbol{\mathcal{M}}$, referred to as the RPA secular matrix, is given by

$$\mathcal{M}_{rs,r's'} = \delta_{rr'}\delta_{ss'}(\epsilon_r - \epsilon_s)(\bar{n}_r n_s - n_r \bar{n}_s) - V_{rs'[r's]} \tag{15.5}$$

and $\boldsymbol{S}$ is a diagonal matrix of elements

$$S_{rs,r's'} = \delta_{rr'}\delta_{ss'}(\bar{n}_r n_s - n_r \bar{n}_s) \tag{15.6}$$

Before addressing its physical content, we shall take a look at the mathematical aspects of the RPA.

The partitioning of the RPA configurations into $1p$-$1h$ and $1h$-$1p$ configurations leads to the following block structure of $\boldsymbol{\mathcal{M}}$ and $\boldsymbol{S}$:

$$\boldsymbol{\mathcal{M}} = \begin{pmatrix} \boldsymbol{A} & \boldsymbol{B} \\ \boldsymbol{B}^* & \boldsymbol{A}^* \end{pmatrix}, \quad \boldsymbol{S} = \begin{pmatrix} \mathbf{1} & \mathbf{0} \\ \mathbf{0} & -\mathbf{1} \end{pmatrix} \tag{15.7}$$

Here, the $1p$-$1h$ sub-block $\boldsymbol{A}$ is a (quadratic) matrix of elements

$$A_{rs,r's'} = \delta_{rr'}\delta_{ss'}(\epsilon_r - \epsilon_s) - V_{rs'[r's]}, \quad \bar{n}_r n_s = \bar{n}_{r'} n_{s'} = 1 \tag{15.8}$$

while the sub-block $\boldsymbol{B}$, coupling the $1p$-$1h$ and $1h$-$1p$ configurations, comprises the matrix elements

$$B_{rs,r's'} = -V_{rs'[r's]}, \quad \bar{n}_r n_s = n_{r'} \bar{n}_{s'} = 1 \tag{15.9}$$

$\boldsymbol{A}^*$ and $\boldsymbol{B}^*$ denote the complex conjugates of $\boldsymbol{A}$ and $\boldsymbol{B}$, respectively. Obviously, the sub-block $\boldsymbol{A}$ is hermitian, and thus, $\boldsymbol{A}^*$ is hermitian as well; the $\boldsymbol{B}$ matrix is symmetric, that is, $B_{ak,jb} = B_{bj,ka}$. According to these properties of the sub-blocks, the entire matrix $\boldsymbol{\mathcal{M}}$ is itself a hermitian matrix – which of course could have been deduced directly from Eq. (15.5).

As will be shown below, the inversion of the ω-dependent matrix $(\omega \boldsymbol{S} - \boldsymbol{\mathcal{M}})$ in Eq. (15.4) is equivalent to solving a modified eigenvalue problem for the matrix $\boldsymbol{\mathcal{M}}$, which is referred to as RPA **pseudo-eigenvalue problem**. Here, the characteristic eigenvalue equation reads

$$\mathcal{M}\begin{pmatrix} \boldsymbol{x}_m \\ \boldsymbol{y}_m \end{pmatrix} = \omega_m \begin{pmatrix} \boldsymbol{x}_m \\ -\boldsymbol{y}_m \end{pmatrix} \quad (15.10)$$

where ω_m denotes the eigenvalue, and the eigenvector consists of $1p$-$1h$ components, $\boldsymbol{x}_m$, and $1h$-$1p$ components, $\boldsymbol{y}_m$:

$$\boldsymbol{X}_m = \begin{pmatrix} \boldsymbol{x}_m \\ \boldsymbol{y}_m \end{pmatrix}$$

The difference to the usual eigenvalue problem is, of course, the minus sign in front of the $1h$-$1p$ components on the right-hand side of the eigenvalue equation. Using a compact matrix notation, the individual equations, $m = 1, 2, \ldots$, can be aggregated as

$$\mathcal{M}\boldsymbol{X} = \boldsymbol{S}\boldsymbol{X}\boldsymbol{\Omega} \quad (15.11)$$

where $\boldsymbol{\Omega}$ and $\boldsymbol{X}$ denote the diagonal matrix of eigenvalues and the eigenvector matrix, respectively.

The structure of the RPA matrices $\mathcal{M}$ and $\boldsymbol{S}$ according to Eq. (15.7) has some specific implications. There are two related sets of solutions referred to as excitation solutions, $m \in \{+\}$, and de-excitation solutions, $m \in \{-\}$. The excitation solutions relate to the upper left block, $\boldsymbol{A}$, which is identical with the TDA (or ADC(1)) secular matrix, presented in Eq. (14.30) of Sect. 14.2. Supposing for a moment that the coupling block can be disregarded, $\boldsymbol{B} = \boldsymbol{0}$, the excitation solutions simply become the TDA solutions. Likewise, the de-excitation solutions are related to the lower right block, $\boldsymbol{A}^*$, of the RPA secular matrix. However, they merely repeat the excitation solutions in a somewhat different shape. Let

$$\omega_+, \ \boldsymbol{X}_+ = \begin{pmatrix} \boldsymbol{x} \\ \boldsymbol{y} \end{pmatrix} \quad (15.12)$$

be the eigenvalue and eigenvector of an excitation solution. It is easily shown that there is a corresponding de-excitation solution with the eigenvector

$$\boldsymbol{X}_- = \begin{pmatrix} \boldsymbol{y}^* \\ \boldsymbol{x}^* \end{pmatrix} \quad (15.13)$$

and an eigenvalue ω_-, being (if real) the negative of ω_+:

$$\omega_- = -\omega_+^* \quad (15.14)$$

What about normalization of the pseudo-eigenvectors? As the reader should check, two eigenvectors belonging to different eigenvalues obey the following pseudo-orthogonality relation

$$\boldsymbol{X}_m^\dagger \boldsymbol{S}\boldsymbol{X}_{m'} = (\boldsymbol{x}_m^\dagger, \boldsymbol{y}_m^\dagger)\begin{pmatrix} \boldsymbol{x}_{m'} \\ -\boldsymbol{y}_{m'} \end{pmatrix} = \boldsymbol{x}_m^\dagger \boldsymbol{x}_{m'} - \boldsymbol{y}_m^\dagger \boldsymbol{y}_{m'} = 0 \ \text{ for } m \neq m' \quad (15.15)$$

Compatible with the orthogonalization, the pseudo-eigenvectors can be normalized according to

$$\boldsymbol{X}_m^\dagger \boldsymbol{X}_m = |\boldsymbol{x}_m|^2 - |\boldsymbol{y}_m|^2 = \begin{cases} 1, & \text{for } m \in \{+\} \\ -1, & \text{for } m \in \{-\} \end{cases} \tag{15.16}$$

Here, one has to distinguish the excitation and de-excitation solutions as in the first case (usually) the norm of the $1p$-$1h$ components is larger than that of the $1h$-$1p$ components, whereas the opposite applies to the de-excitation solutions. The full pseudo-orthonormalization conditions can be comprised in the following compact matrix expression

$$\boldsymbol{X}^\dagger \boldsymbol{S} \boldsymbol{X} = \boldsymbol{S} \tag{15.17}$$

As a simple consequence, $\boldsymbol{S}\boldsymbol{X}^\dagger \boldsymbol{S}\boldsymbol{X} = \mathbf{1}$, and the inverse of $\boldsymbol{X}$ is given by

$$\boldsymbol{X}^{-1} = \boldsymbol{S}\boldsymbol{X}^\dagger \boldsymbol{S} \tag{15.18}$$

Now we may come back to the original form (15.4) of the RPA propagator. The solution of the RPA pseudo-eigenvalue problem established by Eqs. (15.11, 15.17) allows us to write the RPA secular matrix as

$$\boldsymbol{\mathcal{M}} = \boldsymbol{S}\boldsymbol{X}\,\boldsymbol{\Omega}\boldsymbol{X}^{-1} = \boldsymbol{S}\boldsymbol{X}\,\boldsymbol{\Omega}\boldsymbol{S}\,\boldsymbol{X}^\dagger \boldsymbol{S} \tag{15.19}$$

where Eq. (15.18) has been used to arrive at the final expression. Inserting this expression for $\boldsymbol{\mathcal{M}}$ in Eq. (15.4) leads to the form

$$\boldsymbol{\Pi}^{RPA}(\omega) = \boldsymbol{X}\boldsymbol{S}\,(\omega\mathbf{1} - \boldsymbol{\Omega})^{-1}\,\boldsymbol{X}^\dagger \tag{15.20}$$

in which the RPA propagator $\boldsymbol{\Pi}^{RPA}(\omega)$ is expressed entirely in terms of the RPA (pseudo-) eigenvalues and eigenvectors. In that sense, the inversion of the ω-dependent matrix in Eq. (15.4) is equivalent to solving the RPA pseudo-eigenvalue problem.

The RPA pseudo-eigenvalue problem (15.11) can be reformulated as an ordinary eigenvalue problem of the non-hermitian matrix $\boldsymbol{\mathcal{M}}' = \boldsymbol{S}\boldsymbol{\mathcal{M}}$. Being eigenvalues of a non-hermitian matrix, the RPA energies ω_m are not necessarily real numbers. A complex eigenvalue has of course no physical meaning and would indicate a failure of the RPA in the treatment of the concerned excitation.

The possibility of complex eigenvalues as well as other features of the RPA mathematics can nicely be demonstrated by means of a simple model. We shall consider the 2×2 RPA-type matrix

$$\mathcal{M} = \begin{pmatrix} a & b \\ b & a \end{pmatrix} \tag{15.21}$$

where a and b are real numbers. The RPA pseudo-eigenvalue problem

$$\mathcal{M}\begin{pmatrix} x \\ y \end{pmatrix} = \omega \begin{pmatrix} x \\ -y \end{pmatrix} \tag{15.22}$$

has the two eigenvalues

$$\omega_{\pm} = \pm\sqrt{a^2 - b^2} = \pm|a|\sqrt{(1 - b^2/a^2)} \tag{15.23}$$

The corresponding RPA eigenvectors can easily be determined too (see Exercise 15.2). Obviously, complex eigenvalues emerge if $|b| > |a|$, that is, if the magnitude of the coupling matrix element surpasses that of the diagonal element. Supposing $|b|/|a| < 1$ the eigenvalues are real, and the square root can be expanded in a perturbation series, which, for ω_+ and $a > 0$, assumes the form

$$\omega_+ = a - \frac{b^2}{2a} + \ldots \tag{15.24}$$

Note that the denominator is given by the sum (rather than the difference) of the diagonal elements.

The partitioning of the RPA eigenvector matrix in an excitation and de-excitation part,

$$\boldsymbol{X} = (\boldsymbol{X}^+, \boldsymbol{X}^-) \tag{15.25}$$

and the corresponding partitioning of the eigenvalue matrix,

$$\boldsymbol{\Omega} = \begin{pmatrix} \boldsymbol{\Omega}^+ & \mathbf{0} \\ \mathbf{0} & \boldsymbol{\Omega}^- \end{pmatrix} \tag{15.26}$$

allows us to write the RPA propagator (15.20) in the form

$$\boldsymbol{\Pi}^{RPA}(\omega) = \boldsymbol{X}^+ \left(\omega\mathbf{1} - \boldsymbol{\Omega}^+\right)^{-1} (\boldsymbol{X}^+)^\dagger - \boldsymbol{X}^- \left(\omega\mathbf{1} - \boldsymbol{\Omega}^-\right)^{-1} (\boldsymbol{X}^-)^\dagger \tag{15.27}$$

This constitutes, in matrix notation, the spectral representation of $\boldsymbol{\Pi}^{RPA}(\omega)$, that is, the RPA approximation to the spectral representation (13.1) of the exact polarization propagator. More explicitly, the first (excitation) part of Eq. (15.27) reads

$$\Pi^{RPA+}_{rs,r's'}(\omega) = \sum_{m\in\{+\}} \frac{X_{rs,m}X^*_{r's',m}}{\omega - \omega_m} \tag{15.28}$$

In this form, the physical content of the RPA propagator becomes manifest. Evidently, the RPA (pseudo-) eigenvalues can be identified with the excitation energies,

$$\Delta E^{RPA}_m = \omega_m, \quad m \in \{+\} \tag{15.29}$$

The transition moments derive from the RPA eigenvector components in an analogous manner to Eq. (13.6),

$$T_m^{RPA} = \sum_{r,s} X^*_{rs,m} d_{rs} (\bar{n}_r n_s + n_r \bar{n}_s), \; m \in \{+\} \tag{15.30}$$

where d_{rs} are the one-particle matrix elements of a transition operator $\hat{D}$. This expression may be written more compactly as

$$T_m^{RPA} = \boldsymbol{X}_m^\dagger \boldsymbol{d} \tag{15.31}$$

where

$$\boldsymbol{d} = \begin{pmatrix} \boldsymbol{d}_{ph} \\ \boldsymbol{d}_{hp} \end{pmatrix} = \begin{pmatrix} \boldsymbol{d}_{ph} \\ \boldsymbol{d}^*_{ph} \end{pmatrix} \tag{15.32}$$

denotes the (column) vector of p-h and h-p matrix elements d_{rs}.

Perturbation Theoretical Analysis of the RPA

As already mentioned, the RPA energies and transition moments are consistent with those obtained at the TDA (or CIS and ADC(1)) level, since the $1p$-$1h$ sub-block $\boldsymbol{A}$ of the RPA secular matrix is identical with the TDA secular matrix. However, the RPA treatment goes beyond the TDA, the differences beginning at second order, as a result of the off-diagonal blocks $\boldsymbol{B}$ and $\boldsymbol{B}^*$, which couple the (physical) $1p$-$1h$ excitations and the (unphysical) $1h$-$1p$ de-excitations. How can this curious coupling be understood and what is the quality of the results thereby obtained? As a way to better understand the situation, we shall analyze, like in Sect. 8.3, the RPA results using perturbation theory, which, moreover, will allow us to discuss the relevant physics of the excitation process.

Let us consider a single excitation $a \leftarrow k$ from the occupied orbital k to the virtual (unoccupied) orbital a, represented by the zeroth-order (HF) state $|\Phi_{ak}\rangle = c_a^\dagger c_k |\Phi_0\rangle$. The formal PT expansion of the excitation energy, $\Delta E_{ak} = E_{ak} - E_0$, through second order can be written as

$$\Delta E_{ak} = \epsilon_a - \epsilon_k - V_{ak[ak]} \\ + U_{ak}^{(2)}(1p\text{-}1h) + U_{ak}^{(2)}(2p\text{-}2h) + U_{ak}^{(2)}(3p\text{-}3h) - E_0^{(2)} + O(3) \tag{15.33}$$

Here, the three second-order terms, $U_{ak}^{(2)}(\nu p\text{-}\nu h)$, $\nu = 1, 2, 3$, are due to the (first-order) coupling of $|\Phi_{ak}\rangle$ with $1p$-$1h$, $2p$-$2h$, and $3p$-$3h$ excitations, respectively, and $E_0^{(2)}$ is the second-order ground-state energy, as in Eq. (A.1.18).

In zeroth order, the excitation energy is just the difference of the HF orbital energies, $\epsilon_a - \epsilon_k$. The HF orbitals, though, reflect the N-electron charge distribution in the HF ground state. This means, the orbital energy ϵ_a does not account for the instance that there is an electron vacancy in orbital k; nor does ϵ_k account for the presence of an electron in the virtual orbital a. The necessary correction is established at the first-order level: Here, the Coulomb repulsion between electrons in orbitals k and a, $J_{ak} = V_{akak}$ is subtracted from the zeroth-order result. The other first-order term is the exchange integral, $K_{ak} = V_{akka}$, accounting for the energy difference between the singlet and triplet excitations (see Exercise 1.6).

At second order, one goes beyond the static one-particle picture and begins to incorporate electron correlation and dynamic effects accompanying the excitation. The first of the second-order terms in Eq. (15.33), $U_{ak}^{(2)}(1p\text{-}1h)$, is due to the admixture of other $1p\text{-}1h$ excitations. The explicit PT expression is of little interest, as the $1p\text{-}1h$ (single) configuration interaction (CI) is rigorously treated at the RPA level via the $\boldsymbol{A}$ block of the secular matrix. The original HF orbital k and even more so the virtual orbital a may not afford an adequate representation of the final excited state, and that is corrected by means of the $1p\text{-}1h$ admixtures.

More important is the $U_{ak}^{(2)}(2p\text{-}2h)$ contribution, resulting from the coupling with $2p\text{-}2h$ excitations. Via the admixture of $2p\text{-}2h$ configurations of the type $c_b^\dagger c_j c_a^\dagger c_k|\Phi_0\rangle$, $b \neq a$, $j \neq k$ the response of the so far unaffected electrons comes into play. Here, we may distinguish **relaxation** and **polarization**: The electrons "relax" as a response to the removal of an electron from the orbital k; and the electron promoted to the virtual orbital a "polarizes" the charge distribution of the ionic core. The relaxation and polarization response leads to a substantial lowering of the first-order (static) excitation energy. It is interesting to note that these two effects compensate each other to a certain extent, roughly as in $\Delta E_{RP} = -(R-P)^2$, the energy lowerings being $-R^2$ and $-P^2$ separately for relaxation and polarization, respectively. For a more detailled discussion of the PT analysis of the relaxation and polarization energies, the reader is referred to Ref. [5].

The third term, $U_{ak}^{(2)}(3p\text{-}3h)$, accounts for correlation in the excited state. The only $3p\text{-}3h$ excitations that can couple with $|\Phi_{ak}\rangle$ are of the type $c_b^\dagger c_c^\dagger c_i c_j c_a^\dagger c_k|\Phi_0\rangle$, that is, double excitations on the $|\Phi_{ak}\rangle$ state. The explicit PT expression,

$$U_{ak}^{(2)}(3p\text{-}3h) = -\sum_{\substack{b<c\neq a\\ i<j\neq k}} \frac{|V_{bc[ij]}|^2}{\epsilon_b+\epsilon_c-\epsilon_i-\epsilon_j} \tag{15.34}$$

is of the same form as that for $E_0^{(2)}$ (see Eq. A.1.18) and differs only because of the restrictions in the summation indices. Subtracting $E_0^{(2)}$ from $U_{ak}^{(2)}(3p\text{-}3h)$ yields

$$U_{ak}^{(2)}(3p\text{-}3h) - E_0^{(2)} = \sum_{b,i<j} \frac{|V_{ab[ij]}|^2}{\epsilon_a+\epsilon_b-\epsilon_i-\epsilon_j} + \sum_{j,b<c} \frac{|V_{bc[kj]}|^2}{\epsilon_b+\epsilon_c-\epsilon_k-\epsilon_j} - \sum_{j,b} \frac{|V_{ab[kj]}|^2}{\epsilon_a+\epsilon_b-\epsilon_k-\epsilon_j} \tag{15.35}$$

As is to be expected, the correlation energy is somewhat larger in the ground state than in the excited state, and as a result, electron correlation increases the excitation energy. The first two terms in Eq. (15.35) are positive; the negative third term is contained as a partial sum both in the first and the second term.

The corresponding PT expansion of the RPA excitation energy is given by

$$\Delta E_{ak}^{RPA} = \epsilon_a - \epsilon_k - V_{ak[ak]} + U_{ak}^{(2)}(1p\text{-}1h) + U_{ak}^{(2)}(1h\text{-}1p) + O(3) \tag{15.36}$$

As expected, the RPA result reproduces the exact excitation energy through first order and, moreover, correctly accounts for the CI singles contribution, $U_{ak}^{(2)}(1p\text{-}1h)$. As a specific RPA feature, the other second-order term, $U_{ak}^{(2)}(1h\text{-}1p)$, results from the coupling of the considered excitation $a \leftarrow k$ with de-excitation configurations of block $\boldsymbol{A}^*$ via the coupling matrix elements of $\boldsymbol{B}$. What is its physical content? The explicit expression reads

$$U_{ak}^{(2)}(1h\text{-}1p) = -\sum_{j,b} \frac{|V_{ab[kj]}|^2}{\epsilon_a + \epsilon_b - \epsilon_k - \epsilon_j} \tag{15.37}$$

This result can be obtained, for example, using ordinary matrix perturbation theory for the eigenvalue problem of $\boldsymbol{\mathcal{M}}' = \boldsymbol{S}\boldsymbol{\mathcal{M}}$. Alternatively, one can resort to the ADC formulation of the RPA presented in Sect. 15.2 and identify $U_{ak}^{(2)}(1h\text{-}1p)$ with the diagonal element $M_{ak,ak}^{RPA\,(2)}$ of the RPA-ADC secular matrix in second order (see Eqs. 15.61 and 14.39). The comparison with Eq. (15.35) shows that $U_{ak}^{(2)}(1h\text{-}1p)$ is identical with the third term in the latter equation. This means that the RPA treatment recovers one of the three correlation contributions to the excitation energy in second order, however, the one lowering the excitation energy, whereas, as we just have argued, the full second-order correlation increases the excitation energy.

Obviously, the gravest deficiency is the absence of the $U_{ak}^{(2)}(2p\text{-}2h)$ contribution, indicating that in the RPA model the $2p$-$2h$ (and higher) excitations are excluded from the outset. Accordingly, the physically important effects of relaxation and polarization are beyond the RPA description. Therefore, the RPA cannot be seen as a satisfactory approach to the treatment of excitation energies.

In a similar way, one may analyze the RPA transition moments, where the signature of the $1h$-$1p$ de-excitations can already be seen at first order. The exact transition moment for the $a \leftarrow k$ excitation can be expanded as

$$T_{ak} = \langle \Psi_{ak}|\hat{D}|\Psi_0\rangle = \langle \Phi_{ak}|\hat{D}|\Phi_0\rangle + \langle \Psi_{ak}^{(1)}|\hat{D}|\Phi_0\rangle + \langle \Phi_{ak}|\hat{D}|\Psi_0^{(1)}\rangle + O(2) \tag{15.38}$$

There are two first-order terms, related to the first-order excited state and the first-order ground state, respectively. They are matched by corresponding terms in the RPA expansion, reading

$$T_{ak}^{RPA} = d_{ak} + \sum_{b,j} X_{bj,ak}^{(1)} d_{bj} + \sum_{j,b} X_{jb,ak}^{(1)} d_{jb} + O(2) \tag{15.39}$$

where $X_{rs,ak}^{(1)}$ are the components of the first-order RPA eigenvector, $\boldsymbol{X}_{ak}^{(1)}$. In the second term, the sum runs over the $1p$-$1h$ components of $\boldsymbol{X}_{ak}^{(1)}$, reflecting the (first-order) CI singles mixing in $|\Psi_{ak}^{(1)}\rangle$. The third term, to be identified with $\langle \Phi_{ak}|\hat{D}|\Psi_0^{(1)}\rangle$, features the $1h$-$1p$ de-excitation components of the RPA eigenvector, for which the following PT expressions can be established:

$$X^{(1)}_{jb,ak} = \frac{V_{ab[jk]}}{\epsilon_a + \epsilon_b - \epsilon_k - \epsilon_j} \tag{15.40}$$

This means that the RPA transition moments are consistent through first order and, in particular, account for the contributions arising from (first-order) ground-state correlation. In this respect, the RPA performs like the ADC(1) approximation analyzed in Sect. 14.2.

Spectral Properties

In Sect. 14.3, we have discussed the dipole sum rule (Eqs. 14.55 and 14.59) and the equivalence of the length and velocity forms of the transition moments (Eq. 14.64). These are characteristic properties of the exact excitation spectrum. Most notably, the RPA treatment complies with both equivalences, although the excitation energies and transition moments are rather poor approximations to the exact results. So it is interesting to inspect how this remarkable RPA feature comes about. As in Sect. 14.3, we briefly dispense from using atomic units.

At the RPA level, the first spectral moment takes on the form

$$S_1^{RPA}(Z) = \sum_{m\in\{+\}} \Delta E_m^{RPA} |T_m^{RPA}|^2 = \sum_{m\in\{+\}} \omega_m \, \boldsymbol{d}^\dagger \boldsymbol{X}_m \boldsymbol{X}_m^\dagger \boldsymbol{d} \tag{15.41}$$

Equivalently, we may use the de-excitation solutions to obtain

$$S_1^{RPA}(Z) = -\sum_{m\in\{-\}} \omega_m \, \boldsymbol{d}^\dagger \boldsymbol{X}_m \boldsymbol{X}_m^\dagger \boldsymbol{d} \tag{15.42}$$

Combining these two expressions allows us to write $S_1^{RPA}(Z)$ in a compact matrix form as follows,

$$S_1^{RPA}(Z) = \tfrac{1}{2} \boldsymbol{d}^\dagger \boldsymbol{X} \boldsymbol{S} \boldsymbol{\Omega} \boldsymbol{X}^\dagger \boldsymbol{d} \tag{15.43}$$

Now, Eq. (15.19) can be used to replace the eigenvector and eigenvalue matrices by the original RPA secular matrix:

$$S_1^{RPA}(Z) = \frac{1}{2} \boldsymbol{d}^\dagger \boldsymbol{S} \boldsymbol{\mathcal{M}} \boldsymbol{S} \boldsymbol{d} \tag{15.44}$$

This is a remarkable result. It shows that the PT expansion of $S_1^{RPA}(Z)$ terminates after first order, as the secular matrix elements are linear expressions of HF orbital energies and Coulomb integrals. On the other hand, the RPA results are correct through first order so that the identification

$$S_1^{RPA}(Z) = S_1^{(0)}(Z) + S_1^{(1)}(Z) \tag{15.45}$$

with the zeroth- and first-order terms in the PT expansion (14.82) of the exact spectral moment is valid. Recalling the result of Eq. (14.84), this proves the dipole sum rule

$$S_1^{RPA}(Z) = \tfrac{1}{2} N\, \hbar^2 / m_e \tag{15.46}$$

for the RPA excitation spectrum.

For RPA excitations, the identity (14.64) applying to exact states would read

$$\omega_m \boldsymbol{X}_m^\dagger \boldsymbol{d} = -i \boldsymbol{X}_m^\dagger \boldsymbol{p}_z\, \hbar / m_e \tag{15.47}$$

where $\boldsymbol{p}_z$ is the vector of the $1p$-$1h$ and $1h$-$1p$ matrix elements of the momentum operator $\hat{p}_z$, analogous to $\boldsymbol{d}$, being the vector of matrix elements of $\hat{z}$. This equation, so far only asserted, can be replaced by a more abstract version not relating to a particular excitation. To this end, we insert the $\boldsymbol{S}$ matrix twice on the left side of Eq. (15.47), and make use of the RPA secular equation (15.10), which yields

$$\omega_m \boldsymbol{X}_m^\dagger \boldsymbol{d} = \omega_m \boldsymbol{S} \boldsymbol{X}_m^\dagger \boldsymbol{S} \boldsymbol{d} = \boldsymbol{X}_m^\dagger \boldsymbol{\mathcal{M}} \boldsymbol{S} \boldsymbol{d} \tag{15.48}$$

Abstracting the eigenvector $\boldsymbol{X}_m^\dagger$ both on the left- and right-hand side of Eq. (15.47) leads to the “global” identity

$$\boldsymbol{\mathcal{M}} \boldsymbol{S} \boldsymbol{d} = -i\, \boldsymbol{p}_z\, \hbar / m_e \tag{15.49}$$

In fact, the RPA secular matrices $\boldsymbol{\mathcal{M}}$ and $\boldsymbol{S}$ fulfill that identity [6, 7], thus ensuring the equivalence of the length and velocity forms of the RPA transition moments. The identity (15.49) can be shown in a straightforward, if somewhat tedious way (see Exercise 15.4), using the general length–velocity relation at the level of one-particle operators,

$$[\hat{f}, \hat{z}] = [\hat{w}, \hat{z}] - i\, \hat{p}_z\, \hbar / m_e \tag{15.50}$$

according to Eqs. (14.73)–(14.75).

15.2 ADC Formulation of the RPA

The RPA version of the polarization propagator is constituted by a specific class of diagrams, namely the RPA diagrams shown in Fig. 15.1. Obviously, the ADC procedure can be confined to that particular class of diagrams. This allows one to convert the original RPA scheme into ADC-type secular equations [8], as will briefly be discussed in this section.

As in the general case, we start with the spectral representation (15.27):

$$\boldsymbol{\Pi}^{RPA+}(\omega) = \boldsymbol{X}^+ (\omega \boldsymbol{1} - \boldsymbol{\Omega}^+)^{-1} (\boldsymbol{X}^+)^\dagger$$

The corresponding (non-diagonal) ADC form reads

$$\boldsymbol{\Pi}^{RPA+}(\omega) = (\boldsymbol{f}^{RPA})^{\dagger}(\omega\mathbf{1} - \boldsymbol{M}^{RPA})^{-1}\boldsymbol{f}^{RPA} \tag{15.51}$$

Here, $\boldsymbol{M}^{RPA}$ and $\boldsymbol{f}^{RPA}$ are the RPA-ADC secular matrix and the matrix of RPA-ADC transition amplitudes, which are to be constructed by the comparison of the ADC form with the RPA diagrams through successively higher order. $\boldsymbol{M}^{RPA}$ is related to the diagonal matrix $\boldsymbol{\Omega}^{+}$ of RPA excitation energies via a unitary transformation,

$$\boldsymbol{M}^{RPA} = \boldsymbol{Y}\boldsymbol{\Omega}^{+}\boldsymbol{Y}^{\dagger} \tag{15.52}$$

$$\boldsymbol{f}^{RPA} = \boldsymbol{Y}(\boldsymbol{X}^{+})^{\dagger} \tag{15.53}$$

So far, these relations are only formal since the RPA-ADC matrices are still to be constructed. Of course, the construction is such that the eigenvalues of the final (exact) $\boldsymbol{M}^{RPA}$ matrix are the RPA energies of $\boldsymbol{\Omega}^{+}$ and the unitary matrix $\boldsymbol{Y}$ is the corresponding eigenvector matrix (see Eq. 15.57). Note that according to Eq. (15.52) $\boldsymbol{M}^{RPA}$ is a hermitian matrix provided all of the RPA eigenvalues ω_m are real, which will be supposed in the following.

The RPA energies ω_m comprised by $\boldsymbol{\Omega}^{+}$ derive from the $1p$-$1h$ HF excitations $|\Phi_{ak}\rangle = c_a^{\dagger}c_k|\Phi_0\rangle$ (cf. Eq. 15.8). Accordingly, $\boldsymbol{M}^{RPA}$ is a matrix of elements M_{IJ}^{RPA}, where the indices I and J label $1p$-$1h$ configurations (class 1), such as $I \equiv ak$; $\boldsymbol{f}$ is a matrix of elements $f_{I,rs}$ where the first index I labels $1p$-$1h$ configurations, and the index pair (rs) denotes p-h and h-p components.

The matrix elements of $\boldsymbol{M}^{RPA}$ and $\boldsymbol{f}^{RPA}$ are subject to perturbation expansions

$$\boldsymbol{M}^{RPA} = \boldsymbol{M}_{11}^{(0)} + \boldsymbol{M}_{11}^{(1)} + \boldsymbol{M}_{11}^{(2)} + \dots \tag{15.54}$$

$$\boldsymbol{f}^{RPA} = \boldsymbol{f}_{1}^{(0)} + \boldsymbol{f}_{1}^{(1)} + \boldsymbol{f}_{1}^{(2)} + \dots \tag{15.55}$$

Here the subscripts, as in $\boldsymbol{M}_{11}^{(1)}$, indicate that, in contrast to the ADC procedure for the full polarization propagator, the configuration space is restricted to class 1, that is, the $1p$-$1h$ excitations.

Now the formal PT expansion of the ADC form (15.51) can be compared with the RPA diagrams, more specifically with the Goldstone diagrams contributing to $\boldsymbol{\Pi}^{RPA+}(\omega)$, through successively higher order n. This yields a hierarchy of RPA-ADC(n) schemes, $n = 0, 1, 2, \dots$, consisting of the nth-order expansions

$$\boldsymbol{M}^{RPA}(n) = \sum_{\nu=0}^{n} \boldsymbol{M}_{11}^{(\nu)}$$

$$\boldsymbol{f}^{RPA}(n) = \sum_{\nu=0}^{n} \boldsymbol{f}_{1}^{(\nu)} \tag{15.56}$$

of the secular matrix and the transition amplitude matrix.

For given approximations of $\boldsymbol{M}^{RPA}$ and $\boldsymbol{f}^{RPA}$, the spectral information is obtained by solving the hermitian eigenvalue problem (ADC secular equation)

$$\boldsymbol{M}^{RPA}\boldsymbol{Y} = \boldsymbol{Y}\boldsymbol{\Omega}, \quad \boldsymbol{Y}^{\dagger}\boldsymbol{Y} = \mathbf{1} \tag{15.57}$$

where $\boldsymbol{Y}$ and $\boldsymbol{\Omega}$ denote the eigenvector matrix and the diagonal matrix of eigenvalues, respectively.
The RPA spectroscopic amplitudes are obtained according to

$$(\boldsymbol{X}^{+})^{\dagger} = \boldsymbol{Y}^{\dagger}\boldsymbol{f}^{RPA} \tag{15.58}$$

so that the RPA-ADC transition moments are given by

$$T_m^{RPA} = \boldsymbol{Y}_m^{\dagger}\boldsymbol{f}^{RPA}\boldsymbol{d} \tag{15.59}$$

Here, $\boldsymbol{Y}_m$ denotes the mth eigenvector of $\boldsymbol{M}^{RPA}$.

At the RPA-ADC(n) level, the results of Eqs. (15.57) and (15.58) provide approximations to the full RPA energies and transition moments, being consistent through nth order and expected to converge to the RPA results in the limit $n \to \infty$.

Finally, we take a look at the explicit RPA-ADC expressions in the first- and second-order schemes. Through first order, the RPA diagrams are complete so that the derivation of the first-order RPA-ADC scheme does not differ from that for the polarization propagator in Sect. 14.2. According to Eqs. (14.25, 14.26), the ADC(1) expressions read

$$\begin{aligned} M_{ak,a'k'}^{RPA} &= (\epsilon_a - \epsilon_k)\delta_{aa'}\delta_{kk'} - V_{ak'[a'k]} \\ f_{ak,bl}^{RPA} &= \delta_{ab}\delta_{kl} \\ f_{ak,lb}^{RPA} &= \frac{V_{ab[lk]}}{\epsilon_a + \epsilon_b - \epsilon_k - \epsilon_l} \end{aligned} \tag{15.60}$$

In second order, the RPA diagram (2C) in Fig. 13.3 is one of altogether five Feynman diagrams for the full propagator. There are 24 Goldstone diagrams deriving from (C), of which the 12 diagrams contributing to $\boldsymbol{\Pi}^{+}(\omega)$ are shown in Fig. 14.1. In Sect. 14.2, we have addressed the ADC(2) contributions deriving from these 12 diagrams, and the findings here can directly be transferred to the case of the second-order RPA-ADC scheme. Accordingly, the second-order contribution to $\boldsymbol{M}^{RPA}$ is given by the expression $\boldsymbol{C}^{(C)}$ of Eq. (14.39):

$$\boldsymbol{M}_{ak,a'k'}^{RPA\,(2)} = \boldsymbol{C}_{ak,a'k'}^{(C)} \tag{15.61}$$

The second-order part of $\boldsymbol{f}^{RPA}$ consists of three distinct contributions

$$\boldsymbol{f}^{RPA} = \boldsymbol{f}_1^{(C)} + \boldsymbol{f}_1^{(2,5)} + \boldsymbol{f}_1^{(2,6)} \tag{15.62}$$

where $\boldsymbol{f}_1^{(C)}$ derives from the diagrams (C7)–(C10), while the other two contributions are related to diagrams (C5) and (C11), respectively. The explicit expressions are listed in Appendix A.9.

The unitary transformation $\boldsymbol{Y}$ may be viewed as transforming the excited "RPA states" $|\Psi_m^{RPA}\rangle$ to associated intermediate $1p$-$1h$ states $|\widetilde{\Psi}_{ak}^{RPA}\rangle$, so that

$$Y_{ak,m} = \langle \widetilde{\Psi}_{ak}^{RPA} | \Psi_m^{RPA} \rangle \tag{15.63}$$

Could one possibly construct those intermediate RPA states from CE states $c_a^\dagger c_k |\Psi_0^{RPA}\rangle$ based on a RPA ground state $|\Psi_0^{RPA}\rangle$ yet to be determined? What would the presumed RPA ground state look like, and would it be of any use? In Sect. 16.3, we come back to this issue.

Exercises

15.1 Formulate in the time representation the RPA recursion relation according to Fig. 15.2 for the RPA polarization propagator, $\Pi_{rs,r's'}^{RPA}(t,t')$. Derive Eqs. (15.1)–(15.3) using appropriate Fourier transforms.

15.2 Determine the two eigenvectors of the 2×2 RPA model according to Eqs. (15.21,15.22); verify that the pseudo-orthonormalization scheme (15.17) applies.

15.3 Schematic model:
(a) TDA variant: Consider a manifold of m "elementary excitations" $|\Phi_k\rangle, k = 1, \ldots, m$ with "unperturbed" energies ϵ_k, interacting via a uniform "perturbation," $w = \langle \Phi_k | \hat{U} | \Phi_l \rangle, k, l = 1, \ldots, m$. Inspect the eigenvalue equation for the secular matrix and design a graphical scheme to determine the eigenvalues (excitation energies) e_k. The graphical solution allows for an obvious distinction of $m - 1$ ordinary solutions and a particular ("plasmon") solution.
(b) The model can be simplified even further by supposing that the unperturbed energies are degenerate: $\epsilon_k = \overline{e}, k = 1, \ldots, m$. Determine and analyze the analytical solutions for the eigenvalues and eigenvectors of the particular and ordinary solutions.
(c) RPA variant: Augment the TDA excitations with a corresponding set of "de-excitations," $|\overline{\Phi}_k\rangle, k = 1, \ldots, m$, and suppose the same type of uniform coupling, $w = \langle \Phi_k | \hat{U} | \overline{\Phi}_l \rangle, k, l = 1, \ldots, m,$ between excitations and de-excitations (sub-block $\boldsymbol{B}$ of the RPA secular matrix). Transform the RPA eigenvalue equations in a form allowing for graphical solutions and analyze the various solutions.

15.4 Prove the identity (15.49) for a specific $1p$-$1h$ vector component, say $a\alpha k\alpha$ (where a and k denote spatial orbitals). Bring the left side of Eq. (15.49) into spin-free form and replace the matrix element p_{ak} of $\hat{p}_z$ with $d_{ak} = \langle \phi_a | \hat{z} | \phi_k \rangle$ using the relation $-ip_{ak}\hbar/m_e = (\epsilon_a - \epsilon_k) d_{ak} + \langle \phi_a | [\hat{w}, \hat{z}] | \phi_k \rangle$ (see Eqs. (14.73)–(14.75) and Exercise 14.3).

15.5 In a similar way like in Exercise 15.4, show that the dipole sum rule (15.46) derives directly from the RPA expression (15.44).

References

1. Bohm D, Pines D (1951) Phys Rev 82:625
2. Bohm D, Pines D (1953) Phys Rev 92:609
3. Thouless DJ (1961) Nucl Phys 22:78
4. Dirac PAM (1930) J Proc Camb Philos Soc 26:376
5. Schirmer J, Braunstein M, Lee M-T, McKoy V (2001) In: Becker U, Shirley DA (eds) VUV and soft X-ray photoionization. Plenum Press, New York, p 105
6. Harris RA (1969) Chem Phys 50:3947
7. Amusia MY, Cherepkov NA (1975) Case Stud At Phys 5:47
8. Schirmer J, Mertins F (1996b) J Phys B 29:3559

Part V
A Look at Related Methods

We set out from diagrammatic perturbation theory for propagators, developed a practical approach for deriving systematic higher-order approximation schemes, and finally established a relationship to a wave-function description by identifying the direct ADC schemes as intermediate state representations (ISR) based on correlated excited (CE) states. In the final two chapters we will make contact with two important concepts, both of which can be understood as particular ISR variants: The equation-of-motion (EOM) method establishing an algebraic approach to the electron and other propagators (Chap. 16) and the coupled-cluster (CC) theory for generalized electronic excitations (Chap.17). The intuitive idea underlying the use of CE states is that the excitation, removal or attachment of an electron will affect the initial N-electron ground state to a certain extent, but not totally alter it. Therefore the use of CE basis states conveying the ground-state information should afford some advantage over the simple CI treatment in which every excited state must be constructed from the scratch as an expansion in terms of the uncorrelated HF excitations. However, as we have seen, a proper orthonormalization of the CE states is crucial for these advantages to materialize. So it will be interesting to see how the orthonormalization of the CE states is handled in the context of the EOM and CC methods.

Chapter 16
Algebraic Propagator Methods

Algebraic propagator methods, as opposed to methods based on diagrammatic perturbation theory, have been derived within the so-called superoperator formulation [1–3], in which the respective propagator is written and evaluated in the form of a superoperator resolvent [4]. The resulting secular equations are fully equivalent with those obtained in the context of the EOM method [5, 6], originally devised by Rowe [7] in the field of nuclear physics. To some extent, the EOM formulation is more general than the algebraic propagator approach since the former provides a genuine wave-function representation of the (generalized) excited states in terms of an extended set of CE states. In this chapter, we briefly review the EOM method, emphasizing here in particular its ISR characteristics. For a more comprehensive presentation as well as references to the original literature, we refer to the cited review articles and to Ref. [8]. A direct connection to the algebraic propagator methods is established in Sect. 16.2, while the superoperator formulation is reviewed in Appendix A.8.

16.1 Equations-of-Motion (EOM) Method for $N \pm 1$ Electrons

Depending on the choice of the excitation operators, EOM schemes can be derived for neutral (N-electron) excitations, as well as for generalized excitations of $N \pm 1, N \pm 2, \ldots$ electrons. Exemplarily, we will consider the $(N \pm 1)$-electron EOM scheme in the following. The N-electron case, featuring some particularities, will be addressed in Sect. 16.3.

In the EOM formulation, a general cationic state (not necessarily an energy eigenstate) $|\Psi_n^{N-1}\rangle$ is written in the form

J. Schirmer, *Many-Body Methods for Atoms, Molecules and Clusters*, Lecture Notes in Chemistry 94, https://doi.org/10.1007/978-3-319-93602-4_16

$$|\Psi_n^{N-1}\rangle = \hat{\Omega}_n|\Psi_0\rangle \tag{16.1}$$

where $\hat{\Omega}_n$ denotes an associated excitation operator yet to be determined, and $|\Psi_0\rangle$ is the exact ground state of the N-electron system. In addition to Eq. (16.1), the operator $\hat{\Omega}_n$ is subject to the requirement

$$\hat{\Omega}_n^\dagger|\Psi_0\rangle = 0 \tag{16.2}$$

which is commonly referred to as "killer condition" (KC). The excitation operators are assumed to be linear expansions

$$\hat{\Omega}_n = \sum_J X_{Jn}\hat{O}_J \tag{16.3}$$

with regard to a manifold of "basis" operators $\hat{O}_J$. Apparently, a suitable choice of such basis operators is the set (11.2) of physical $(N-1)$-electron operators,

$$\{\hat{C}_J^{N-1}\} = \left\{c_k;\ c_a^\dagger c_k c_l, k < l;\ c_a^\dagger c_b^\dagger c_j c_k c_l,\ a < b, j < k < l;\ \ldots\right\} \tag{16.4}$$

As discussed in Sect. 11.1, the CE states $\hat{C}_J^{N-1}|\Psi_0\rangle$ obtained by applying these operators to the exact ground state form a complete set of $(N-1)$-electron states so that Eq. (16.1) can be fulfilled. However, the operator set (16.4) is not sufficient to solve Eq. (16.2) as well. To satisfy both EOM conditions, the operator set has to be augmented with "unphysical" operators

$$\{\hat{C}_J^{N+1\dagger}\} = \left\{c_a;\ c_i^\dagger c_b c_a, a < b;\ c_j^\dagger c_i^\dagger c_c c_b c_a,\ i < j, a < b < c;\ \ldots\right\} \tag{16.5}$$

obtained as the hermitian conjugates of physical $(N+1)$-electron excitation operators $\hat{C}_J^{N+1}$. In the combined operator set,

$$\{\hat{O}_I\} \equiv \{\hat{C}_J^{N-1}\} \cup \{\hat{C}_J^{N+1\dagger}\} \tag{16.6}$$

also referred to as Manne–Dalgaard basis, the capital letters used as subscripts denote both $(N-1)$- and $(N+1)$-electron configurations, $I \in \{N-1, N+1\}$. As first shown by Dalgaard [9], there is a unique solution of Eqs. (16.1), (16.2) in the form of an expansion (16.3) based on the Manne–Dalgaard operators (16.6).

Supplied with a suitable operator basis, we may now turn to the secular equations needed to determine the expansion coefficients X_{Jn} for energy eigenstates. The Schrödinger equation for the ionic state (16.1) can be combined with that for the ground state as follows:

$$[\hat{H}, \hat{\Omega}_n]|\Psi_0\rangle = \omega_n\hat{\Omega}_n|\Psi_0\rangle \tag{16.7}$$

where $\omega_n = E^{N-1} - E_0$ denotes the excitation (ionization) energy of the cationic energy eigenstate $\hat{\Omega}_n|\Psi_0\rangle$. In a related way, the killer condition can be written in the form

$$\langle\Psi_0|[\hat{H}, \hat{\Omega}_n] = 0 \tag{16.8}$$

Now we "multiply" (form a scalar product) the first equation from the left with $\langle\Psi_0|\hat{O}_I^\dagger$ and the second equation from the right with $\hat{O}_I^\dagger|\Psi_0\rangle$,

$$\langle\Psi_0|\hat{O}_I^\dagger[\hat{H}, \hat{\Omega}_n]|\Psi_0\rangle = \omega_n\langle\Psi_0|\hat{O}_I^\dagger\hat{\Omega}_n|\Psi_0\rangle \tag{16.9}$$

$$\langle\Psi_0|[\hat{H}, \hat{\Omega}_n]\hat{O}_I^\dagger|\Psi_0\rangle = 0 \tag{16.10}$$

These two equations can be added to give

$$\langle\Psi_0|\{\hat{O}_I^\dagger, [\hat{H}, \hat{\Omega}_n]\}|\Psi_0\rangle = \omega_n\langle\Psi_0|\{\hat{O}_I^\dagger, \hat{\Omega}_n\}|\Psi_0\rangle \tag{16.11}$$

where $\{\hat{A}, \hat{B}\} = \hat{A}\hat{B} + \hat{B}\hat{A}$ denotes the anticommutator of the operators $\hat{A}$, $\hat{B}$. Note that the killer condition has been used on the right-hand side to replace $\langle\Psi_0|\hat{O}_I^\dagger\hat{\Omega}_n|\Psi_0\rangle$ with $\langle\Psi_0|\{\hat{O}_I^\dagger, \hat{\Omega}_n\}|\Psi_0\rangle$. Inserting the expansion (16.3) for $\hat{\Omega}_n$ yields the secular equations,

$$\sum_J (\mathsf{M}_{IJ} - \omega_n S_{IJ})X_{Jn} = 0 \tag{16.12}$$

where

$$\mathsf{M}_{IJ} = \langle\Psi_0|\{\hat{O}_I^\dagger, [\hat{H}, \hat{O}_J]\}|\Psi_0\rangle \tag{16.13}$$

$$S_{IJ} = \langle\Psi_0|\{\hat{O}_I^\dagger, \hat{O}_J\}|\Psi_0\rangle \tag{16.14}$$

are the elements of the EOM secular matrix $\mathbf{M}$ and the metric (or overlap) matrix $\boldsymbol{S}$, respectively. Using matrix notation, Eq. (16.12) takes on the form

$$\mathbf{M}\boldsymbol{X} = \boldsymbol{S}\boldsymbol{X}\boldsymbol{\Omega} \tag{16.15}$$

The normalization of the eigenvectors is obtained according to

$$\boldsymbol{X}^\dagger\boldsymbol{S}\boldsymbol{X} = \mathbf{1} \tag{16.16}$$

which follows from the orthonormalization of the final states,

$$\langle\Psi_0|\{\hat{\Omega}_n^\dagger, \hat{\Omega}_m\}|\Psi_0\rangle = \delta_{nm} \tag{16.17}$$

invoking here again the killer condition.

In the EOM secular equations, the operators $\hat{O}_I$ and $\hat{O}_I^\dagger$ are treated on an equal footing, and since the latter generate $(N+1)$-electron states when acting on $|\Psi_0\rangle$, one will expect that the set of EOM solutions comprises also the energy eigenstates of the $(N+1)$-electron system. In fact, our derivation could have set out from $(N+1)$-electron states. Using here the hermitian adjoint operator basis

$$\{\hat{O}_J^\dagger\} \equiv \{\hat{C}_J^{N+1}\} \cup \{\hat{C}_J^{N-1\dagger}\} \tag{16.18}$$

leads again to Eqs. (16.12)–(16.14). This shows that the EOM secular problem provides a unified treatment of both $(N-1)$- and $(N+1)$-electron excitations. The EOM eigenvalues relate to electron affinities and ionization energies,

$$\omega_n = \begin{cases} E_0 - E_n^{N+1}, & n \in \{N+1\} \\ E_n^{N-1} - E_0, & n \in \{N-1\} \end{cases} \tag{16.19}$$

The general orthonormalization conditions (16.17) apply to the entire set of $(N \pm 1)$-electron solutions. One may note that the anticommutators $\{\hat{O}_I^\dagger, \hat{O}_J\}$ in Eq. (16.14) vanish if $I \in \{N+1\}$ and $J \in \{N-1\}$ (or vice versa). Accordingly, $\boldsymbol{S}$ has an obvious block structure,

$$\boldsymbol{S} = \begin{pmatrix} \boldsymbol{S}^+ & \boldsymbol{0} \\ \boldsymbol{0} & \boldsymbol{S}^- \end{pmatrix} \tag{16.20}$$

where $\boldsymbol{S}^+$ and $\boldsymbol{S}^-$ denote the sub-blocks associated with $I, J \in \{N+1\}$ and $I, J \in \{N-1\}$, respectively.

The state representations of the $(N \pm 1)$-electron solutions are given by

$$\langle \Psi_m^{N+1}| = \sum_J \langle \Psi_0|\hat{O}_J\, X_{Jm}^*, \quad m \in \{N+1\} \tag{16.21}$$

$$|\Psi_n^{N-1}\rangle = \sum_J X_{Jn}\, \hat{O}_J|\Psi_0\rangle, \quad n \in \{N-1\} \tag{16.22}$$

while the associated KC relations can be written as

$$\sum_J X_{Jm}\hat{O}_J|\Psi_0\rangle = 0, \quad m \in \{N+1\} \tag{16.23}$$

$$\sum_J \langle \Psi_0|\hat{O}_J X_{Jn}^* = 0, \quad n \in \{N-1\} \tag{16.24}$$

Transition moments, such as the spectroscopic factors (3.20), are obtained according to

$$x_p^{(n)} = \langle \Psi^{N-1}|c_p|\Psi_0\rangle = \sum_J X_{Jn}^* \langle \Psi_0|\{\hat{O}_J^\dagger, c_p\}|\Psi_0\rangle, \; n \in \{N-1\} \tag{16.25}$$

as the scalar product of the respective eigenvector and a vector of of transition moments for the basis operators, $\langle \Psi_0 | \{ \hat{O}_J^\dagger, c_p \} | \Psi_0 \rangle$. Here anticommutators can be used as a result of the KC relation (16.24).

The explicit forms (16.21), (16.22) of the final states make apparent that the EOM qualifies as a particular ISR approach, in which the treatment of the $(N-1)$- and $(N+1)$-electron systems is combined. In fact, the EOM secular equations can be obtained explicitly as a Fock-space state representation of a generalized hamiltonian, as will be demonstrated in the ensuing Sect. 16.2.

Orthonormalization and Approximation Schemes

We note that the EOM secular matrix $\mathbf{M}$ is hermitian, and the likewise hermitian overlap matrix $\boldsymbol{S}$ consists of two positive definite sub-blocks. Accordingly, the eigenvalue problem can be transformed into a regular hermitian one. This of course touches upon the issue of the appropriate orthonormalization. As we have seen in Sect. 12.1, the excitation class Gram-Schmidt orthonormalization of the CE states is crucial with regard to the canonical order relations and the separability of the secular matrices. This finding applies also to the EOM schemes. In the given form, based on non-orthonormal CE states, the EOM secular matrices are neither "canonical" nor separable. This can be seen by inspecting, for example, the matrix elements of the $1h$/$3h$-$2p$ coupling blocks of $\boldsymbol{S}$ and $\mathbf{M}$, displaying non-vanishing first-order contributions (Exercise 16.1),

$$\mathbf{M}^{(1)}_{i,abjkl} = \epsilon_i \left(-\delta_{ij} \frac{V_{kl[ab]}}{\epsilon_a + \epsilon_b - \epsilon_k - \epsilon_l} + \delta_{ik} \frac{V_{jl[ab]}}{\epsilon_a + \epsilon_b - \epsilon_j - \epsilon_l} - \delta_{il} \frac{V_{jk[ab]}}{\epsilon_a + \epsilon_b - \epsilon_j - \epsilon_k} \right) \tag{16.26}$$

Similar first-order contributions arise in the overlap matrix element, $S^{(1)}_{i,abjkl}$, which indicates that the problem can be traced to the fact that the EOM basis states are not orthonormal. Again, symmetric orthonormalization based on $\boldsymbol{S}^{-1/2}$ is not expedient, but the ECO-Gram-Schmidt procedure described in Chap. 11 can be extended to the EOM case [8, 10], which procures the desired properties.

In its general form, the EOM method is an exact approach to $(N \pm 1)$-electron states and energies, though obviously not a practicable one. To devise viable approximation schemes, one has to truncate the operator expansion manifold in appropriate ways and replace the exact ground state with consistent finite PT expansions in the expressions (16.13), (16.14) for the secular matrix elements. An example is the consistent third-order EOM approximation scheme discussed in Ref. [6]. Here the explicit operator manifold comprises the $1h$, $1p$, $2h$-$1p$, and $2p$-$1h$ excitation classes; the PT expansions of the secular matrix elements in the respective sub-blocks are such that the energies of the $1h$ and $1p$ main states are treated consistently through third order.

Like the Dyson equation discussed in Chap. 8, the EOM method entangles the treatment of $N-1$ electrons with that of $N+1$ electrons. Obviously, this procedure is to be seen as a mathematical device rather than being physically motivated. The advantages and disadvantages of the $(N \pm 1)$-electron coupling with regard to comparable separated procedures have been addressed in Chap. 10. The coupled schemes require lower-order PT expansions for the secular matrix elements than in comparable separate schemes. However, this is contrasted by the unfavorably large configuration spaces, the size being the sums of the $(N-1)$- and $(N+1)$-electron subspaces. Moreover, the EOM solutions of interest lie in the middle of the spectrum, which is not expedient for the iterative diagonalization routines used in the large matrix eigenvalue problem.

16.2 A State Representation of the EOM Secular Equations

An alternative derivation of the EOM secular equations based explicitly on a state representation of a generalized hamiltonian was presented in Ref. [8], and a brief review of that approach should be instructive. For a convenient notation, we here suppose that the underlying one-particle basis functions (spin-orbitals) are real functions. This is hardly a restriction, since, in the absence of magnetic fields, the hamiltonian is real so that the HF orbitals can always be chosen as real functions. An extension to the case of complex orbitals [11] is addressed in Exercises 16.2 and 16.3.

We consider $(N\pm1)$-electron Fock-space states defined as

$$|\Theta_I\rangle = (\hat{O}_I^\dagger + \hat{O}_I)|\Psi_0\rangle, \quad I \in \{N+1, N-1\} \tag{16.27}$$

where $\hat{O}_I$ are the Manne–Dalgaard operators (16.6). By construction, these states are superpositions of states of $N+1$ and $N-1$ electrons,

$$|\Theta_I\rangle = |\Theta_I^{N+1}\rangle + |\Theta_I^{N-1}\rangle \tag{16.28}$$

First we establish that the overlap matrix elements $\langle\Theta_I|\Theta_J\rangle$ can be identified with the EOM overlap matrix elements S_{IJ}:

$$\begin{aligned}\langle\Theta_I|\Theta_J\rangle &= \langle\Psi_0|(\hat{O}_I^\dagger + \hat{O}_I)(\hat{O}_J^\dagger + \hat{O}_J)|\Psi_0\rangle \\ &= \langle\Psi_0|\hat{O}_I^\dagger\hat{O}_J + \hat{O}_I\hat{O}_J^\dagger|\Psi_0\rangle \\ &= \langle\Psi_0|\{\hat{O}_I^\dagger, \hat{O}_J\}|\Psi_0\rangle = S_{IJ}\end{aligned} \tag{16.29}$$

In the second line, terms of the type $\langle\Theta_I^{N+1}|\Theta_J^{N-1}\rangle$ have been discarded, as they vanish according to the Fock-space extension of the scalar product to states of different particle numbers (see Eq. 2.2). Moreover, here we use the identity $\langle\Psi_0|\hat{O}_I\hat{O}_J^\dagger|\Psi_0\rangle = \langle\Psi_0|\hat{O}_J\hat{O}_I^\dagger|\Psi_0\rangle$, which is valid under the assumption that the spin-orbitals underlying the operators are real functions.

The states (16.27) form a set of linear independent, complete states in the $(N\pm1)$-subspace of the Fock space. This is a consequence of the linear independence and completeness of the constituents $|\Theta_I^{N+1}\rangle$ and $|\Theta_I^{N-1}\rangle$ in the respective subspaces. For example, the CE states $|\Psi_J^0\rangle = \hat{O}_J|\Psi_0\rangle$, $J \in \{N-1\}$ are linear independent (and complete) in the $(N-1)$-electron Hilbert space (see Sect. 11.1). Consequently, the subset of Θ-states, $|\Theta_I\rangle$, $I \in \{N-1\}$, is linear independent in the Fock space too. The linear independence of the full set of Θ-states, $I \in \{N+1, N-1\}$, follows from the orthogonality of the two subsets according to Eq. (16.20).

We now may expand an arbitrary $(N-1)$-electron state, $|\Psi_u^{N-1}\rangle$, in terms of the $|\Theta_I\rangle$-states as follows:

$$|\Psi_u^{N-1}\rangle = \sum_{I,J} |\Theta_I\rangle (\boldsymbol{S}^{-1})_{IJ} \langle\Theta_J|\Psi_u^{N-1}\rangle = \sum_I X_{Iu}\, |\Theta_I\rangle \tag{16.30}$$

The expansion coefficients used in the second equation are given by

$$X_{Iu} = \sum_J (\boldsymbol{S}^{-1})_{IJ} \langle\Psi_0|\hat{O}_J^\dagger|\Psi_u^{N-1}\rangle \tag{16.31}$$

Since $|\Psi_u^{N-1}\rangle$ is an $(N-1)$-electron state, Eq. (16.30) decomposes into two separate equations,

$$|\Psi_u^{N-1}\rangle = \sum_I X_{Iu}\, \hat{O}_I|\Psi_0\rangle \tag{16.32}$$

$$0 = \sum_I X_{Iu}\, \hat{O}_I^\dagger|\Psi_0\rangle \tag{16.33}$$

This can be seen as a direct proof for the existence of a unique solution of Eqs. (16.1), (16.2) based on the operator manifold (16.6).

Next we show that the EOM secular matrix $\mathbf{M}$ can be derived as the representation of a modified hamiltonian in terms of the $|\Theta_I\rangle$ states. We consider the operator

$$\hat{\mathcal{H}} = (\hat{H} - E_0)(N - \hat{N}) \tag{16.34}$$

where $\hat{N} = \sum_p c_p^\dagger c_p$ is the particle number operator. Obviously, the factor $(N - \hat{N})$ does not affect the eigenstates of $\hat{H} - E_0$, but merely changes the sign of the eigenvalues for the $(N+1)$-electron solutions. The matrix elements of $\hat{\mathcal{H}}$ can be evaluated as follows:

$$\begin{aligned}
\mathcal{H}_{IJ} = \langle\Theta_I|\hat{\mathcal{H}}|\Theta_J\rangle =& \langle\Theta_I|(\hat{H} - E_0)(N - \hat{N})|\Theta_J\rangle \\
=& \langle\Psi_0|\hat{O}_I^\dagger(\hat{H} - E_0)\hat{O}_J|\Psi_0\rangle - \langle\Psi_0|\hat{O}_I(\hat{H} - E_0)\hat{O}_J^\dagger|\Psi_0\rangle \\
=& \langle\Psi_0|\hat{O}_I^\dagger[\hat{H}, \hat{O}_J]|\Psi_0\rangle + \langle\Psi_0|[\hat{H}, \hat{O}_J]\hat{O}_I^\dagger|\Psi_0\rangle \\
=& \langle\Psi_0|\{\hat{O}_I^\dagger, [\hat{H}, \hat{O}_J]\}|\Psi_0\rangle = \mathsf{M}_{IJ}
\end{aligned} \tag{16.35}$$

In the second line, contributions of the type $\langle N-1|N+1\rangle$ and $\langle N+1|N-1\rangle$ have been discarded; in the third line, the identity

$$\langle\Psi_0|\hat{O}_J(\hat{H} - E_0)\hat{O}_I^\dagger|\Psi_0\rangle = \langle\Psi_0|\hat{O}_I(\hat{H} - E_0)\hat{O}_J^\dagger|\Psi_0\rangle$$

has been used, supposing again that the underlying one-particle states (and $\hat{H}$) are real. The latter result, together with the overlap relations (16.29), establishes that the EOM secular equations are identical with the eigenvalue equations of the modified hamiltonian (16.34) in the $(N\pm1)$-electron subspace of the Fock space.

An eigenstate of $\hat{\mathcal{H}}$ can be written as

$$|\Psi_n\rangle = \sum_I X_{In}|\Theta_I\rangle = \sum_I X_{In}(\hat{O}_I^\dagger + \hat{O}_I)|\Psi_0\rangle \tag{16.36}$$

where X_{In} are the components of the corresponding eigenvector of $\mathbf{M}$. Since $|\Psi_n\rangle$ is either an $(N+1)$-electron or an $(N-1)$-electron state, the latter equation splits into two separate equations, reading for an $(N-1)$-electron solution, $n \in \{N-1\}$,

$$|\Psi_n\rangle = |\Psi_n^{N-1}\rangle = \sum_I X_{In}\hat{O}_I|\Psi_0\rangle \tag{16.37}$$

$$0 = \sum_I X_{In}\hat{O}_I^\dagger|\Psi_0\rangle \tag{16.38}$$

which of course just reproduces the corresponding Eqs. (16.22), (16.24).

Relation to the Electron Propagator

The formulation of the EOM secular problem in terms of a state representation of the modified hamiltonian $\hat{\mathcal{H}}$ allows for a direct approach to the electron propagator in the form of Eqs. (3.24), (3.25). This is established by observing that the propagator matrix elements $G_{pq}(\omega)$ can be written as

$$G_{pq}(\omega) = \langle\Theta_p|(\omega + \hat{\mathcal{H}})^{-1}|\Theta_q\rangle \tag{16.39}$$

that is, as matrix elements of the resolvent operator $(\omega + \hat{\mathcal{H}})^{-1}$ in terms of the $1h/1p$ states, $|\Theta_p\rangle = (c_p^\dagger + c_p)|\Psi_0\rangle$. This can be verified as follows:

$$\begin{aligned}\langle\Theta_p|(\omega + \hat{\mathcal{H}})^{-1}|\Theta_q\rangle =&\langle\Psi_0|c_p(\omega + \hat{\mathcal{H}})^{-1}c_q^\dagger|\Psi_0\rangle + \langle\Psi_0|c_q^\dagger(\omega + \hat{\mathcal{H}})^{-1}c_p|\Psi_0\rangle\\ =&\langle\Psi_0|c_p(\omega - \hat{H} + E_0)^{-1}c_q^\dagger|\Psi_0\rangle + \langle\Psi_0|c_q^\dagger(\omega + \hat{H} - E_0)^{-1}c_p|\Psi_0\rangle\\ =&G_{pq}^+(\omega) + G_{pq}^-(\omega)\end{aligned} \tag{16.40}$$

As in the derivation of Eq. (16.29), only pure $(N+1)$- or $(N-1)$-electron matrix elements need to be retained; in the first line, we have used that

$$\langle\Psi_0|c_p^\dagger(\omega + \mathcal{H})^{-1}c_q|\Psi_0\rangle = \langle\Psi_0|c_q^\dagger(\omega + \hat{\mathcal{H}})^{-1}c_p|\Psi_0\rangle$$

which supposes a real-valued one-particle representation. Note that the operator $\hat{\mathcal{H}}$ and the corresponding resolvent simplify if applied to an $(N+1)$- or $(N-1)$-electron state, e.g., $\hat{\mathcal{H}}c_q^\dagger|\Psi_0\rangle = (-\hat{H} + E_0)c_q^\dagger|\Psi_0\rangle$. The representation (16.39) means that the EOM secular equations, which according to Eqs. (16.29), (16.35) represent secular equations for $\hat{\mathcal{H}}$ and, in extension, the resolvent operator $(\omega + \hat{\mathcal{H}})^{-1}$, allow one to determine the poles and residues of the electron propagator. Another way of relating the EOM secular problem to algebraic propagator equations is the superoperator formalism reviewed in Appendix A.8.

16.3 EOM Treatment of N-Electron Excitations

While the derivation of the EOM method for N-electron excitations is essentially analogous to the $(N \pm 1)$-case, there emerge some distinct features which need to be addressed.

Excited states $|\Psi_n\rangle$ of the N-electron system are represented in the form of an operator expansion

$$\hat{\Omega}_n = \sum_J X_{Jn}\hat{O}_J \tag{16.41}$$

acting on the exact ground state,

$$|\Psi_n\rangle = \hat{\Omega}_n|\Psi_0\rangle \tag{16.42}$$

In addition, there is the killer condition,

$$\hat{\Omega}_n^\dagger|\Psi_0\rangle = 0 \tag{16.43}$$

which here also implies that the excited state is orthogonal to the ground state, $\langle\Psi_n|\Psi_0\rangle = 0$.

The Manne–Dalgaard operator basis required for a unique solution of Eqs. (16.42), (16.43),

$$\{\hat{O}_J\} = \left\{c_a^\dagger c_k;\, c_a^\dagger c_b^\dagger c_k c_l,\; a < b, k < l;\; \dots\right\} \cup \left\{c_k^\dagger c_a;\, c_l^\dagger c_k^\dagger c_b c_a,\; a < b, k < l;\; \dots\right\} \tag{16.44}$$

is formed by the set (14.10) of physical N-electron excitation operators, $\{\hat{C}_I\}$, and the set $\{\hat{C}_I^\dagger\}$ of their hermitian adjoints, also referred to as de-excitation operators. Like the $1p$-$1h, 2p$-$2h, \dots,$ classes of physical excitations $(I \in \{+\})$, one may distinguish classes $1h$-$1p$, $2h$-$2p, \dots$ of de-excitation operators, $(I \in \{-\})$.

In the derivation of the secular equations one proceeds along Eqs. (16.7)–(16.10), but then uses commutators rather than anticommutators in Eq. (16.11), as a consequence of subtracting Eq. (16.10) from Eq. (16.9) The resulting secular equations read

$$\mathbf{M}X = SX\Omega \tag{16.45}$$

with the secular matrices given by

$$\mathbf{M}_{IJ} = \langle \Psi_0 | [\hat{O}_I^\dagger, [\hat{H}, \hat{O}_J]] | \Psi_0 \rangle \tag{16.46}$$

$$S_{IJ} = \langle \Psi_0 | [\hat{O}_I^\dagger, \hat{O}_J] | \Psi_0 \rangle \tag{16.47}$$

Formally, Eq. (16.45) looks like the secular equation (16.15) in the $(N \pm 1)$-case, but, in fact, there is a fundamental difference. The partitioning of the secular matrices $\mathbf{M}$ and S with regard to excitation ($\{+\}$) and de-excitation ($\{-\}$) configurations reveals an RPA-type structure,

$$\mathbf{M} = \begin{pmatrix} A & B \\ B^\dagger & A^* \end{pmatrix}, \quad S = \begin{pmatrix} S^+ & \mathbf{0} \\ \mathbf{0} & -(S^+)^* \end{pmatrix} \tag{16.48}$$

While $\mathbf{M}$ itself is hermitian, the sign structure of the S matrix indicates that the N-electron secular equation (16.45) represents a pseudo-eigenvalue problem of RPA-type. In Sect. 15.1, the mathematical features of the RPA pseudo-eigenvalue problem have been discussed at length, and many of the above findings also apply to the more general N-electron EOM secular problem.

The solution manifold of Eq. (16.45) comprises two interrelated sets, namely the excitation solutions, $m \in \{+\}$, and de-excitation solutions, $m \in \{-\}$. For any excitation solution, $n \in \{+\}$, with the eigenvalue

$$\omega_n = E_n - E_0, \tag{16.49}$$

there is a corresponding de-excitation solution, $\overline{n} \in \{-\}$, of negative energy,

$$\omega_{\overline{n}} = -\omega_n, \tag{16.50}$$

The respective eigenvectors, X_n and $X_{\overline{n}}$, are interrelated as in Eqs. (15.12), (15.13). Obviously, the de-excitation solutions can be regarded as redundant since they convey the same physical information as the excitation solutions.

The EOM eigenvectors satisfy the pseudo-orthonormalization relations

$$X^\dagger S X = \begin{pmatrix} \mathbf{1} & \mathbf{0} \\ \mathbf{0} & -\mathbf{1} \end{pmatrix} \tag{16.51}$$

For the excitation solutions, $n \in \{+\}$, this is consistent with the usual orthonormalization of the energy eigenstates,

$$\langle \Psi_n | \Psi_m \rangle = \langle \Psi_0 | [\hat{\Omega}_n^\dagger, \hat{\Omega}_m] | \Psi_0 \rangle = \delta_{nm}, \quad n, m \in \{+\} \tag{16.52}$$

where the first equation is a consequence of the KC relation (16.43). In the case of the de-excitation solutions, $n \in \{-\}$, where $\hat{\Omega}_n = \sum X_{Jn}\hat{O}_J^\dagger$, the KC relation eliminates the first term of the $[\hat{\Omega}_n^\dagger, \hat{\Omega}_m]$ commutator,

$$\langle\Psi_0|[\hat{\Omega}_n^\dagger, \hat{\Omega}_m]|\Psi_0\rangle = -\langle\Psi_0|\hat{\Omega}_m\hat{\Omega}_n^\dagger|\Psi_0\rangle = -\delta_{nm}, \ \ n, m \in \{-\} \tag{16.53}$$

which explains the minus sign in the "normalization" of the de-excitation eigenvectors.

The state representation of an excitation solution, $n \in \{+\}$, explicitly reads

$$|\Psi_n\rangle = \sum_I X_{In}\hat{O}_I|\Psi_0\rangle = \sum_{I\in\{+\}} X_{In}\hat{C}_I|\Psi_0\rangle + \sum_{I\in\{-\}} X_{In}\hat{C}_{\overline{I}}^\dagger|\Psi_0\rangle \tag{16.54}$$

The two terms in the second equation relate to excitation and de-excitation components, respectively, of the eigenvector $\boldsymbol{X}_n$; note that in the second term the subscripts $\overline{I}$ are related to $I \in \{-\}$ as implied in (16.44), e.g., $\overline{ka} \equiv ak$. In a similar way, the KC relation (16.43) can be written as

$$0 = \sum_I X_{In}^*\hat{O}_I^\dagger|\Psi_0\rangle = \sum_{I\in\{+\}} X_{In}^*\hat{C}_I^\dagger|\Psi_0\rangle + \sum_{I\in\{-\}} X_{In}^*\hat{C}_{\overline{I}}|\Psi_0\rangle \tag{16.55}$$

Transition moments (for excitation solutions) are obtained according to

$$T_m = \langle\Psi_m|\hat{D}|\Psi_0\rangle = \langle\Psi_0|[\hat{\Omega}_m^\dagger, \hat{D}]|\Psi_0\rangle = \sum_I X_{Im}^* D_I \tag{16.56}$$

where the use of the commutator is justified because of the KC relation (16.55); D_I are transition matrix elements for the basis states,

$$D_I = \langle\Psi_0|[\hat{O}_I^\dagger, \hat{D}]|\Psi_0\rangle \tag{16.57}$$

In a similar way, properties of excited states and, more general, transition moments for different excited states can be obtained according to

$$T_{mn} = \langle\Psi_m|\hat{D}|\Psi_n\rangle = \langle\Psi_0|[\hat{\Omega}_m^\dagger, \hat{D}\hat{\Omega}_n]|\Psi_0\rangle = \sum_{I,J} X_{Im}^* X_{Jn} D_{IJ} \tag{16.58}$$

Here the basis state matrix elements are given by

$$D_{IJ} = \langle\Psi_0|[\hat{O}_I^\dagger, \hat{D}\hat{O}_J]|\Psi_0\rangle \tag{16.59}$$

The RPA and Other EOM Approximations

Approximation schemes to the (in principle) exact EOM equations can be devised by truncating the excitation operator manifold and adopting finite PT expansions for the ground state in a consistent way. Of special interest is the RPA scheme, which we have extensively reviewed in Chap. 15. In fact, the RPA secular equations (15.5), (15.6) result from Eqs. (16.46), (16.47) upon restricting the EOM operator manifold to the $1p$-$1h$ and $1h$-$1p$ configurations and replacing $|\Psi_0\rangle$ with the HF ground state $|\Phi_0\rangle$ (see Exercise 16.4). The derivation as an EOM approximation supplies the RPA solutions with state representations according to Eqs. (16.54), (16.55); a brief inspection of these RPA states should be of interest.

As in Sect. 15.1, we consider the RPA solution for a single excitation, $a \leftarrow k$, According to Eq. (16.54), the related state representation derives from

$$|\Psi_{ak}^{RPA}\rangle \sim \sum_{b,l} \left(X_{bl,ak} c_b^\dagger c_l + X_{lb,ak} c_l^\dagger c_b \right) |\Psi_0\rangle \tag{16.60}$$

where $X_{rs,ak}$ denote the components of the RPA eigenvector, the zeroth-order contribution being $X_{rs,ak}^{(0)} = \delta_{ra}\delta_{sk}$. This expression combines an approximate excitation operator, $\hat{\Omega}_{ak}^{RPA}$, with the exact ground state, $|\Psi_0\rangle$, and one may ask whether there is more consistent approximation for the latter. To address this question, it is useful to inspect the KC relations (16.55) for the RPA excitation,

$$0 \sim \sum_{b,l} \left(X_{bl,ak}^* c_l^\dagger c_b + X_{lb,ak}^* c_b^\dagger c_l \right) |\Psi_0\rangle \tag{16.61}$$

Using first-order PT expansions, $|\Psi_0\rangle = |\Phi_0\rangle + |\Psi_0^{(1)}\rangle + \ldots$, and $X_{rs,ak} = \delta_{ra}\delta_{sk} + X_{rs,ak}^{(1)} + \ldots$, the KC expression can be expanded through first order,

$$c_k^\dagger c_a |\Psi_0^{(1)}\rangle + \sum_{b,l} X_{lb,ak}^{(1)*} c_b^\dagger c_l |\Phi_0\rangle + O(2) = 0 + O(2) \tag{16.62}$$

In zeroth order, the KC relation is trivially fulfilled, and also the two first-order terms cancel each other, as can be seen by comparing $X_{lb,ak}^{(1)}$ (see Eq. 15.40) and the $2p$-$2h$ coefficients in $|\Psi_0^{(1)}\rangle$. Obviously, the RPA excitations satisfy the KC relations consistently through first order, which also shows that the first-order ground state, $|\Phi_0\rangle + |\Psi_0^{(1)}\rangle$, is consistent with the RPA level of theory. One may try to go beyond the first-order ground state and devise an RPA ground state, $|\Psi_0^{RPA}\rangle$, being consistent at second and higher order, but since the RPA errors are of first and second order in the states and excitation energies, respectively, the relevance of such constructs is questionable.

For the RPA state (16.60), the PT expansion through first order reads

$$|\Psi_{ak}^{RPA}\rangle = c_a^\dagger c_k |\Phi_0\rangle + \sum_{b,l} X_{bl,ak}^{(1)} c_b^\dagger c_l |\Phi_0\rangle + c_a^\dagger c_k |\Psi_0^{(1)}\rangle + O(2) \tag{16.63}$$

There are two first-order terms, of which the former simply accounts for the mixing with other $1p$-$1h$ excitations. The second term correctly describes the (first-order) admixture of $3p$-$3h$ configurations in the exact final state. This makes explicit what was inferred in Sect. 15.1 from the PT analysis of the excitation energy, namely that the RPA accounts for the effect of the $3p$-$3h$ admixtures.

Furthermore, the first-order state expansion reveals the main shortcoming of the RPA, that is, the neglect of $2p$-$2h$ admixtures. As a result, the RPA does not account for the important relaxation and polarization effects (see Sect. 15.1). For a more appropriate treatment of excitations in the N-electron system, one has to include the $2p$-$2h$ excitation operators together with their $2h$-$2p$ de-excitation counterparts, as is the case in the second-order polarization propagator approximation (SOPPA) scheme [12, 13].

The considerations in Sect. 16.1 concerning the order relations and the separability of the EOM secular matrices apply to the N-electron case as well. In the form of Eqs. (16.46), (16.47), $\mathbf{M}$ is neither canonical nor separable, and it requires again an orthonormalization procedure of ECO-Gram-Schmidt type to afford the latter properties [8].

The N-electron EOM treatment features a non-hermitian secular problem combining physical excited configurations and an equally large manifold of unphysical de-excitation configurations. As in the $(N \pm 1)$-electron case above, one may weigh advantages and disadvantages of the EOM concept as compared to the plain ISR schemes of the type presented in Chap. 14. Besides the size of the EOM secular matrix, being here exactly twice the size of comparable plain ISR schemes, the fact that the EOM treatment involves a non-hermitian RPA-type secular problem constitutes another disadvantage.

Exercises

16.1 Verify the expression (16.26) for the first-order contribution to the EOM secular matrix element $M_{i,abjkl}$; evaluate the analogous contribution, $S^{(1)}_{i,abjkl}$, for the EOM overlap matrix.

16.2 Field operators in coordinate space representation:
(a) Revisit the abstract definition (2.3) of the creation operator $c_q^\dagger$ and "translate" it to the coordinate space representation. Let $\Psi(\boldsymbol{\xi}_1, \ldots, \boldsymbol{\xi}_N)$ denote an antisymmetrized and normalized wave function of N electrons (N even). Write the $(N+1)$-electron wave functions $c_q^\dagger \Psi(\boldsymbol{\xi}_1, \ldots, \boldsymbol{\xi}_N)$ in terms of a properly antisymmetrized linear combination of products of the type $\Psi\phi_q$.
(b) Write $c_q \Psi(\boldsymbol{\xi}_1, \ldots, \boldsymbol{\xi}_N)$ as an expansion of $(N-1)$-electron basis states Φ_n in analogy to Eq. (2.7).
(c) Let ϕ_q be a complex-valued orbital. The coordinate space representation according to (a) and (b) allows one to define "complex" field operators, $c_q^{*\dagger}$ and c_q^*. Verify that $c_q^{*\dagger}\Psi = (c_q^\dagger\Psi)^*$ and $c_q^*\Psi = (c_q\Psi)^*$ for a real-valued wave function Ψ.

16.3 EOM state representation in case of complex-valued orbitals:
Suppose a complex one-particle representation and modify the definition (16.27) of Fock-space states according to $|\Theta_I\rangle = (\hat{O}_I^{*\dagger} + \hat{O}_I)|\Psi_0\rangle$, where the

complex operators $\hat{O}_I$ are defined via the coordinate space representation (as in Exercise 16.2).

(a) Consider the $1p/1h$ states $|\Theta_p\rangle = (c_p^{*\dagger} + c_p)|\Psi_0\rangle$ and show that

$$\langle\Theta_p|\Theta_q\rangle = \langle\Psi_0|\{c_p^\dagger, c_q\}|\Psi_0\rangle =: S_{pq} \ (= \delta_{pq}\ )$$

and

$$\langle\Theta_p|(\hat{H} - E_0)(N - \hat{N})|\Theta_q\rangle = \langle\Psi_0|\{c_p^\dagger, [\hat{H}, c_q]\}|\Psi_0\rangle =: \mathsf{M}_{pq}$$

Here the hamiltonian is assumed to be real, and so is Ψ_0 (possibly up to an irrelevant phase factor).

(b) Establish analogous results for arbitrary $|\Theta_I\rangle$ states.

16.4 Derive the RPA equations (15.5), (15.6) as an approximation to the EOM expressions (16.46), (16.47) obtained by restricting the EOM operator manifold to the $1p$-$1h$ and $1h$-$1p$ operators and replacing $|\Psi_0\rangle$ with the HF ground state $|\Phi_0\rangle$. Note that the relative phase of the $1h$-$1p$ operators implied by the definition (16.44) differs from that underlying Eq. (15.5). Verify that this is consistent with the transition moments $D_{rs}^{RPA} = \langle\Phi_0|[\hat{O}_{rs}^\dagger, \hat{D}]|\Phi_0\rangle$ deriving from Eq. (16.57).

References

1. Pickup BT, Goscinski O (1973) Mol Phys 26:1013
2. Öhrn Y, Born G (1981) Adv Quantum Chem 13:1
3. Linderberg J, Öhrn Y (2004) Propagators in quantum chemistry. Wiley, New York
4. Goscinski D, Lukman B (1970) Chem. Phys Lett 7:573
5. McCurdy CW, Rescigno TN, Yeager DL, McKoy V (1977) In: Schaefer HF (ed) Methods of electronic structure theory. Plenum, New York, p 339
6. Herman MF, Freed KF, Yeager DL (1981) Adv Chem Phys 48:1
7. Rowe DJ (1968) Rev Mod Phys 40:153
8. Mertins F, Schirmer J, Tarantelli A (1996) Phys Rev A 53:2153
9. Dalgaard E (1979) Int J Quantum Chem 15:169
10. Herman MF, Freed KF, Yeager DL (1980) J Chem Phys 72:602
11. Mertins F (1995) Dissertation, Universität Heidelberg
12. Nielsen ES, Jørgensen P, Oddershede J (1980) J Chem Phys 73:6238
13. Oddershede J, Jørgensen P, Yeager DL (1984) Comput Phys Rep 2:33

Chapter 17
Coupled-Cluster Methods for Generalized Excitations

The extension of the coupled-cluster (CC) method, originally devised for ground-states [1, 2], to the treatment of (generalized) electronic excitations is based on CE states in which the CC ground-state parametrization is used. To deal with the non-orthogonality of the CE states and, even more importantly, to obtain tractable secular equations, one introduces, as a second expansion manifold, the set of associated biorthogonal states. The corresponding mixed or biorthogonal (B) representation of the (shifted) hamiltonian gives rise to the non-hermitian BCC secular matrix.

This BCC secular problem lies at the core of the excited-state CC methodology, which comprises two historically independent developments or brands, namely the CCLR (coupled-cluster linear response) theory (see [3] and references therein) and the EOM- CC (equation-of-motion coupled-cluster) approach (see [4] and references therein). A closely related method is the SAC- CI (symmetry-adapted cluster configuration interaction) scheme [5]. For comprehensive presentations of ground- and excited-state CC theory, the reader is referred to textbooks [6, 7] and review articles [8–10].

In this chapter, we give an introduction into the BCC concept and review in particular the order relations and separability properties of the BCC secular equations. We here refer to the extensive analysis given in Ref. [11] which may be consulted for further details.

17.1 Ground-State Coupled-Cluster Formulation

In the CC approach, the ground state is represented in the form of an exponential operator acting on the HF ground state $|\Phi_0\rangle$:

$$|\Psi_0^{cc}\rangle = e^{\hat{T}}|\Phi_0\rangle \tag{17.1}$$

J. Schirmer, *Many-Body Methods for Atoms, Molecules and Clusters*, Lecture Notes in Chemistry 94, https://doi.org/10.1007/978-3-319-93602-4_17

Here, the cluster operator $\hat{T}$ in the exponential is given by the expansion

$$\hat{T} = \sum_I t_I \hat{C}_I \tag{17.2}$$

in terms of the physical excitation operators, $\hat{C}_I$, as specified in Eq. (14.10), and the CC amplitudes (or coefficients) t_I. We note that the exponential operator can be simplified according to

$$e^{\hat{T}} = \prod_I e^{t_I \hat{C}_I} = \prod_I (1 + t_I \hat{C}_I) \tag{17.3}$$

since the $\hat{C}_I$ operators commute, and their powers vanish: $\hat{C}_I^m \equiv 0, m \geq 2$.

We recall some notations introduced in Chap. 11. The successive excitation classes will be numbered $\nu = 1, 2, 3, \ldots$, that is, class μ is formed by the μp-μh excitations. The notation $[J]$ is used to specify the excitation class to which configuration J belongs: $[J] = \mu$ if J labels a μp-μh excitation. The classification of the excitations allows one to write $\hat{T}$ as the sum

$$\hat{T} = \sum_{\nu=1}^{N} \hat{T}_\nu \tag{17.4}$$

of class-specific operators

$$\hat{T}_\nu = \sum_{[I]=\nu} t_I \hat{C}_I \tag{17.5}$$

For example, $\hat{T}_2$ is given by

$$\hat{T}_2 = \sum_{a<b,k<l} t_{abkl}\, c_a^\dagger c_b^\dagger c_k c_l \tag{17.6}$$

The CC amplitudes obey the (nonlinear) ground-state CC equations obtained by projecting the Schrödinger equation

$$\hat{H}|\Psi_0^{cc}\rangle = E_0|\Psi_0^{cc}\rangle \tag{17.7}$$

onto the space spanned by the CI states, $|\Phi_I\rangle = \hat{C}_I|\Phi_0\rangle$, including the HF ground state $|\Phi_0\rangle$. Practical CC approximation schemes are obtained by truncating the operator manifold at successively higher excitation classes, such as in the CCSD approximation, in which the operator manifold comprises the single (S) and double (D) excitation operators. Here, we will not discuss the actual derivation of the respective CC equations and computational aspects, being well documented in the CC literature cited above.

It should be noted that the CC equations do not constitute a variational approach to the CC amplitudes. This has implications to the performance of the truncated CC schemes, which deteriorates when the ground state is no longer adequately described by a single dominant (reference) configuration. To cope with such situations, multi-reference (MR) CC schemes have been developed (see Refs. [9, 12]), which are much more complicated though.

As a point of particular interest, the CC amplitudes fulfill the order relations

$$t_I \sim O([I]-1),\ [I] > 1 \tag{17.8}$$

which are a consequence of the exponential ansatz [13]. The CC order relations can be contrasted with the CI order relations (A.1.24),

$$x_I \sim \begin{cases} \frac{1}{2}O([I]),\ [I] \text{ even} \\ \frac{1}{2}O([I]+1),\ [I] \text{ odd}, > 1 \end{cases} \tag{17.9}$$

Here, x_I are the coefficients in the ground-state CI expansion (see Appendix A.1),

$$|\Psi_0\rangle = |\Phi_0\rangle + \sum_I x_I \hat{C}_I\,|\Phi_0\rangle \tag{17.10}$$

The $2p$-$2h$ and $3p$-$3h$ amplitudes are of first and second order, respectively, both in the CC and CI expansion; but beginning with the $4p$-$4h$ excitations, where $x_I \sim O(2)$ and $t_I \sim O(3)$, the orders of the CC amplitudes are increasingly higher than those of their CC counterparts.

As analyzed by Hubbard [13], the exponential ansatz (17.1) rests on a generalized linked-cluster theorem, which implies that, in contrast to the CI coefficients, the t-amplitudes do not exhibit "non-linking" PT contributions, that is, terms involving coupling matrix elements of the type H_{IJ}, where I and J differ by a double excitation, $[I]-[J]=\pm 2$. An elaboration of these findings is given in the first section of Appendix A.6.

17.2 Biorthogonal Coupled-Cluster Representation

Now we can turn to the BCC approach to electronic excitations, where we consider specifically the case of N-electron excitations. Of course, the BCC concept is quite general and can easily be adapted to ionization (IP- EOM- CC), electron attachment (EA- EOM- CC), and other cases of interest.

The BCC secular matrix $\boldsymbol{M}^{cc}$ is obtained as a non-hermitian representation of the (shifted) hamiltonian $\hat{H} - E_0$,

$$M_{IJ}^{cc} = \langle \overline{\Phi}_I | \hat{H} - E_0 | \Psi_J^0 \rangle \tag{17.11}$$

in terms of two distinct sets of states:

(i) the CC states (right expansion manifold $\{R\}$)

$$|\Psi_J^0\rangle = \hat{C}_J|\Psi_0^{cc}\rangle = \hat{C}_J e^{\hat{T}}|\Phi_0\rangle \tag{17.12}$$

(ii) the associated biorthogonal states (left expansion manifold $\{L\}$)

$$\langle\overline{\Phi}_I| = \langle\Phi_0|\hat{C}_I^\dagger e^{-\hat{T}} \tag{17.13}$$

Here, the operators $\hat{C}_I$ are the physical excitation operators already used in the ground-state CC expansion.

The biorthonormality of the two sets of states

$$\langle\overline{\Phi}_I|\Psi_J^0\rangle = \langle\Phi_I|e^{-\hat{T}}e^{\hat{T}}|\Phi_J\rangle = \delta_{IJ} \tag{17.14}$$

is an obvious consequence of the orthonormalization of the CI configurations $|\Phi_I\rangle = \hat{C}_I|\Phi_0\rangle$. Since the operator $\hat{T}$ is formed entirely of physical excitation operators, it commutes with any operator $\hat{C}_J$. Thus, the CC states of Eq. (17.12) can likewise be obtained by applying the exponential operator to the CI states,

$$|\Psi_J^0\rangle = e^{\hat{T}}\hat{C}_J|\Phi_0\rangle = e^{\hat{T}}|\Phi_J\rangle \tag{17.15}$$

The matrix elements (17.11) may also be written in the form

$$M_{IJ}^{cc} = \langle\overline{\Phi}_I|(\hat{H} - E_0)\hat{C}_J|\Psi_0^{cc}\rangle = \langle\Phi_0|\hat{C}_I^\dagger e^{-\hat{T}}[\hat{H}, \hat{C}_J]e^{\hat{T}}|\Phi_0\rangle. \tag{17.16}$$

where the ground-state energy E_0 no longer appears explicitly.

The two expansion manifolds establishing the BCC representation are of quite different quality. The states of $\{R\}$ are essentially the "correlated excited" (CE) states (11.1), (14.9) underlying the ISR construction presented in Chaps. 11 and 14. As may be anticipated, the CE states are superior to the biorthogonal $\{L\}$ states, if more complex. Obviously, a CC state of class $[I]$ can be written according to

$$|\Psi_I^0\rangle = e^{\hat{T}}|\Phi_I\rangle = |\Phi_I\rangle + \sum_{K,[K]>[I]} z_K^{(I)}|\Phi_K\rangle \tag{17.17}$$

as a linear combination of $|\Phi_I\rangle$ and CI configurations of *higher* excitation classes, $[K] > [I]$, extending through N-tuple excitations. By contrast, the CI expansion of a biorthogonal state from the $\{L\}$ set reads

$$\langle\overline{\Phi}_I| = \langle\Phi_I|e^{-\hat{T}} = \langle\Phi_I| + \sum_{K,[K]<[I]} z_K^{(I)}\langle\Phi_K| \tag{17.18}$$

that is, a linear combination of $\langle\Phi_I|$ and CI excitations of *lower* classes, $[K] = 1, \ldots, [I]-1$ (including $|\Phi_0\rangle$ as a zeroth class in the case of N-electron excitations).

As is easily seen, the linear space spanned by the biorthogonal states through a given excitation class μ is identical with the corresponding space of CI configurations:

$$\text{span}\{\langle\Phi_I|e^{-\hat{T}},\ [I] = (0), 1, \ldots, \mu\} = \text{span}\{\langle\Phi_I|,\ [I] = (0), 1, \ldots, \mu\} \tag{17.19}$$

This suggests that expansions in terms of the biorthogonal states of the $\{L\}$ set are essentially of CI type.

The BCC secular matrix is manifestly non-hermitian, so that one has to deal with a right and a left eigenvalue problem, reading as follows:

$$\begin{aligned} \boldsymbol{M}^{cc}\boldsymbol{X} &= \boldsymbol{X}\boldsymbol{\Omega} \\ \boldsymbol{Y}^\dagger\boldsymbol{M}^{cc} &= \boldsymbol{\Omega}\boldsymbol{Y}^\dagger \end{aligned} \tag{17.20}$$

Here, $\boldsymbol{X}$ and $\boldsymbol{Y}$ denote the matrices of the right and left eigenvectors, respectively, and $\boldsymbol{\Omega}$ is the diagonal matrix of eigenvalues ω_n, to be identified with the (vertical) electronic excitation energies,

$$\omega_n = E_n - E_0. \tag{17.21}$$

The two sets of secular equations can be combined,

$$\boldsymbol{Y}^\dagger\boldsymbol{M}^{cc}\,\boldsymbol{X} = \boldsymbol{\Omega}, \quad \boldsymbol{Y}^\dagger\boldsymbol{X} = \boldsymbol{1} \tag{17.22}$$

where the right and left eigenvectors are mutually biorthonormal. Accordingly, the corresponding right and left excited states,

$$|\Psi_n^{(r)}\rangle = \sum_I X_{In}|\Psi_I^0\rangle \tag{17.23}$$

$$\langle\Psi_m^{(l)}| = \sum_I Y_{Im}^*\langle\overline{\Phi}_I| \tag{17.24}$$

are biorthonormal as well, $\langle\Psi_m^{(l)}|\Psi_n^{(r)}\rangle = \delta_{mn}$.

Extended BCC Expansion

In general, the right excited states $|\Psi_n^{(r)}\rangle$ are not yet eigenstates of $\hat{H}$, because the expansion manifold $\{R\}$ of Eq. (17.12) is incomplete as long as the ground state $|\Psi_0^{cc}\rangle$ is not taken into account. Using the extended CC expansion manifold $\{R^x\} = \{|\Psi_0^{cc}\rangle, |\Psi_I^0\rangle\}$ on the right-hand side, and, likewise, the extended biorthogonal manifold $\{L^x\} = \{\langle\Phi_0|, \langle\overline{\Phi}_I|\}$ on the left side, one obtains an extended BCC representation of $\hat{H} - E_0$:

$$\boldsymbol{M}^x = \begin{pmatrix} 0 & \boldsymbol{v} \\ \boldsymbol{0} & \boldsymbol{M}^{cc} \end{pmatrix}. \tag{17.25}$$

Here, $\boldsymbol{v} = (v_1, v_2, \ldots)$ is the (row) vector of the elements

$$v_I = \langle \Phi_0 | \hat{H} | \Psi_I^0 \rangle \tag{17.26}$$

that is, the coupling matrix elements between the HF ground state and the excited CC states. We note that in the usage of the EOM-CC approach the extended matrix $\boldsymbol{M}^x$ is denoted by $\overline{\boldsymbol{H}}$; in the CCLR context, on the other hand, the original BCC secular matrix $\boldsymbol{M}^{cc}$ is referred to as the CC Jacobian, $\boldsymbol{A}$, and the coupling vector $\boldsymbol{v}$ is denoted as $\boldsymbol{\eta}$.

The extension of the BCC secular problem has some implications. Obviously, there is one more eigenvalue, namely $\omega_0 = 0$, associated with the ground state. The corresponding right eigenvector,

$$\boldsymbol{X}'_0 = \begin{pmatrix} 1 \\ \boldsymbol{0} \end{pmatrix} \tag{17.27}$$

is trivial, reconfirming that $|\Psi_0^{cc}\rangle$ is the exact ground state. The left eigenvector takes on the form

$$\boldsymbol{Y}_0^{\prime\dagger} = (1, \boldsymbol{Y}_0^\dagger) \tag{17.28}$$

where $\boldsymbol{Y}_0^\dagger$ is a row vector given by

$$\boldsymbol{Y}_0^\dagger = -\boldsymbol{v}(\boldsymbol{M}^{cc})^{-1} \tag{17.29}$$

The corresponding state

$$\langle \overline{\Psi}_0 | = \langle \Phi_0 | + \sum_I Y_{I0}^* \langle \overline{\Phi}_I | \tag{17.30}$$

is referred to as "dual" ground state (denoted $\langle \Lambda |$ in the CCLR literature). In view of the nature of the left expansion manifold, the dual ground state can be viewed essentially as a CI-type representation of the ground state.

As the structure of $\boldsymbol{M}^x$ shows, the excited state eigenvalues are given by the eigenvalues ω_n of the $\boldsymbol{M}^{cc}$ sub-block. Here, the left eigenvectors of $\boldsymbol{M}^x$ are obtained as the simple extensions

$$\boldsymbol{Y}_n^{\prime\dagger} = (0, \ \boldsymbol{Y}_n^\dagger) \tag{17.31}$$

of the left eigenvectors of $\boldsymbol{M}^{cc}$. Accordingly, the excited states $\langle \Psi_n^{(l)} |$ of Eq. (17.24) are proper eigenstates of $\hat{H}$. In particular, they are orthogonal to the exact ground state,

$$\langle \Psi_n^{(l)} | \Psi_0^{cc} \rangle = 0 \tag{17.32}$$

which follows from $\langle \overline{\Phi}_I | \Psi_0^{cc} \rangle = 0$. By contrast, the extended right-hand eigenvectors may have a non-vanishing zeroth component,

$$\boldsymbol{X}'_n = \begin{pmatrix} x_n \\ \boldsymbol{X}_n \end{pmatrix} \tag{17.33}$$

Here, $\boldsymbol{X}_n$ denotes the nth right eigenvector of $\boldsymbol{M}^{cc}$, and the x_n component is given by

$$x_n = \omega_n^{-1} \boldsymbol{v}\, \boldsymbol{X}_n = \boldsymbol{v}\,(\boldsymbol{M}^{cc})^{-1} \boldsymbol{X}_n = -\boldsymbol{Y}_0^\dagger \boldsymbol{X}_n \tag{17.34}$$

using here the eigenvalue equation $\boldsymbol{M}^{cc}\boldsymbol{X}_n = \omega_n \boldsymbol{X}_n$ and, moreover, Eq. (17.29). As a consequence, the right expansion of an excited eigenstate takes on the form

$$|\Psi_n^{(x)}\rangle = x_n |\Psi_0^{cc}\rangle + |\Psi_n^{(r)}\rangle \tag{17.35}$$

where $|\Psi_n^{(r)}\rangle$ is given by Eq. (17.23). We note the relation

$$x_n = \langle \Phi_0 | \Psi_n^{(x)} \rangle \tag{17.36}$$

which follows from $\langle \Phi_0 | \Psi_n^{(r)} \rangle = 0$. The excited eigenstates $|\Psi_n^{(x)}\rangle$ are orthogonal to the dual ground state,

$$\langle \overline{\Psi}_0 | \Psi_n^{(x)} \rangle = x_n + \boldsymbol{Y}_0^\dagger \boldsymbol{X}_n = 0 \tag{17.37}$$

which follows from the relations $\langle \overline{\Phi}_I | \Psi_n^{(r)} \rangle = X_{In}$ and Eq. (17.34).

Transition Moments

For spectral intensities, the squared moduli $|T_n|^2$ of the transition moments (13.6) are required, involving normalized ground and excited states. In the BCC representation, a properly normalized expression for $|T_n|^2$ is obtained according to

$$|T_n|^2 = T_n^{(l)} T_n^{(r)} \tag{17.38}$$

where

$$T_n^{(l)} = \langle \Psi_n^{(l)} | \hat{D} | \Psi_0^{cc} \rangle \tag{17.39}$$

$$T_n^{(r)} = \langle \overline{\Psi}_0 | \hat{D} | \Psi_n^{(x)} \rangle \tag{17.40}$$

denote transition moments associated with the left and right forms of the excited states. The left transition moment can be written as the scalar product,

$$T_n^{(l)} = \boldsymbol{Y}_n^\dagger \boldsymbol{F}^{(l)} \tag{17.41}$$

of the left eigenvector, $\boldsymbol{Y}_n$, and a vector $\boldsymbol{F}^{(l)}$ of basis set transition moments,

$$F_I^{(l)} = \langle \overline{\Phi}_I | \hat{D} | \Psi_0^{cc} \rangle \tag{17.42}$$

In a similar way, one may define transition moments for the right expansion manifold,

$$F_I^{(r)} = \langle\overline{\Psi}_0|\hat{D}|\Psi_I^0\rangle \tag{17.43}$$

and write the right transition moments as

$$T_n^{(r)} = x_n\langle\overline{\Psi}_0|\hat{D}|\Psi_0^{cc}\rangle + \boldsymbol{F}^{(r)\dagger}\boldsymbol{X}_n \tag{17.44}$$

Here, the first term derives from a possible ground-state admixture in $|\Psi_n^{(x)}\rangle$ according to Eq. (17.35). We note that the biorthonormality relations $\langle\overline{\Psi}_0|\Psi_0^{cc}\rangle = 1$ and $\langle\Psi_n^{(l)}|\Psi_n^{(x)}\rangle = 1$ ensure the proper normalization of the product (17.38).

17.3 Order Relations and Separability Properties

The different quality of the two expansion manifolds becomes apparent in the order structure of the BCC secular matrix $\boldsymbol{M}^{cc}$ shown in Fig. 17.1. As explained in Chap. 12, the entries in the $\boldsymbol{M}_{\mu\nu}^{cc}$ sub-blocks associated with the excitation-class partitioning denote the lowest (non-vanishing) PT orders of the matrix elements in these blocks. The upper right (UR) triangular part of $\boldsymbol{M}^{cc}$ shows the characteristic CI order structure as in Fig. 12.5 (there for the case of $(N-1)$-electron excitations). By contrast, the lower left (LL) triangular part ($\mu > \nu$) displays the canonical order relations (see Eqs. (12.1), (14.44))

$$\boldsymbol{M}_{\mu\nu}^{cc} \sim O(\mu - \nu) \tag{17.45}$$

The CI structure of the UR part can be inferred by using the expansions (17.17), (17.18) in the BCC matrix elements M_{IJ}^{cc}, $[I] < [J]$:

$$\langle\overline{\Phi}_I|\hat{H} - E_0|\Psi_J^0\rangle = \langle\Phi_I|\hat{H}|\Phi_J\rangle + \sum_{[K]<[I]}\sum_{[L]>[J]} z_K^{(I)} z_L^{(J)}\langle\Phi_K|\hat{H}|\Phi_L\rangle \tag{17.46}$$

Obviously, there are no contributions from the double sum, because the excitation classes of the states K and L differ at least by a triple excitation, that is, $[L] - [K] \geq 3$, so that all matrix elements $\langle\Phi_K|\hat{H}|\Phi_L\rangle$ vanish. In fact, this shows that the BCC and CI secular matrix elements in the UR blocks are identical:

$$M_{IJ}^{cc} = H_{IJ}, \text{ for } [I] < [J] \tag{17.47}$$

For matrix elements M_{IJ}^{cc} of a diagonal block, $[I] = [J]$, we obtain

$$M_{IJ}^{cc} = \langle\Phi_I|\hat{H} - E_0|\Phi_J\rangle + \sum_{[K]=[I]-1}\sum_{[L]=[I]+1} z_K^{(I)} z_L^{(J)}\langle\Phi_K|\hat{H}|\Phi_L\rangle \tag{17.48}$$

Obviously, the BCC and CI expressions differ beyond first order, which of course does not affect the trivial order relation, $\boldsymbol{M}_{\mu\mu}^{cc} \sim O(0)$, in the diagonal blocks.

Fig. 17.1 Order structure of the BCC secular matrix $\boldsymbol{M}^{cc}$

	$1p$-$1h$	$2p$-$2h$	$3p$-$3h$	$4p$-$4h$	$5p$-$5h$	...
$1p$-$1h$	0	1	1	-	-	...
$2h$-$2p$	1	0	1	1	-	...
$3p$-$3h$	2	1	0	1	1	-
$4p$-$4h$	3	2	1	0	1	1
$5p$-$5h$	4	3	2	1	0	1
⋮	⋮	⋮	⋮	⋮	⋮	⋱

The less obvious (canonical) order relations for the LL part of $\boldsymbol{M}^{cc}$ are addressed in Appendix A.6.

As discussed in Sect. 12.1 and Appendix A.6, the order structure of the secular matrix determines the truncation errors inherent to approximations obtained by limiting the configuration space. As a consequence of the CI structure in the UR matrix blocks of the BBC secular matrix, the truncation properties are somewhat weaker than those in a full canonical order structure. For example, neglecting the triple excitations in the BCC expansion causes a third-order error (as can be seen from Fig. 17.1), whereas in the ISR-ADC scheme, the corresponding error is of fourth order. For the energies of $1p$-$1h$ (single) excitations, the BCC truncation error orders (TEO) are given by (see Ref. [11])

$$O_{TE}(\mu) = \begin{cases} \frac{3}{2}\mu, & \mu \text{ even} \\ \frac{1}{2}(3\mu + 1), & \mu \text{ odd} \end{cases} \tag{17.49}$$

where μ denotes the highest excitation class included in the secular expansion. This formula can be compared to Eq. (14.45) specifying the truncation errors in the ISR-ADC case. In Table 17.1, the explicit BCC, ISR-ADC, and CI truncation errors are listed for successively larger configuration spaces.

With regard to the transition moments, we may refer to Ref. [11], where a comprehensive analysis has been given of the truncation errors in the left and right transition moments. For the single ($1p$-$1h$) excitations, the left and right transition moments have the same truncation error characteristics, shown in Table 17.1 for the lowest 6 expansion levels.

Table 17.1 Truncation errors (PT order) for excitation energies and transition moments of singly excited states: comparison of CI, BCC, and ADC approaches for the lowest six truncation levels

Truncation level	Excitation energies			Transition moments		
	CI	BCC	ADC	CI	BCC	ADC
1	2	2	2	1	1	2
2	2	3	4	2	3	4
3	4	5	6	3	4	6
4	4	6	8	4	6	8
5	6	8	10	5	7	10
6	6	9	12	6	9	12

Separability

An analysis of the separability properties of the BCC secular equations and the size-consistency of the resulting excitation energies and transition moments can easily be performed along the lines of Sect. 12.2. We confine ourselves to a brief sketch of the essential features.

The hybrid character of the BCC secular matrix is reflected in the block structure associated with the separate fragment model shown in Fig. 17.2. The LL part is separable, whereas the UR part displays the non-separable CI structure.

According to the generalized linked-cluster theorem (see Sect. A.6.1), the $\hat{T}$ operator can be written as the sum of fragment operators,

$$\hat{T} = \hat{T}_A + \hat{T}_B \tag{17.50}$$

and the exponential of the $\hat{T}$ operator factorizes,

$$e^{\hat{T}} = e^{\hat{T}_A} e^{\hat{T}_B} \tag{17.51}$$

$\boldsymbol{M}^{cc}_{AA}$	0	$\boldsymbol{M}^{cc}_{A,AB}$
0	$\boldsymbol{M}^{cc}_{BB}$	$\boldsymbol{M}^{cc}_{B,AB}$
0	0	$\boldsymbol{M}^{cc}_{AB,AB}$

Fig. 17.2 Block structure of the BCC secular matrix $\boldsymbol{M}^{cc}$ with respect to the separate fragment model

As a consequence, both the CC states and the biorthogonal states can be written as products of fragment states. For a non-local excitation, $J \equiv J_{AB}$, the CC state takes on the form

$$|\Psi^0_{J_{AB}}\rangle = |\Psi^A_{J_A}\rangle|\Psi^B_{J_B}\rangle \tag{17.52}$$

where

$$|\Psi^A_{J_A}\rangle = \hat{C}_{J_A} e^{\hat{T}_A}|\Phi^A_0\rangle \tag{17.53}$$

denotes a CC state for fragment A. The corresponding biorthogonal state reads

$$\langle\overline{\Phi}_{J_{AB}}| = \langle\overline{\Phi}^A_{J_A}|\langle\overline{\Phi}^B_{J_B}| \tag{17.54}$$

where the fragment biorthogonal state, say for A, is given by

$$\langle\overline{\Phi}^A_{J_A}| = \langle\Phi^A_0|\hat{C}^\dagger_{J_A} e^{-\hat{T}_A} \tag{17.55}$$

Using the factorization of the left and right BCC basis states, the block structure of $\boldsymbol{M}^{cc}$ can readily be established (Exercise 17.4). As an example, we consider a matrix element in the $\boldsymbol{M}^{cc}_{A,AB}$ block:

$$\begin{aligned} M^{cc}_{I_A,J_{AB}} =& \langle\overline{\Phi}_{I_A}|\hat{H}_A + \hat{H}_B|\Psi^0_{J_{AB}}\rangle \\ =& \langle\overline{\Phi}^A_{I_A}|\hat{H}_A|\Psi^A_{J_A}\rangle\langle\Phi^B_0|\Psi^B_{J_B}\rangle + \langle\overline{\Phi}^A_{I_A}|\Psi^A_{J_A}\rangle\langle\Phi^B_0|\hat{H}_B|\Psi^B_{J_B}\rangle \end{aligned}$$

While the first term in the last line vanishes due to the orthogonality of $|\Phi^B_0\rangle$ and $|\Psi^B_{J_B}\rangle$, the second term may not vanish for $I_A = J_A$, since $|\Phi^B_0\rangle$ is not an eigenstate of $\hat{H}_B$. In a similar way, one may verify that

$$\boldsymbol{M}^{cc}_{AA} = \boldsymbol{M}^{cc}(A) \tag{17.56}$$

where $\boldsymbol{M}^{cc}(A)$ denotes the BCC secular matrix of fragment A.

What are the consequences of the semi-separable structure of the BCC secular matrix? There are no ramifications for the excitation energies. The characteristic polynomial for the fragment secular matrix, $\boldsymbol{M}^{cc}_{AA}$, is a factor of the characteristic polynomial of the full $\boldsymbol{M}^{cc}$ matrix, so that the eigenvalues of $\boldsymbol{M}^{cc}_{AA}$ are a subset of all eigenvalues. This means that the energies are size-consistent: the BCC results for local excitations do not depend on whether the method is applied to the respective fragment or to the composite.

For the transition moments, the situation is more nuanced. The left transition moments are size-consistent, although the left eigenvectors are non-separable: The eigenvector $\boldsymbol{Y}_n$ for a local excitation in one of the fragments has non-vanishing non-local components, $Y_{J_{AB},n} \neq 0$. However, this does not matter because in the vector of the left basis set transition moments, $\boldsymbol{F}^{(l)}$, all non-local components vanish: $F^{(l)}_{J_{AB}} = 0$. The right transition moments (17.40) are not size-consistent, even though the right eigenvectors are separable. The problem is due to the use of the dual ground state (17.30) in the right basis set transition moments (17.43). The dual ground state, being essentially a CI expansion, is not separable, that is, a factorization

$$\langle\overline{\Psi}_0| = \langle\overline{\Psi}_0^A|\langle\overline{\Psi}_0^B| \tag{17.57}$$

of the composite ground state is attained only in the exact (full) BCC treatment. It should be noted that in the CCLR formulation this shortcoming is avoided, as here a separable, if more elaborate expression is employed for the right transition moments [14, 15].

Exercises

17.1 (a) Compare the CC expansion $|\Psi_0\rangle = e^{\hat{T}_1+\hat{T}_2+\hat{T}_3+\dots}|\Phi_0\rangle$ for the ground state with the CI expansion $|\Psi_0\rangle = (1 + \hat{S}_1 + \hat{S}_2 + \hat{S}_3 + \dots)|\Phi_0\rangle$, where $\hat{S}_\nu = \sum_{[I]=\nu} x_I \hat{C}_I$ denotes the excitations of class ν in the CI expansion. Express the four lowest CI operators $\hat{S}_1, \dots, \hat{S}_4$, in terms of the CC operators $\hat{T}_1, \dots, \hat{T}_4$.
(b) Deduce the order relations for $\hat{T}_1$, $\hat{T}_2$, and $\hat{T}_3$ from those for $\hat{S}_1$, $\hat{S}_2$, and $\hat{S}_3$; what is the situation for $\hat{T}_4$?

17.2 (a) Consider the excitation operator $\hat{C}_K$ for a quadruple excitation $K \equiv abcd; ijkl$ and establish that this operator can be written in 18 distinct ways as a product of two "disjunct" double excitations, such as $\hat{C}_{abcd;ijkl} = \hat{C}_{acil}\hat{C}_{bdjk}$.
(b) Inspect the PT expression for the ground-state amplitude $x^{(2)}_{abcd;ijkl} = \langle\Phi_{abcd;ijkl}|\Psi_0^{(2)}\rangle$ in second order (see Appendix A.1), and show that $x^{(2)}_{abcd;ijkl}$ can be written as a sum over products $x_I^{(1)}x_J^{(1)}$ of first-order amplitudes for (disjunct) double excitations, I, J, where $\hat{C}_{abcd;ijkl} = \hat{C}_I\hat{C}_J$.
(c) Use the results of (b) together with Exercise 17.1 to show that the CC amplitude $t_{abcd;ijkl}$ is (at least) of third order.

17.3 Consider the BCC representation for an $(N-1)$-electron system (IP- EOM-CC) and expand the coupling matrix elements $M^{cc}_{i,abjkl}$ and $M^{cc}_{abjkl,i}$ for a $1h$ state and a $3h$-$2p$ state through first order.

17.4 Establish the block structure of the BCC secular matrix with respect to separate fragment partitioning (Fig. 17.2).

17.5 Apply the ground-state CC concept to the 2E-2O model considered in Exercise 2.4. Here, the CC operator consists of a single double excitation, $\hat{T} = t\, c^\dagger_{u\alpha}c^\dagger_{u\beta}c_{g\beta}c_{g\alpha}$ so that $|\Psi_0^{cc}\rangle = e^{\hat{T}}|\Phi_0\rangle = |\Phi_0\rangle + t|\Phi_1\rangle$. Project the Schrödinger equation for $|\Psi_0^{cc}\rangle$ onto $|\Phi_0\rangle$ and $|\Phi_1\rangle$ and determine t and E_0.

17.6 CCD(doubles) treatment of the multiple 2E-2O system (see Exercise 12.2): Consider the CC operator $\hat{T} = \sum_i t_i\hat{C}_i$ where $\hat{C}_i = c^\dagger_{iu\alpha}c^\dagger_{iu\beta}c_{ig\beta}c_{ig\alpha}$, $i = 1, \dots, M$, are double excitation operators for the M sub-systems. Derive the ground-state CC equations and verify that the amplitudes are all equal, $t_i = t$. Determine t, and establish that the CCD results reproduce the exact ground-state energy and wave function.

References

1. Coester F (1958) Nucl Phys 7:421
2. Cízek J (1966) J Phys Chem 45:4256
3. Koch H, Jensen HJA, Jørgensen P, Helgaker T (1990) J Chem Phys 93:3345
4. Stanton JF, Bartlett RJ (1993) J Chem Phys 98:7029
5. Nakatsuji H, Hirao K (1977) Chem Phys Lett 47:569
6. Helgaker T, Jørgensen P, Olsen J (2000) Molecular electronic structure theory. J. Wiley, New York
7. Bartlett RJ, Shavitt I (2009) Many-body methods in chemistry and physics: MBPT and coupled-cluster theory. Cambridge University Press, Cambridge
8. Christiansen O (2006) Theor Chem Acc 116:106
9. Bartlett RJ, Musial M (2007) Rev Mod Phys 79:291
10. Krylov A (2008) Ann Rev Phys Chem 59:433
11. Schirmer J, Mertins F (2009) Theor Chem Acc 125:145
12. Mukherjee D, Pal S (1989) Adv Quantum Chem 20:561
13. Hubbard J (1957) Proc R Soc A 240:539
14. Koch H, Jørgensen P (1990) J Chem Phys 93:3333
15. Koch H, Kobayashi R, de Merás AS, Jørgensen P (1994) J Chem Phys 100:4393

Appendix

A.1 Basic Tools

A.1.1 Perturbation Theory for the Ground State

In the following, a brief recapitulation is given of Rayleigh–Schrödinger perturbation theory (RSPT) for the N-electron ground state. Here, we adopt the elegant general derivation of ground-state perturbation theory presented by March, Young, and Sampanthar in Chap. 1 of their textbook [1].

As usual, the starting point is the decomposition

$$\hat{H} = \hat{H}_0 + \hat{H}_I \tag{A.1.1}$$

of the Hamiltonian into an unperturbed part $\hat{H}_0$ and an interaction part $\hat{H}_I$. Let $|\Phi_0\rangle$ denote the ground state of $\hat{H}_0$ with the energy $E_0^{(0)}$, and

$$\hat{Q}_0 = \hat{\mathbb{1}} - |\Phi_0\rangle\langle\Phi_0| \tag{A.1.2}$$

the projector onto the orthogonal complement of $|\Phi_0\rangle$. The exact ground state $|\Psi_0\rangle$ can be written as

$$|\Psi_0\rangle = |\Phi_0\rangle + \hat{Q}_0|\Psi_0\rangle \tag{A.1.3}$$

which implies *intermediate normalization*, $\langle\Phi_0|\Psi_0\rangle = 1$. The Schrödinger equation for $|\Psi_0\rangle$ allows us to write

$$(\epsilon - \hat{H})|\Psi_0\rangle = (\epsilon - E_0)|\Psi_0\rangle \tag{A.1.4}$$

where ϵ is an arbitrary parameter to be specified later. Applying $\hat{Q}_0$ to the latter equation gives

$$(\epsilon - \hat{H}_0)\hat{Q}_0|\Psi_0\rangle = \hat{Q}_0(\hat{H}_I + \epsilon - E_0)|\Psi_0\rangle \tag{A.1.5}$$

J. Schirmer, *Many-Body Methods for Atoms, Molecules and Clusters*, Lecture Notes in Chemistry 94, https://doi.org/10.1007/978-3-319-93602-4

and

$$\hat{Q}_0|\Psi_0\rangle = (\epsilon - \hat{H}_0)^{-1}\hat{Q}_0(\hat{H}_I + \epsilon - E_0)|\Psi_0\rangle \tag{A.1.6}$$

The latter expression for $\hat{Q}_0|\Psi_0\rangle$ can be used in Eq. (A.1.3), yielding an implicit equation

$$|\Psi_0\rangle = |\Phi_0\rangle + \frac{\hat{Q}_0}{\epsilon - \hat{H}_0}(\hat{H}_I + \epsilon - E_0)|\Psi_0\rangle \tag{A.1.7}$$

for $|\Psi_0\rangle$, which can be solved formally by iteration:

$$|\Psi_0\rangle = |\Phi_0\rangle + \sum_{\nu=1}^{\infty}\left[\frac{\hat{Q}_0}{\epsilon - \hat{H}_0}(\hat{H}_I + \epsilon - E_0)\right]^{\nu}|\Phi_0\rangle \tag{A.1.8}$$

The associated expansion of E_0 deriving from the energy expression

$$E_0 = \langle\Phi_0|\hat{H}_0 + \hat{H}_I|\Psi_0\rangle \tag{A.1.9}$$

is given by

$$E_0 = E_0^{(0)} + \langle\Phi_0|\hat{H}_I|\Phi_0\rangle + \langle\Phi_0|\hat{H}_I\sum_{\nu=1}^{\infty}\left[\frac{\hat{Q}_0}{\epsilon - \hat{H}_0}(\hat{H}_I + \epsilon - E_0)\right]^{\nu}|\Phi_0\rangle \tag{A.1.10}$$

The closed-form expansions (A.1.8), (A.1.10) still contain the exact energy E_0. To obtain explicit perturbation series, the expansion

$$E_0 = E_0^{(0)} + E_0^{(1)} + E_0^{(2)} + \ldots \tag{A.1.11}$$

has to be used in an appropriate way. Here, the individual terms can be determined successively from Eq. (A.1.10).

The familiar Rayleigh–Schrödinger (RS) perturbation theory results from Eqs. (A.1.8) and (A.1.10) by setting $\epsilon = E_0^{(0)}$. Another obvious choice, namely $\epsilon = E_0$, leads to the so-called Brillouin–Wigner (BW) perturbation theory which will be briefly addressed at the end of this section.

Note that the formal development presented so far is completely general and can easily be transferred to the case of a one-particle system, essentially by adapting the notations accordingly: Write the hamiltonian as $\hat{h} = \hat{h}_0 + \hat{h}_1$ and let ϕ_0 and ψ_0 denote the unperturbed and exact ground states, respectively, and e_0 and $e_0^{(0)}$ the corresponding energies.

Let us construct the actual first- and second-order terms in the expansion of $|\Psi_0\rangle$. The explicit RS perturbation expansion through second order reads

$$|\Psi_0\rangle = |\Phi_0\rangle + \frac{\hat{Q}_0}{E_0^{(0)} - \hat{H}_0}\hat{H}_I|\Phi_0\rangle + \frac{\hat{Q}_0}{E_0^{(0)} - \hat{H}_0}(\hat{H}_I - E_0^{(1)})\frac{\hat{Q}_0}{E_0^{(0)} - \hat{H}_0}\hat{H}_I|\Phi_0\rangle + O(3) \tag{A.1.12}$$

where in the second-order term the last numerator $(\hat{H}_I - E_0^{(1)})$ has been simplified because $\hat{Q}_0|\Phi_0\rangle = 0$.

To further evaluate the terms in the PT expansion, one may insert the resolution of the identity

$$\hat{\mathbb{1}} = \sum_I |\Phi_I\rangle\langle\Phi_I| \tag{A.1.13}$$

to the left of each $\hat{H}_I$ operator. Here, $|\Phi_I\rangle$ denote the ground and excited HF states (or, more general, eigenstates of $\hat{H}_0$) as specified by Eq. (2.24).

The first-order term thus becomes

$$|\Psi_0^{(1)}\rangle = \sum_{J\neq 0} \frac{1}{E_0^{(0)} - E_J^{(0)}}|\Phi_J\rangle\langle\Phi_J|\hat{H}_I|\Phi_0\rangle \tag{A.1.14}$$

Here, the states $|\Phi_J\rangle$ are restricted to the $1p$-$1h$ and $2p$-$2h$ excitations, since the matrix element $\langle\Phi_J|\hat{H}_I|\Phi_0\rangle$ vanishes for higher excitations. Supposing the Møller-Plesset (MP) partitioning of the hamiltonian (see Eqs. 4.2–4.4), there are no $1p$-$1h$ components either, because the matrix elements

$$\langle\Phi_{ak}|\hat{H}_I|\Phi_0\rangle = w_{ak} + \sum_r V_{ar[kr]}n_r = 0 \tag{A.1.15}$$

vanish according to Eq. (4.6). This is often referred to as *Brillouin's theorem*. Accordingly, the first-order ground state is a linear combination of double excitations,

$$|\Psi_0^{(1)}\rangle = \sum_{a<b,k<l} x_{abkl}^{(1)}|\Phi_{abkl}\rangle \tag{A.1.16}$$

where the coefficients are given by the characteristic PT fractions

$$x_{abkl}^{(1)} = \frac{V_{ab[kl]}}{\epsilon_a + \epsilon_b - \epsilon_k - \epsilon_l} \tag{A.1.17}$$

With the help of Eq. (A.1.10), the second-order ground-state energy is given by

$$E_0^{(2)} = \langle \Phi_0 | \hat{H}_I | \Psi_0^{(1)} \rangle = - \sum_{a<b,k<l} \frac{|V_{ab[kl]}|^2}{\epsilon_a + \epsilon_b - \epsilon_k - \epsilon_l} \tag{A.1.18}$$

The second-order ground state consists of two terms,

$$|\Psi_0^{(2)}\rangle = \frac{\hat{Q}_0}{E_0^{(0)} - \hat{H}_0} \hat{H}_I \frac{\hat{Q}_0}{E_0^{(0)} - \hat{H}_0} \hat{H}_I |\Phi_0\rangle - E_0^{(1)} \frac{\hat{Q}_0}{(E_0^{(0)} - \hat{H}_0)^2} \hat{H}_I |\Phi_0\rangle \tag{A.1.19}$$

where $E_0^{(1)} = \langle \Phi_0 | \hat{H}_I | \Phi_0 \rangle$. Applying again the resolution of the identity (twice in the first term) and using the first-order result, $|\Psi_0^{(2)}\rangle$ may be written as

$$\begin{aligned} |\Psi_0^{(2)}\rangle = & \sum_{J \neq 0} \sum_{c<d,i<j} \frac{1}{E_0^{(0)} - E_J^{(0)}} |\Phi_J\rangle \langle \Phi_J | \hat{H}_I | \Phi_{cdij} \rangle \frac{V_{cd[ij]}}{\epsilon_c + \epsilon_d - \epsilon_i - \epsilon_j} \\ & + E_0^{(1)} \sum_{a<b,k<l} |\Phi_{abkl}\rangle \frac{V_{ab[kl]}}{(\epsilon_a + \epsilon_b - \epsilon_k - \epsilon_l)^2} \end{aligned} \tag{A.1.20}$$

Here, the sum over the states $|\Phi_J\rangle$ in the first term comprises single, double, triple, and quadruple excitations (νp-νh, $\nu = 1, 2, 3, 4$).

Let us consider the $|\Phi_{abkl}\rangle$ component in $|\Psi_0^{(2)}\rangle$. The corresponding amplitude $x_{abkl}^{(2)}$ can be written as

$$\begin{aligned} x_{abkl}^{(2)} = \langle \Phi_{abkl} | \Psi_0^{(2)} \rangle = & - \sum_{\substack{c<d,i<j \\ \neq (abkl)}} \frac{\langle \Phi_{abkl} | \hat{H}_I | \Phi_{cdij} \rangle V_{cd[ij]}}{(\epsilon_a + \epsilon_b - \epsilon_k - \epsilon_l)(\epsilon_c + \epsilon_d - \epsilon_i - \epsilon_j)} \\ & - \frac{\langle \Phi_{abkl} | \hat{H}_I - E_0^{(1)} | \Phi_{abkl} \rangle V_{ab[kl]}}{(\epsilon_a + \epsilon_b - \epsilon_k - \epsilon_l)^2} \end{aligned} \tag{A.1.21}$$

Here, we have separated the diagonal contribution in the sum over the $2p$-$2h$ states which can be combined with the second term of Eq. (A.1.20), effecting the modified diagonal matrix element $\langle \Phi_{abkl} | \hat{H}_I - E_0^{(1)} | \Phi_{abkl} \rangle$ in the numerator.

As another example, the $1p$-$1h$ excitations in $|\Psi_0^{(2)}\rangle$ contribute with the amplitudes

$$x_{ak}^{(2)} = \langle \Phi_{ak} | \Psi_0^{(2)} \rangle = - \frac{1}{\epsilon_a - \epsilon_k} \sum_{c<d,i<j} \langle \Phi_{ak} | \hat{H}_I | \Phi_{cdij} \rangle \frac{V_{cd[ij]}}{\epsilon_c + \epsilon_d - \epsilon_i - \epsilon_j} \tag{A.1.22}$$

As these examples show, the general PT expansion (A.1.8) for $|\Psi_0\rangle$ establishes individual PT expansions for the amplitudes (or CI coefficients)

$$x_J = \langle \Phi_J | \Psi_0 \rangle \tag{A.1.23}$$

The onset of these expansions, that is, the PT order $O(x_J)$ of the first non-vanishing contribution, depends on the excitation class $[J]$ of the respective excitation J. The amplitudes x_{abkl} are of first order (and that would apply to x_{ak} as well were it not for Brillouin's theorem). The x-amplitudes for the triple and quadruple excitations (excitation classes 3 and 4, respectively) begin in second order; in third order, the next two excitation classes, $\nu = 5$ and 6, come into play, and so forth. The general order relations read

$$O(x_J) = \begin{cases} \frac{1}{2}[J], & [J] \text{ even} \\ \frac{1}{2}([J]+1), & [J] \text{ odd}, > 1 \end{cases} \tag{A.1.24}$$

They follow from the structure of the expansion (A.1.8) and the instance that via the two-body Coulomb operator contained in $\hat{H}_I$ there is a (non-vanishing) coupling of states of excitation class ν to states of the two higher classes, $\nu + 1$ and $\nu + 2$.

A.1.2 Matrix Algebra

There are some basic matrix algebra techniques which are generally useful and apply, in particular, to some of the topics treated in this book. While these algebra tools are simple, they cannot necessarily be considered common knowledge so that the following brief inspection may be helpful.

Inverse of a Partitioned Matrix

Consider a square matrix

$$\boldsymbol{M} = \begin{pmatrix} \boldsymbol{A} & \boldsymbol{B} \\ \boldsymbol{C} & \boldsymbol{D} \end{pmatrix} \tag{A.1.25}$$

consisting as indicated of sub-blocks, where $\boldsymbol{A}$ and $\boldsymbol{D}$ are square matrices, and $\boldsymbol{B}$ and $\boldsymbol{C}$ are rectangular matrices of corresponding size. The inverse of the partitioned matrix

$$\begin{pmatrix} \boldsymbol{A} & \boldsymbol{B} \\ \boldsymbol{C} & \boldsymbol{D} \end{pmatrix}^{-1} = \begin{pmatrix} \boldsymbol{E} & \boldsymbol{F} \\ \boldsymbol{G} & \boldsymbol{H} \end{pmatrix} \tag{A.1.26}$$

has the same block form, and the sub-blocks of the inverse can be directly written in terms of the original blocks according to

$$\begin{aligned} \boldsymbol{E} &= (\boldsymbol{A} - \boldsymbol{B}\boldsymbol{D}^{-1}\boldsymbol{C})^{-1}, & \boldsymbol{F} &= -\boldsymbol{A}^{-1}\boldsymbol{B}\boldsymbol{H} \\ \boldsymbol{G} &= -\,\boldsymbol{D}^{-1}\boldsymbol{C}\boldsymbol{E}, & \boldsymbol{H} &= (\boldsymbol{D} - \boldsymbol{C}\boldsymbol{A}^{-1}\boldsymbol{B})^{-1} \end{aligned} \tag{A.1.27}$$

To derive these equations, one may break down the general inversion product,

$$\begin{pmatrix} \boldsymbol{A} & \boldsymbol{B} \\ \boldsymbol{C} & \boldsymbol{D} \end{pmatrix} \begin{pmatrix} \boldsymbol{E} & \boldsymbol{F} \\ \boldsymbol{G} & \boldsymbol{H} \end{pmatrix} = \begin{pmatrix} \mathbf{1} & \mathbf{0} \\ \mathbf{0} & \mathbf{1} \end{pmatrix}$$

into the following four sub-block equations:

$$\begin{aligned} \boldsymbol{A}\boldsymbol{E} + \boldsymbol{B}\boldsymbol{G} &= \mathbf{1} \\ \boldsymbol{A}\boldsymbol{F} + \boldsymbol{B}\boldsymbol{H} &= \mathbf{0} \\ \boldsymbol{C}\boldsymbol{E} + \boldsymbol{D}\boldsymbol{G} &= \mathbf{0} \\ \boldsymbol{C}\boldsymbol{F} + \boldsymbol{D}\boldsymbol{H} &= \mathbf{1} \end{aligned}$$

Now one may, for example, solve the third equation for $\boldsymbol{G}$. The result $\boldsymbol{G} = -\boldsymbol{D}^{-1}\boldsymbol{C}\boldsymbol{E}$ can be used to replace $\boldsymbol{G}$ in the first equation, which then can easily be solved for $\boldsymbol{E}$ in the form given by Eq. (A.1.27).

A side remark: The simple expression

$$\begin{pmatrix} a & b \\ c & d \end{pmatrix}^{-1} = \frac{1}{ad - bc} \begin{pmatrix} d & -b \\ -c & a \end{pmatrix} \tag{A.1.28}$$

for the inverse of a 2×2 matrix is useful in dealing with small matrices. Obviously, the inversion of a 3×3 matrix or a 4×4 matrix can be accomplished in a straightforward way (leading to closed-form expressions) by using a (1-2) or (2-2) partitioning scheme, respectively, and combining Eq. (A.1.28) with the partitioning formulas (A.1.27). Such an approach is particularly helpful if one deals with matrices where the matrix elements are functions of one or several variables.

Partitioning of an Eigenvalue Problem

The partitioning technique can also be applied to the matrix eigenvalue problem. Consider the eigenvalue equation for (a hermitian) matrix $\boldsymbol{M}$ given in the form of Eq. (A.1.25):

$$\begin{pmatrix} \boldsymbol{A} & \boldsymbol{B} \\ \boldsymbol{C} & \boldsymbol{D} \end{pmatrix} \begin{pmatrix} \boldsymbol{x} \\ \boldsymbol{y} \end{pmatrix} = \lambda \begin{pmatrix} \boldsymbol{x} \\ \boldsymbol{y} \end{pmatrix} \tag{A.1.29}$$

Here, λ denotes an eigenvalue of $\boldsymbol{M}$, and the corresponding eigenvector is written in an obvious partitioned form. More explicitly, the latter equation is composed of the two sub-block equations

$$\begin{aligned}\boldsymbol{A}\boldsymbol{x} + \boldsymbol{B}\boldsymbol{y} &= \lambda\boldsymbol{x}\\ \boldsymbol{C}\boldsymbol{x} + \boldsymbol{D}\boldsymbol{y} &= \lambda\boldsymbol{y}\end{aligned}$$

coupling the $\boldsymbol{x}$ and $\boldsymbol{y}$ components of the eigenvector. Solving the second equation for $\boldsymbol{y}$ gives

$$\boldsymbol{y} = (\lambda\mathbf{1} - \boldsymbol{D})^{-1}\boldsymbol{C}\boldsymbol{x} \tag{A.1.30}$$

which can be used to replace $\boldsymbol{y}$ in the first equation. The result is the following pseudo-eigenvalue equation for $\boldsymbol{x}$:

$$\left(\boldsymbol{A} + \boldsymbol{B}(\lambda\mathbf{1} - \boldsymbol{D})^{-1}\boldsymbol{C}\right)\boldsymbol{x} = \lambda\boldsymbol{x} \tag{A.1.31}$$

While the matrix on the left-hand side is of smaller dimension than that of the original matrix $\boldsymbol{M}$, it depends on the respective eigenvalue λ. This means one has to resort to an appropriate iterative procedure in order to solve the pseudo-eigenvalue equation, which of course brings up computational issues such as the convergence of the procedure to selected (or a manifold of) pseudo-eigenpairs λ_k, $\boldsymbol{x}_k$. Once such an eigenpair has been determined, the $\boldsymbol{y}$ components of the full eigenvector of $\boldsymbol{M}$ can be obtained using Eq. (A.1.30).

Note that, in general, two pseudo-eigenvectors $\boldsymbol{x}_k$ and $\boldsymbol{x}_l$ associated with distinct eigenvalues, $\lambda_k \neq \lambda_l$, are not orthogonal. Orthogonality only applies to the full eigenvectors of $\boldsymbol{M}$, of which the $\boldsymbol{x}$ components are only a part. (As pseudo-eigenvectors of Eq. (A.1.31) $\boldsymbol{x}_k$ and $\boldsymbol{x}_l$ derive from distinct sub-block matrices.)

The computational benefit of partitioning the eigenvalue problem depends of course on the specifics of the problem under consideration. Generally, a partitioning scheme will be desirable in which the sub-block $\boldsymbol{A}$ is of small dimension and preferably coupled only weakly to the (large) sub-block $\boldsymbol{D}$ via the off-diagonal blocks $\boldsymbol{B}$ and $\boldsymbol{C}$ $(= \boldsymbol{B}^\dagger)$. The advantage of a low-dimensional pseudo-eigenvalue problem, being of the dimension of $\boldsymbol{A}$, is offset to a certain extent by the need of inverting the large matrix $(\lambda\mathbf{1} - \boldsymbol{D})$ for various values of λ in the iterative procedure. The latter task is addressed in the following sub-section.

The simplest partitioning scheme is one where the sub-block $\boldsymbol{A}$ is one-dimensional:

$$\boldsymbol{m} = \begin{pmatrix} a & b_1 & \dots & b_m \\ c_1 & d_{11} & \dots & d_{1m} \\ \vdots & \vdots & \ddots & \vdots \\ c_m & d_{m1} & \dots & d_{mm} \end{pmatrix} \tag{A.1.32}$$

The corresponding one-dimensional eigenvalue equation reads

$$\lambda - a - \boldsymbol{b}^t(\lambda - \boldsymbol{d})^{-1}\boldsymbol{c} = 0 \tag{A.1.33}$$

where $\boldsymbol{d}$ denotes the matrix of elements d_{ij} and $\boldsymbol{b}$ and $\boldsymbol{c}$ are (column) vectors of the elements b_i and c_j, respectively. If $\boldsymbol{d}$ is already diagonal, $d_{kl} = \delta_{kl}d_k$, and, moreover, supposing $b_i = c_i^*$ for hermiticity of $\boldsymbol{m}$, the eigenvalue equation simplifies to

$$\lambda - a - \sum_{k=1}^{m} \frac{|b_k|^2}{\lambda - d_k} = 0 \tag{A.1.34}$$

This allows for "graphical" solutions obtained as the intersections of the straight line $f(\lambda) = \lambda - a$ and the pole function $g(\lambda) = \sum_k |b_k|^2(\lambda - d_k)^{-1}$.

Inversion of Resolvent-Type Matrices

Let $\boldsymbol{M}$ be a hermitian matrix of dimension n. The task of inverting the matrix $(\omega\mathbf{1} - \boldsymbol{M})$ as a function of a variable ω is encountered in resolvent-type matrices such as

$$\boldsymbol{R}(\omega) = (\omega\mathbf{1} - \boldsymbol{M})^{-1} \tag{A.1.35}$$

This inversion problem is essentially equivalent to solving the eigenvalue problem for $\boldsymbol{M}$,

$$\boldsymbol{M}\boldsymbol{X} = \boldsymbol{X}\boldsymbol{\Omega}, \quad \boldsymbol{X}^\dagger\boldsymbol{X} = \mathbf{1} \tag{A.1.36}$$

Here, $\boldsymbol{\Omega}$ denotes the diagonal matrix of eigenvalues, $\omega_1, \ldots, \omega_n$, and $\boldsymbol{X}$ is the matrix of eigenvectors (represented by columns of $\boldsymbol{X}$). According to the eigenvalue equations, the original matrix $\boldsymbol{M}$ can be written as

$$\boldsymbol{M} = \boldsymbol{X}\boldsymbol{\Omega}\boldsymbol{X}^\dagger \tag{A.1.37}$$

Using this form and the orthonormality relations for $\boldsymbol{X}$ in the resolvent matrix (A.1.35) yields

$$\begin{aligned}\boldsymbol{R}(\omega) &= (\omega\boldsymbol{X}\boldsymbol{X}^\dagger - \boldsymbol{X}\boldsymbol{\Omega}\boldsymbol{X}^\dagger)^{-1} = \left[\boldsymbol{X}(\omega\mathbf{1} - \boldsymbol{\Omega})\boldsymbol{X}^\dagger\right]^{-1} \\ &= \boldsymbol{X}(\omega\mathbf{1} - \boldsymbol{\Omega})^{-1}\boldsymbol{X}^\dagger\end{aligned} \tag{A.1.38}$$

The last line gives an explicit expression of $(\omega\mathbf{1} - \boldsymbol{M})^{-1}$ in terms of the eigenvalues and eigenvectors of $\boldsymbol{M}$; the inversion of the diagonal matrix $(\omega\mathbf{1} - \boldsymbol{\Omega})$ is of course trivial.

A particular matrix element of $\boldsymbol{R}(\omega)$ is given by

$$\boldsymbol{R}_{pq}(\omega) = \sum_{k=1}^{n} X_{pk} \frac{1}{\omega - \omega_k} X_{qk}^* \tag{A.1.39}$$

While the relation (A.1.38) is of obvious theoretical interest, one may wonder about its usefulness in actual computations. Again, this depends on the actual problem under consideration. If $\boldsymbol{M}$ is a large matrix, full diagonalization is certainly not a desirable or feasible option. Often, however, one is not interested in the resolvent matrix as a function of the energy variable, but rather in its particular pole positions, that is, selected eigenvalues of $\boldsymbol{M}$. Then, of course the eigenvalue problem of $\boldsymbol{M}$ is to be dealt with in the first place, and one can resort to the iterative diagonalization methods such as the Davidson [2] or Lanczos procedures [3, 4]. In case one actually needs $\boldsymbol{R}(\omega)$ as a function of ω such as for frequency-dependent polarizabilities (see Appendix A.7), a viable computational scheme can be based on the Lanczos algorithm. There the exact spectral representation according to Eq. (A.1.38) can be approximated by a corresponding representation in terms of a set of L Lanczos eigenvalues and eigenvectors (Lanczos pseudo-spectrum), where L is the number of Lanczos iteration steps.

An Application: Brillouin–Wigner Perturbation Theory

Using the partitioning technique in the eigenvalue problem of the hamiltonian allows one to understand the essence of the Brillouin–Wigner perturbation theory. The BW expansion for the ground-state energy is obtained from Eq. (A.1.10) with the choice $\epsilon = E_0$. For notational ease, we consider the case of a one-particle system, where that expansion takes on the form

$$\begin{aligned} e_0 =& e_0^{(0)} + \langle\phi_0|\hat{h}_1|\phi_0\rangle + \langle\phi_0|\hat{h}_1 \sum_{\nu=1}^{\infty} \left(\frac{\hat{q}_0\,\hat{h}_1}{e_0 - \hat{h}_0}\right)^{\nu} |\phi_0\rangle \\ =& \langle\phi_0|\hat{h}|\phi_0\rangle + \langle\phi_0|\hat{h}_1 \frac{\hat{q}_0}{e_0 - \hat{h}_0}\hat{h}_1|\phi_0\rangle + \langle\phi_0|\hat{h}_1 \frac{\hat{q}_0}{e_0 - \hat{h}_0}\hat{h}_1 \frac{\hat{q}_0}{e_0 - \hat{h}_0}\hat{h}_1|\phi_0\rangle + \ldots \end{aligned} \tag{A.1.40}$$

Here, $\hat{q}_0 = \hat{1} - |\phi_0\rangle\langle\phi_0|$ is the one-particle analogue to the projector (A.1.2). Truncating the expansion after the nth order term gives rise to an implicit equation for the energy e_0 of the type

$$e_0 = f_n(e_0) \tag{A.1.41}$$

Obviously, the BW procedure does not lead to a usual PT expansion for e_0. Rather, it will be seen to be a specific way of applying a CI treatment to the ground state.

Representing the hamiltonian in terms of the eigenstates ϕ_k, $k = 0, 1, \ldots$ of $\hat{h}_0$,

$$\boldsymbol{h} = \begin{pmatrix} h_{00} & h_{01} & h_{02} & \ldots \\ h_{10} & h_{11} & h_{12} & \ldots \\ h_{20} & h_{21} & h_{22} & \ldots \\ \vdots & \vdots & \vdots & \ddots \end{pmatrix} \tag{A.1.42}$$

the Schrödinger equation takes on the form

$$\boldsymbol{h}\boldsymbol{x} = e_0 \boldsymbol{x} \tag{A.1.43}$$

where $\boldsymbol{x}$ denotes the ground-state eigenvector. Partitioning of the Schrödinger equation as in Eq. (A.1.32), one obtains the one-dimensional secular equation

$$e_0 - h_{00} - \boldsymbol{v}^\dagger (e_0 - \tilde{\boldsymbol{h}})^{-1} \boldsymbol{v} = 0 \tag{A.1.44}$$

Here, $\boldsymbol{v}$ is the (column) vector of the matrix elements

$$v_k = h_{k0} = \langle \phi_k | \hat{h}_1 | \phi_0 \rangle, \; k \geq 1 \tag{A.1.45}$$

and $\tilde{\boldsymbol{h}}$ denotes the sub-block of $\boldsymbol{h}$ with matrix elements $h_{kl}, k, l \geq 1$. Writing h_{00} more explicitly as $h_{00} = e_0^{(0)} + \langle \phi_0 | \hat{h}_1 | \phi_0 \rangle$ and, correspondingly, $\tilde{\boldsymbol{h}}$ as

$$\tilde{\boldsymbol{h}} = \boldsymbol{e}^{(0)} + \tilde{\boldsymbol{h}}_1 \tag{A.1.46}$$

where $\boldsymbol{e}^{(0)}$ is the diagonal matrix of eigenvalues $e_k^{(0)}$ of $\hat{h}_0, k \geq 1$, the one-dimensional eigenvalue equation (A.1.44) takes on the form

$$e_0 = e_0^{(0)} + \langle \phi_0 | \hat{h}_1 | \phi_0 \rangle + \boldsymbol{v}^\dagger (e_0 \mathbf{1} - \boldsymbol{e}^{(0)} - \tilde{\boldsymbol{h}}_1)^{-1} \boldsymbol{v} \tag{A.1.47}$$

Expanding the matrix inverse in a geometrical series according to

$$\begin{aligned}
(e_0 \mathbf{1} - \boldsymbol{e}^{(0)} - \tilde{\boldsymbol{h}}_1)^{-1} =& (e_0 \mathbf{1} - \boldsymbol{e}^{(0)})^{-1} \left(\mathbf{1} - \frac{\tilde{\boldsymbol{h}}_1}{e_0 \mathbf{1} - \boldsymbol{e}^{(0)}} \right)^{-1} \\
=& (e_0 \mathbf{1} - \boldsymbol{e}^{(0)})^{-1} + (e_0 \mathbf{1} - \boldsymbol{e}^{(0)})^{-1} \tilde{\boldsymbol{h}}_1 (e_0 \mathbf{1} - \boldsymbol{e}^{(0)})^{-1} + \ldots
\end{aligned}$$

the right-hand of side Eq. (A.1.47) can be directly identified with the BW expansion (A.1.40). To make the BW expressions more explicit, one may replace each $\hat{q}_0$ operator with $\sum_{k \neq 0} |\phi_k\rangle \langle \phi_k|$.

To conclude, the BW approach to the ground-state energy is essentially equivalent to a CI treatment, in which the CI eigenvalue problem is recast via partitioning into a one-dimensional iterative eigenvalue equation. The energy-dependent inversion of the (large) residual matrix is handled by truncating the associated geometric series at successively higher orders.

A.2 Proof of the Gell-Mann and Low Theorem

The proof of the Gell-Mann and Low theorem presented in the following is essentially based on the version given in the textbook by Fetter and Walecka [5].

Theorem:

If the state

$$|\Psi\rangle = \lim_{\epsilon\to 0} \frac{\hat{U}_\epsilon(0,-\infty)|\Phi_0\rangle}{\langle\Phi_0|\hat{U}_\epsilon(0,-\infty)|\Phi_0\rangle} \quad \text{(A.2.1)}$$

exists in all orders of perturbation theory, then it is an eigenstate of $\hat{H}$ with the eigenvalue

$$E_0 = E_0^{(0)} + \lim_{\epsilon\to 0} \frac{\langle\Phi_0|\hat{H}_I\hat{U}_\epsilon(0,-\infty)|\Phi_0\rangle}{\langle\Phi_0|\hat{U}_\epsilon(0,-\infty)|\Phi_0\rangle} \quad \text{(A.2.2)}$$

Proof:
For brevity, we use the notation

$$|\Psi_0(\epsilon)\rangle = \hat{U}_\epsilon(0,-\infty)|\Phi_0\rangle \quad \text{(A.2.3)}$$

where $\hat{U}_\epsilon(0,-\infty)$ is given as in Eq. (4.41).
The *first step* is to derive an expression for $\hat{H}|\Psi_0(\epsilon)\rangle$.
Consider the commutator

$$[\hat{H}_0, \hat{U}_\epsilon(0,-\infty)]|\Phi_0\rangle = (\hat{H}_0 - E_0^{(0)})|\Psi_0(\epsilon)\rangle \quad \text{(A.2.4)}$$

To evaluate the commutator on the left side, consider the nth order term in $\hat{U}_\epsilon(0,-\infty)$:

$$\begin{aligned}
[\hat{H}_0, \hat{H}_I(t_i)\hat{H}_I(t_j)\ldots\hat{H}_I(t_k)] &= [\hat{H}_0, \hat{H}_I(t_i)]\hat{H}_I(t_j)\ldots\hat{H}_I(t_k) \\
&+ \hat{H}_I(t_i)[\hat{H}_0, \hat{H}_I(t_j)]\ldots\hat{H}_I(t_k) + \ldots \\
&+ \hat{H}_I(t_i)\hat{H}_I(t_j)\ldots[\hat{H}_0, \hat{H}_I(t_k)]
\end{aligned}$$

where $t_i > t_j > \cdots > t_k$ is a specific ordering of the n time arguments. Using Eq. (4.18), the commutators may be replaced by time derivatives,

$$[\hat{H}_0, \hat{H}_I(t)] = -i\frac{\partial}{\partial t}\hat{H}_I(t)$$

which yields

$$[\hat{H}_0, \hat{H}_I(t_i)\hat{H}_I(t_j)\ldots\hat{H}_I(t_k)] = (-i)\left(\frac{\partial}{\partial t_1} + \cdots + \frac{\partial}{\partial t_n}\right)\hat{H}_I(t_i)\hat{H}_I(t_j)\ldots\hat{H}_I(t_k)$$

This allows us to write the right-hand side of Eq. (A.2.4) in the form

$$(\hat{H}_0 - E_0^{(0)})|\Psi_0(\epsilon)\rangle = -\sum_{n=1}^{\infty} \frac{(-i)^{n-1}}{n!} \int_{-\infty}^{0} \mathrm{d}t_1\, e^{\epsilon t_1} \dots \int_{-\infty}^{0} \mathrm{d}t_n\, e^{\epsilon t_n}$$
$$\hat{\mathcal{T}}\left[\left(\sum_{k=1}^{n} \frac{\partial}{\partial t_k}\right) \hat{H}_I(t_1)\hat{H}_I(t_2)\dots \hat{H}_I(t_n)\right]|\Phi_0\rangle \qquad \text{(A.2.5)}$$

Next, the time derivatives can be placed before the time-ordering operator. This is possible since

$$\left(\sum_{k=1}^{n} \frac{\partial}{\partial t_k}\right) \theta(t_{P(1)} - t_{P(2)})\theta(t_{P(2)} - t_{P(3)})\dots\theta(t_{P(n-1)} - t_{P(n)}) \equiv 0$$

where $P(i)$ denotes a permutation of the integers $i = 1, \dots, n$. The latter identity holds because each factor $\theta(t_{P(j)} - t_{P(j+1)})$ is differentiated twice, and the respective two time arguments have a plus and a minus sign, respectively. In the simple case $n = 2$, for example, we find

$$(\frac{\partial}{\partial t_1} + \frac{\partial}{\partial t_2})\theta(t_1 - t_2) \equiv \delta(t_1 - t_2) - \delta(t_1 - t_2) \equiv 0$$

In consequence, Eq. (A.2.5) takes on the form

$$(\hat{H}_0 - E_0^{(0)})|\Psi_0(\epsilon)\rangle = -\sum_{n=1}^{\infty} \frac{(-i)^{n-1}}{n!} \int_{-\infty}^{0} \mathrm{d}t_1\, e^{\epsilon t_1} \dots \int_{-\infty}^{0} \mathrm{d}t_n\, e^{\epsilon t_n}$$
$$\left(\sum_{k=1}^{n} \frac{\partial}{\partial t_k}\right) \hat{\mathcal{T}}[\hat{H}_I(t_1)\hat{H}_I(t_2)\dots \hat{H}_I(t_n)]|\Phi_0\rangle \qquad \text{(A.2.6)}$$

To proceed, let us note that

(i) each time derivative on the right-hand side of Eq. (A.2.6) yields the same contribution; that is, we may replace $(\sum_{k=1}^{n} \frac{\partial}{\partial t_k})$ by $n \frac{\partial}{\partial t_1}$.
(ii) we may then use partial integration for the t_1 integration according to

$$e^{\epsilon t_1} \frac{\partial}{\partial t_1}(\dots) = \frac{\partial}{\partial t_1}(e^{\epsilon t_1}(\dots)) - \epsilon e^{\epsilon t_1}(\dots)$$

This allows us to write Eq. (A.2.6) as

$$(\hat{H}_0 - E_0^{(0)})|\Psi_0(\epsilon)\rangle =$$
$$= -\hat{H}_I \sum_{n=1}^{\infty} \frac{(-i)^{n-1}}{(n-1)!} \int_{-\infty}^{0} \mathrm{d}t_2\, e^{\epsilon t_2} \dots \int_{-\infty}^{0} \mathrm{d}t_n\, e^{\epsilon t_n} \hat{\mathcal{T}}\left[\hat{H}_I(t_2)\dots \hat{H}_I(t_n)\right]|\Phi_0\rangle$$
$$+ \epsilon \sum_{n=1}^{\infty} \frac{(-i)^{n-1}}{(n-1)!} \int_{-\infty}^{0} \mathrm{d}t_1\, e^{\epsilon t_1} \dots \int_{-\infty}^{0} \mathrm{d}t_n\, e^{\epsilon t_n} \hat{\mathcal{T}}[\hat{H}_I(t_1)\dots \hat{H}_I(t_n)]|\Phi_0\rangle \qquad \text{(A.2.7)}$$

Note that in the first line on the right-hand side the operator $\hat{H}_I = \hat{H}_I(0)$ could be placed to the left of the time-ordering operator, because $t_1 = 0$ is the maximal value in any of the remaining time arguments. Equation (A.2.7) can be written in a more compact form as follows:

$$(\hat{H}_0 - E_0^{(0)})|\Psi_0(\epsilon)\rangle = -\hat{H}_I|\Psi_0(\epsilon)\rangle + i\epsilon g \frac{\partial}{\partial g}|\Psi_0(\epsilon)\rangle\bigg|_{g=1} \tag{A.2.8}$$

Here, the first term on the right-hand side is readily identified with the first term on the right-hand side of Eq. (A.2.7). The second term on the right-hand side can be understood as the result of introducing a coupling strength parameter associated with the interaction part, i.e., $\hat{H}_I \to g\hat{H}_I$, so that in nth order the relation

$$g\frac{\partial}{\partial g}g^n = ng^n$$

holds. The aim of the first step is thus achieved, the result being of the form

$$(\hat{H} - E_0^{(0)})|\Psi_0(\epsilon)\rangle = i\epsilon g \frac{\partial}{\partial g}|\Psi_0(\epsilon)\rangle\bigg|_{g=1} \tag{A.2.9}$$

At this point, however, the limit $\epsilon \to 0$ cannot be carried out, because $|\Psi_0(\epsilon)\rangle$ contains diverging contributions.

Second step: Bring Eq. (A.2.9) into a form complying with the premise of the theorem, so that the adiabatic limit can be taken.

Multiplying Eq. (A.2.9) from the left by $\frac{\langle\Phi_0|}{\langle\Phi_0|\Psi_0(\epsilon)\rangle}$ yields

$$E_0(\epsilon) - E_0^{(0)} = i\epsilon g \frac{\partial}{\partial g}\log\langle\Phi_0|\Psi_0(\epsilon)\rangle \tag{A.2.10}$$

where $E_0(\epsilon)$ is defined as in Eq. (A.2.2) without taking the limit $\epsilon \to 0$; the choice $g = 1$ is no longer explicitly indicated. On the other hand, we may multiply Eq. (A.2.9) from the left by $\langle\Phi_0|\Psi_0(\epsilon)\rangle^{-1}$, yielding

$$\begin{aligned}(\hat{H} - E_0^{(0)})\frac{|\Psi_0(\epsilon)\rangle}{\langle\Phi_0|\Psi_0(\epsilon)\rangle} &= i\epsilon g \frac{1}{\langle\Phi_0|\Psi_0(\epsilon)\rangle}\frac{\partial}{\partial g}|\Psi_0(\epsilon)\rangle \\ &= i\epsilon g \frac{\partial}{\partial g}\frac{|\Psi_0(\epsilon)\rangle}{\langle\Phi_0|\Psi_0(\epsilon)\rangle} - i\epsilon g|\Psi_0(\epsilon)\rangle\frac{\partial}{\partial g}\frac{1}{\langle\Phi_0|\Psi_0(\epsilon)\rangle} \\ &= i\epsilon g \frac{\partial}{\partial g}\frac{|\Psi_0(\epsilon)\rangle}{\langle\Phi_0|\Psi_0(\epsilon)\rangle} + \frac{|\Psi_0(\epsilon)\rangle}{\langle\Phi_0|\Psi_0(\epsilon)\rangle} i\epsilon g \frac{\partial}{\partial g}\log\langle\Phi_0|\Psi_0(\epsilon)\rangle \end{aligned} \tag{A.2.11}$$

Using Eq. (A.2.10) in the second term of the last line of Eq. (A.2.11), takes us to the final result

$$(\hat{H} - E_0(\epsilon))\frac{|\Psi_0(\epsilon)\rangle}{\langle\Phi_0|\Psi_0(\epsilon)\rangle} = i\epsilon g \frac{\partial}{\partial g}\frac{|\Psi_0(\epsilon)\rangle}{\langle\Phi_0|\Psi_0(\epsilon)\rangle} \tag{A.2.12}$$

which is of a form in which the limit $\epsilon \to 0$ can be taken on both sides. According to the assumption,

$$|\Psi\rangle = \lim_{\epsilon \to 0} \frac{|\Psi_0(\epsilon)\rangle}{\langle \Phi_0 | \Psi_0(\epsilon)\rangle}$$

exists in all orders of perturbation theory, and this property is not affected by applying the operation $g\frac{\partial}{\partial g}$ (in each order n). Then, due to the factor ϵ, the right-hand side vanishes for $\epsilon \to 0$, and we obtain

$$(\hat{H} - E_0) \lim_{\epsilon \to 0} \frac{|\Psi_0(\epsilon)\rangle}{\langle \Phi_0 | \Psi_0(\epsilon)\rangle} = 0$$

which completes the proof.

Combining Operators and the Time-Evolution Operator:

In deriving Eq. (4.56), the following extended transitivity property of the time-evolution operator has been used:

$$\hat{U}_\epsilon(\infty, t)\hat{O}_I(t)\hat{U}_\epsilon(t, -\infty) = \sum_{n=0}^{\infty} \frac{(-i)^n}{n!} \int_{-\infty}^{\infty} \mathrm{d}t_1 e^{-\epsilon|t_1|} \ldots \int_{-\infty}^{\infty} \mathrm{d}t_n\, e^{-\epsilon|t_n|} \hat{\mathcal{T}}\left[\hat{H}_I(t_1) \ldots \hat{H}_I(t_n)\hat{O}_I(t)\right] \tag{A.2.13}$$

To prove this property, let us consider the nth order term on the right-hand side of Eq. (A.2.13) for a given value of t,

$$X^{(n)}(t) = \frac{(-i)^n}{n!} \int_{-\infty}^{\infty} \mathrm{d}t_1 \ldots \int_{-\infty}^{\infty} \mathrm{d}t_n\, \hat{\mathcal{T}}\left[\hat{H}_I(t_1) \ldots \hat{H}_I(t_n)\hat{O}_I(t)\right] \tag{A.2.14}$$

omitting here the switching functions for simplicity. The n-fold time integrations can be broken up into $(n+1)!$ distinct contributions according to the different time-orderings of the $n+1$ time arguments $t_1, \ldots, t_n, t$. These time-orderings can further be classified according to the number $\mu = 0, \ldots, n$, counting the time arguments larger than t; correspondingly, $\nu = n - \mu$ is the number of time arguments smaller than t. For a given μ, there are $\binom{n}{\mu}$ ways to select μ time arguments from the set $t_1, \ldots, t_n$, and for each selection, there are $\mu!\,\nu!$ time-orderings in which the first μ and the last ν time arguments are permuted among themselves. Note that the total number of time-orderings in class μ is $n! = \binom{n}{\mu}\,\mu!\nu!$. Let

$$t_i > t_j \cdots > t_k > t_l > t > t_r > \cdots > t_u$$

denote a specific time-ordering of class μ; that is, $t_i > t_j \cdots > t_k > t_l$ are μ time arguments. The corresponding contribution in the integrand can be written as

$$\theta(t_i - t_j)\dots\theta(t_k - t_l)\theta(t_l - t)\theta(t - t_r)\dots\hat{H}_I(t_i)\hat{H}_I(t_j)\dots\hat{H}_I(t_l)$$
$$\hat{O}_I(t)\hat{H}_I(t_r)\dots\hat{H}_I(t_u)$$

Obviously, the $\mu!$ permuted time-orderings of the first μ time arguments can again be combined according to

$$\hat{\mathcal{T}}\left[\hat{H}_I(t_i)\dots\hat{H}_I(t_l)\right] \tag{A.2.15}$$

in a corresponding $\hat{\mathcal{T}}$ product, and the same holds for the last ν time arguments. Using the fact that the time arguments in the time integrations are dummy variables which can be renamed at will, Eq. (A.2.14) can be written as

$$X^{(n)}(t) = \frac{(-i)^n}{n!}\sum_{\mu=0}^{n}\binom{n}{\mu}\int_t^\infty \mathrm{d}t_1\dots\int_t^\infty \mathrm{d}t_\mu\,\hat{\mathcal{T}}\left[\hat{H}_I(t_1)\dots\hat{H}_I(t_\mu)\right]$$
$$\hat{O}_I(t)\int_{-\infty}^t \mathrm{d}t_1\dots\int_{-\infty}^t \mathrm{d}t_\nu\,\hat{\mathcal{T}}\left[\hat{H}_I(t_1)\dots\hat{H}_I(t_\nu)\right],\quad \nu = n-\mu \tag{A.2.16}$$

Using this result on the right-hand side of Eq. (A.2.13) and taking again the switching functions into account, we obtain

$$\sum_{n=0}^{\infty}\frac{(-i)^n}{n!}\int_{-\infty}^{\infty}\mathrm{d}t_1 e^{-\epsilon|t_1|}\dots\int_{-\infty}^{\infty}\mathrm{d}t_n\, e^{-\epsilon|t_n|}\,\hat{\mathcal{T}}\left[\hat{H}_I(t_1)\dots\hat{H}_I(t_n)\hat{O}_I(t)\right]$$
$$=\sum_{n=0}^{\infty}\frac{(-i)^n}{n!}\sum_{\mu,\nu=0}^{n}\delta_{n,\mu+\nu}\binom{n}{\mu}\int_t^\infty \mathrm{d}t_1 e^{-\epsilon|t_1|}\dots\int_t^\infty \mathrm{d}t_\mu e^{-\epsilon|t_\mu|}\,\hat{\mathcal{T}}\left[\hat{H}_I(t_1)\dots\hat{H}_I(t_\mu)\right]$$
$$\hat{O}_I(t)\int_{-\infty}^t \mathrm{d}t_1 e^{-\epsilon|t_1|}\dots\int_{-\infty}^t \mathrm{d}t_\nu e^{-\epsilon|t_\nu|}\,\hat{\mathcal{T}}\left[\hat{H}_I(t_1)\dots\hat{H}_I(t_\nu)\right] \tag{A.2.17}$$

As in the proof of $e^{x+y} = e^x e^y$, the summations on the right-hand side can be reordered according to

$$\sum_{n=0}^{\infty}\frac{(-i)^n}{n!}\sum_{\mu,\nu=0}^{n}\delta_{n,\mu+\nu}\binom{n}{\mu} \to \sum_{\mu=0}^{\infty}\frac{(-i)^\mu}{\mu!}\sum_{\nu=0}^{\infty}\frac{(-i)^\nu}{\nu!} \tag{A.2.18}$$

which brings us to the left side of Eq. (A.2.13).

A.3 Proof of Wick's Theorem

The key step in the proof of Wick's theorem [6] is the following
Lemma 1: Let $\hat{a}_i\hat{a}_j\ldots\hat{a}_s\hat{a}_t$ be a product of time-dependent fermion operators and $\hat{b}$ an operator with a time argument smaller than those of the factors in the product. Then the following identity holds:

$$\begin{aligned}\hat{\mathcal{N}}\left[\hat{a}_i\hat{a}_j\ldots\hat{a}_s\hat{a}_t\right]\hat{b} = {} & \hat{\mathcal{N}}\left[\hat{a}_i\hat{a}_j\ldots\hat{a}_s\overbracket{\hat{a}_t\hat{b}}\right] \\ & + \hat{\mathcal{N}}\left[\hat{a}_i\hat{a}_j\ldots\overbracket{\hat{a}_s\hat{a}_t\hat{b}}\right] + \cdots + \hat{\mathcal{N}}\left[\overbracket{\hat{a}_i\hat{a}_j\ldots\hat{a}_s\hat{a}_t\hat{b}}\right] \\ & + \hat{\mathcal{N}}\left[\hat{a}_i\hat{a}_j\ldots\hat{a}_s\hat{a}_t\hat{b}\right]\end{aligned} \tag{A.3.1}$$

Proof: The operator $\hat{b}$ can either be a physical operator, $\hat{b}=\hat{v}$, or an unphysical operator, $\hat{b}=\hat{u}$. In the latter case, the lemma is essentially trivial. All contractions on the right-hand side vanish,

$$\hat{a}_l^{\cdot}\hat{u}^{\cdot} = \hat{\mathcal{T}}\left[\hat{a}_l\hat{u}\right] - \hat{\mathcal{N}}\left[\hat{a}_l\hat{u}\right] = 0$$

since each of the products $\hat{a}_l\hat{u}$ is both time- and normal-ordered. Moreover,

$$\hat{\mathcal{N}}\left[\hat{a}_i\ldots\hat{a}_t\hat{u}\right] = \hat{\mathcal{N}}\left[\hat{a}_i\ldots\hat{a}_t\right]\hat{u}$$

as $\hat{u}$ is unphysical, so that Eq. (A.3.1) is fulfilled.

Now let us consider the case, where $\hat{b}=\hat{v}$ is a physical operator. Without loss of generality, we can assume that the factors in the operator product are already normal-ordered (otherwise a corresponding rearrangement could be performed on both sides of Eq. A.3.1). In particular, we may assume that there are μ physical and ν unphysical factors in the original product, so that the left side of Eq. (A.3.1) can be written as

$$\hat{\mathcal{N}}\left[\hat{v}_1\ldots\hat{v}_\mu\hat{u}_1\ldots\hat{u}_\nu\right]\hat{v} = \hat{v}_1\ldots\hat{v}_\mu\hat{u}_1\ldots\hat{u}_\nu\hat{v} \tag{A.3.2}$$

To proceed, we commute $\hat{v}$ successively to the left. The first step gives

$$\begin{aligned}\hat{v}_1\ldots\hat{v}_\mu\hat{u}_1\ldots\hat{u}_\nu\hat{v} = & -\hat{v}_1\ldots\hat{v}_\mu\hat{u}_1\ldots\hat{u}_{\nu-1}\hat{v}\hat{u}_\nu \\ & + \hat{v}_1\ldots\hat{v}_\mu\hat{u}_1\ldots\hat{u}_{\nu-1}\left\{\hat{u}_\nu,\hat{v}\right\}\end{aligned} \tag{A.3.3}$$

Now the anticommutator $\left\{\hat{u}_\nu,\hat{v}\right\}$ on the right-hand side can be replaced by the contraction $\hat{u}_\nu^{\cdot}\hat{v}^{\cdot}$,

$$\left\{\hat{u}_\nu,\hat{v}\right\} = \hat{u}_\nu\hat{v} + \hat{v}\hat{u}_\nu = \hat{\mathcal{T}}\left[\hat{u}_\nu\hat{v}\right] - \hat{\mathcal{N}}\left[\hat{u}_\nu\hat{v}\right] = \hat{u}_\nu^{\cdot}\hat{v}^{\cdot} \tag{A.3.4}$$

since $\hat{\boldsymbol{\mathcal{T}}}\left[\hat{u}_\nu\hat{v}\right] = \hat{u}_\nu\hat{v}$ and $\hat{\boldsymbol{\mathcal{N}}}\left[\hat{u}_\nu\hat{v}\right] = -\hat{v}\hat{u}_\nu$. Accordingly, Eq. (A.3.3) can be written in the form

$$\begin{aligned}\hat{v}_1 \ldots \hat{v}_\mu \hat{u}_1 \ldots \hat{u}_\nu \hat{v} = &- \hat{v}_1 \ldots \hat{v}_\mu \hat{u}_1 \ldots \hat{u}_{\nu-1} \hat{v} \hat{u}_\nu \\ &+ \hat{\boldsymbol{\mathcal{N}}}\left[\hat{v}_1 \ldots \hat{v}_\mu \hat{u}_1 \ldots \hat{u}_{\nu-1} \overbracket{\hat{u}_\nu \hat{v}}\right]\end{aligned} \tag{A.3.5}$$

where the contraction has been inserted in the $\hat{\boldsymbol{\mathcal{N}}}$ product (see Eq. 5.15). Obviously, the second term on the right-hand side reproduces the first contracted $\hat{\boldsymbol{\mathcal{N}}}$ product in Eq. (A.3.1).

In the same way, the anticommutator arising in the second step, $\left\{\hat{u}_{\nu-1}, \hat{v}\right\}$ can be replaced by the contraction $\hat{u}^{\bullet}_{\nu-1}\hat{v}^{\bullet}$:

$$\begin{aligned}\hat{v}_1 \ldots \hat{v}_\mu \hat{u}_1 \ldots \hat{u}_\nu \hat{v} =& \hat{v}_1 \ldots \hat{v}_\mu \hat{u}_1 \ldots \hat{u}_{\nu-2} \hat{v} \hat{u}_{\nu-1} \hat{u}_\nu \\ &+ \hat{\boldsymbol{\mathcal{N}}}\left[\hat{v}_1 \ldots \hat{v}_\mu \hat{u}_1 \ldots \hat{u}_{\nu-1} \overbracket{\hat{u}_\nu \hat{v}}\right] \\ &+ \hat{\boldsymbol{\mathcal{N}}}\left[\hat{v}_1 \ldots \hat{v}_\mu \hat{u}_1 \ldots \hat{u}_{\nu-2} \overbracket{\hat{u}_{\nu-1} \hat{u}_\nu \hat{v}}\right]\end{aligned} \tag{A.3.6}$$

Note that the phase (-1) in the first term on the right-hand side of Eq. (A.3.5) has been accounted for by writing the $\hat{u}^{\bullet}_{\nu-1}\hat{v}^{\bullet}$ contraction in a separated form (see Eq. 5.15) maintaining the original order of the operators. Performing ν commutations in such a way, the original product (A.3.2) can be written as

$$\begin{aligned}\hat{v}_1 \ldots \hat{v}_\mu \hat{u}_1 \ldots \hat{u}_\nu \hat{v} =& (-1)^\nu \hat{v}_1 \ldots \hat{v}_\mu \hat{v} \hat{u}_1 \ldots \hat{u}_\nu + \hat{\boldsymbol{\mathcal{N}}}\left[\hat{v}_1 \ldots \hat{v}_\mu \hat{u}_1 \ldots \hat{u}_{\nu-1} \overbracket{\hat{u}_\nu \hat{v}}\right] \\ &+ \hat{\boldsymbol{\mathcal{N}}}\left[\hat{v}_1 \ldots \hat{v}_\mu \hat{u}_1 \ldots \hat{u}_{\nu-2} \overbracket{\hat{u}_{\nu-1} \hat{u}_\nu \hat{v}}\right] + \ldots\end{aligned} \tag{A.3.7}$$

where the operator $\hat{v}$ is situated to the left of the $\hat{u}$ operators and ν normal-ordered products with one contraction have emerged as a result of commuting $\hat{v}$ to that position.

At this point, we may stop, because a contraction (and anticommutator) of $\hat{v}$ and any of the μ physical operators vanishes. Obviously, the first term on the right-hand side written as

$$(-1)^\nu \hat{v}_1 \ldots \hat{v}_\mu \hat{v} \hat{u}_1 \ldots \hat{u}_\nu = \hat{\boldsymbol{\mathcal{N}}}\left[\hat{v}_1 \ldots \hat{v}_\mu \hat{u}_1 \ldots \hat{u}_\nu \hat{v}\right] \tag{A.3.8}$$

reproduces the contraction-free normal product on the right-hand side of Eq. (A.3.1). This concludes the proof of the lemma.

It should be noted that the lemma can readily be generalized to the case where the $\hat{\boldsymbol{\mathcal{N}}}$ product contains one or more contractions, since the contractions can be taken out of the respective $\hat{\boldsymbol{\mathcal{N}}}$ product.

Using Lemma 1, Wick's theorem follows readily by induction with regard to the number n of operators.

Proof of Wick's theorem: For $n = 2$, Wick's theorem (cf. Eq. 5.16) merely reformulates the definition (5.6) of a contraction:

$$\hat{\mathcal{T}}\left[\hat{a}\hat{b}\right] = \hat{\mathcal{N}}\left[\hat{a}\hat{b}\right] + \hat{a}^{\bullet}\hat{b}^{\bullet}$$

Suppose that the theorem holds for n factors and consider a product of $n + 1$ factors $\hat{a}_1 \ldots \hat{a}_n \hat{a}_{n+1}$. The last operator $\hat{a}_{n+1}$ may be chosen such that its time argument is smaller than those of the other factors, which allows us to write

$$\hat{\mathcal{T}}\left[\hat{a}_1 \ldots \hat{a}_n\right]\hat{a}_{n+1} = \hat{\mathcal{T}}\left[\hat{a}_1 \ldots \hat{a}_n \hat{a}_{n+1}\right] \tag{A.3.9}$$

Multiplying both sides of Wick's Eq. (5.16) on the right by $\hat{a}_{n+1}$ yields

$$\begin{aligned}
\hat{\mathcal{T}}\left[\hat{a}_1 \ldots \hat{a}_n\right]\hat{a}_{n+1} &= \hat{\mathcal{T}}\left[\hat{a}_1 \ldots \hat{a}_n \hat{a}_{n+1}\right] \\
&= \hat{\mathcal{N}}\left[\hat{a}_1 \ldots \hat{a}_n\right]\hat{a}_{n+1} + \hat{\mathcal{N}}\left[\overbracket{\hat{a}_1\hat{a}_2}\hat{a}_3 \ldots\right]\hat{a}_{n+1} \\
&\quad + \hat{\mathcal{N}}\left[\overbracket{\hat{a}_1\hat{a}_2\hat{a}_3} \ldots\right]\hat{a}_{n+1} + \ldots + \hat{\mathcal{N}}\left[\overbracket{\hat{a}_1\hat{a}_2}\overbracket{\hat{a}_3\hat{a}_4} \ldots\right]\hat{a}_{n+1} + \ldots
\end{aligned} \tag{A.3.10}$$

Applying Lemma 1 individually to all the terms on the right-hand side, one obviously reproduces the right-hand side of Wick's operator identity for a product of $n + 1$ operators:

$$\begin{aligned}
\hat{\mathcal{T}}\left[\hat{a}_1 \ldots \hat{a}_n \hat{a}_{n+1}\right] &= \hat{\mathcal{N}}\left[\hat{a}_1 \ldots \hat{a}_n \hat{a}_{n+1}\right] \\
&\quad + \hat{\mathcal{N}}\left[\text{sum over all possible pairs of contractions}\right]
\end{aligned} \tag{A.3.11}$$

Finally, the initial restriction with regard to the time argument of $\hat{a}_{n+1}$ can be lifted, since the factors within the $\hat{\mathcal{T}}$ and $\hat{\mathcal{N}}$ products can be changed at will, resulting only in a common phase on both sides of Eq. (A.3.11).

A.4 Time-Ordered Diagrams: Derivation of Goldstone Rules

To establish the rules (G1)–(G4) of Sect. 7.2 for assigning analytic expressions to the time-ordered Feynman diagrams, one has to perform the required time integrations for an arbitrary nth order diagram in a systematic and generic way. The procedure presented in the following is based on notes by F. Mertins [7].

Let us consider an nth order Feynman diagram, $D = D(t, t')$, where t, t' denote the two external time arguments. The Fourier transformed diagram, $D(\omega)$, can be obtained according to

$$D(\omega) = \int_{-\infty}^{\infty} dt\, e^{i\omega t} D(t, 0) \tag{A.4.1}$$

using here the legitimate choice $t' = 0$. This means that altogether the computation of $D(\omega)$ requires $n + 1$ time integrations:

$$D(\omega) = \int_{-\infty}^{\infty} dt \int_{-\infty}^{\infty} dt_n \int_{-\infty}^{\infty} dt_{n-1} \dots \int_{-\infty}^{\infty} dt_1\, e^{i\omega t} D(t, 0; t_1, \dots t_n) \tag{A.4.2}$$

Here, $D(t, 0; t_1, \dots, t_n)$ denotes the full time-dependent form of D (prior to the n internal time integrations).

The $n + 1$ time integrations in Eq. (A.4.2) can be decomposed with respect to distinct time-orderings of the time arguments. Let $\tau_1, \tau_2, \dots, \tau_{n+1}$ denote a particular permutation P of the time arguments $t, t_1, \dots, t_n$ such that

$$\tau_1 \leq \tau_2 \leq \dots \leq \tau_m \leq 0 \leq \tau_{m+1} \leq \dots \leq \tau_{n+1} \tag{A.4.3}$$

The corresponding contribution to the full time integral (A.4.2) is given by

$$D^P(\omega) = \underbrace{\int_0^{\infty} d\tau_{n+1} \int_0^{\tau_{n+1}} d\tau_n \dots \int_0^{\tau_{m+2}} d\tau_{m+1}}_{(B)} \underbrace{\int_{-\infty}^{0} d\tau_m \int_{-\infty}^{\tau_m} d\tau_{m-1} \dots \int_{-\infty}^{\tau_2} d\tau_1}_{(A)} \mathcal{D}(\tau_1, \dots) \tag{A.4.4}$$

where $\mathcal{D}(\tau_1, \dots)$ is the integrand of Eq. (A.4.2) as a function of the τ-variables. The original Feynman diagram may be redrawn such that the order of the vertices (both internal and external) reflects the particular permutation of the time arguments. This is referred to as a time-ordered or Goldstone diagram (rule G1).

Let us first consider part (A) of the integral, comprising the variables $\tau_i \leq 0$. In analogy to the treatment of the time-evolution operator in Sect. 4.4, we introduce new variables, $x_1, \dots, x_m$:

$$\begin{aligned} x_m &= \tau_m & \tau_m &= x_m \\ x_{m-1} &= \tau_{m-1} - \tau_m & \tau_{m-1} &= x_{m-1} + x_m \\ &\vdots & &\vdots \\ x_1 &= \tau_1 - \tau_2 & \tau_1 &= x_1 + x_2 + \cdots + x_m \end{aligned} \tag{A.4.5}$$

Here, an x_i-variable is assigned to each pair of successive τ-arguments (with $x_m = \tau_m - 0$ for $i = m$). The second column in (A.4.5) specifies the inverse transformation. The Jacobian determinant of the transformation (A.4.5) simply is

$$\left| \left(\frac{\partial \tau_i}{\partial x_j} \right) \right| = 1 \tag{A.4.6}$$

and the integration limits become $(-\infty, 0)$ for any of the x_i-ntegrations. Accordingly, (A) can be written as

$$(A) \equiv \int_{-\infty}^{0} \mathrm{d}x_m \int_{-\infty}^{0} \mathrm{d}x_{m-1} \ldots \int_{-\infty}^{0} \mathrm{d}x_1 \, \mathcal{D}$$

Before applying a similar transformation to part (B),

$$\begin{aligned} (B) &\equiv \int_0^{\infty} \mathrm{d}\tau_{n+1} \int_0^{\tau_{n+1}} \mathrm{d}\tau_n \ldots \int_0^{\tau_{m+2}} \mathrm{d}\tau_{m+1} \ldots \\ &= \int_0^{\infty} \mathrm{d}\tau_{m+1} \int_{\tau_{m+1}}^{\infty} \mathrm{d}\tau_{m+2} \ldots \int_{\tau_n}^{\infty} \mathrm{d}\tau_{n+1} \ldots \end{aligned}$$

let us note that the original integration procedure (first line above) is equivalent to performing the integrations as indicated in the second line. Applying now the transformation

$$\begin{aligned} y_{m+1} &= \tau_{m+1} & \tau_{m+1} &= y_{m+1} \\ y_{m+2} &= \tau_{m+2} - \tau_{m+1} & \tau_{m+2} &= y_{m+1} + y_{m+2} \\ &\vdots & &\vdots \\ y_{n+1} &= \tau_{n+1} - \tau_n & \tau_{n+1} &= y_{m+1} + \cdots + y_n + y_{n+1} \end{aligned} \tag{A.4.7}$$

part (B) becomes

$$(B) \equiv \int_0^{\infty} \mathrm{d}y_{n+1} \int_0^{\infty} \mathrm{d}y_n \ldots \int_0^{\infty} \mathrm{d}y_{m+1} \ldots$$

Altogether, the $(n+1)$-fold integral (A.4.4) takes on the form

$$D^P(\omega) = \int_0^{\infty} \mathrm{d}y_{n+1} \int_0^{\infty} \mathrm{d}y_n \ldots \int_0^{\infty} \mathrm{d}y_{m+1} \int_{-\infty}^{0} \mathrm{d}x_m \int_{-\infty}^{0} \mathrm{d}x_{m-1} \ldots \int_{-\infty}^{0} \mathrm{d}x_1 \, \tilde{\mathcal{D}} \tag{A.4.8}$$

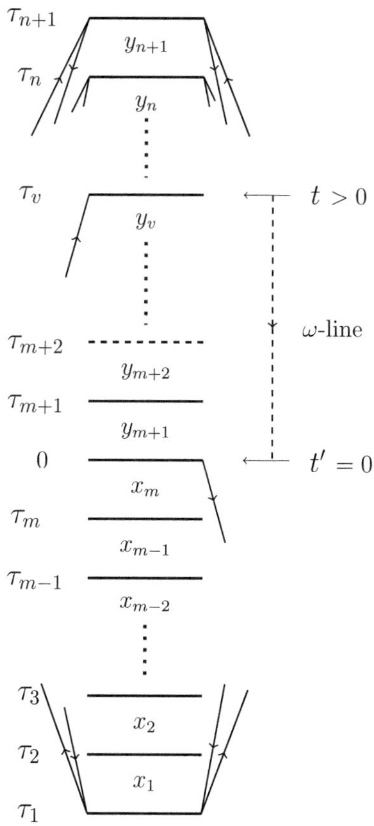

Fig. A.1 Schematic representations of the $n + 2$ vertices of an nth order Feynman diagram associated with a specific time-ordering of the time arguments. Both the internal and external vertices are drawn as horizontal lines. The time arguments τ_ν, $\nu = 1, \ldots, n + 1$ denote a permutation of the original time arguments t_i, $i = 1, \ldots, n$ and t; the time argument of the lower external vertex has been set to $t' = 0$

where $\tilde{\mathcal{D}}$ denotes the integrand as a function of the new variables. As will be seen, the new integration variables, x_i and y_j, can be assigned to "cuts" between consecutive variables of the original set $\tau_1, \ldots, \tau_{n+1}$. The situation is depicted in Fig. A.1 representing the time-ordered diagram D^P in a schematic way. Here, the horizontal lines represent the $n + 2$ general (inner and external) vertices of the time-ordered Feynman diagram, each being associated with one of the $n + 2$ time arguments $\tau_1, \ldots, \tau_m, 0, \tau_{m+1} \ldots, \tau_{n+1}$. There are n internal vertices (normally depicted as wiggly interaction lines), with two outgoing and two incoming free fermion lines each. The two external vertices of the original Feynman diagram have positions according to the respective time-ordering. In Fig. A.1, the upper external vertex is assigned to $t = \tau_v$, with one free fermion line ending here. The other external vertex is assigned to the $t' = 0$ line (between τ_m and τ_{m+1}), where one free fermion line sets out toward a lower lying (inner) vertex.

The time-orderings form two classes, (I) and (II), according to $t > 0$ and $t < 0$, respectively. To be specific, we shall confine us to the case $t > 0$, as assumed in Fig. A.1; the treatment of time-orderings with $t < 0$ is completely analogous.

The integrand is given as a product of free Green's functions and the $e^{i\omega t}$ factor of the Fourier transform. Let us recall the general form (cf. Eq. 3.52)

$$iG_q^0(\tau,\tau') = \begin{cases} \theta(\tau-\tau')e^{-i\epsilon_q(\tau-\tau')}e^{-\eta(\tau-\tau')}\overline{n}_q \\ (-1)\theta(\tau'-\tau)e^{-i\varepsilon_q(\tau-\tau')}e^{\eta(\tau-\tau')}n_q \end{cases} \tag{A.4.9}$$

of a free Green's function, beginning at the vertex τ' and ending at the vertex τ. Here and in the following, the convergence factors $e^{\pm\eta(\tau-\tau')}$ required for the Fourier transforms of the θ-functions are explicitly taken into account.

In a time-ordered diagram, either $\tau > \tau'$ or $\tau < \tau'$. In the former case, G_q^0 is given by the upper expression on the right-hand side, associated with particle states, $\overline{n}_q$. The corresponding G^0-line is referred to as a particle line, its direction arrow pointing upwards (from τ' towards τ). For $\tau < \tau'$, the G^0-line is directed downwards, representing the hole part of G^0 according to the second line in Eq. (A.4.9).

This is summed up in rule (G2): In a time-ordered diagram, the direction arrows of the G^0-lines distinguish particle contributions (arrow upwards) and hole contributions (arrow downwards). Note that each hole line introduces a factor (-1) in the integrand (cf. rule G4).

Now we inspect how a G^0-line running between two vertices, $\tau_s < \tau_r$, contributes to the respective integrations in (A.4.8). Let us first consider a *particle line*, $G_p^0(\tau_r, \tau_s)$, running from τ_s to τ_r; here, the index p stands for "particle." We may distinguish three cases with respect to the extension of the G_p^0-line within the diagram:

(*i*) $r > s \geq m+1$: Since

$$\tau_r - \tau_s = y_r + y_{r-1} + \cdots + y_{s+1}$$

one obtains a factor

$$e^{-i\epsilon_p y_i}e^{-\eta y_i}, \quad i = s+1, s+2, \ldots, r$$

for any y-coordinate encompassed by the τ_r and τ_s vertices. Note that the lower vertex of the particle line could also be $\tau_s = 0$, which would bring the lowest y-coordinate, that is, y_{m+1} into play.

(*ii*) $r > m \geq s$: According to

$$\tau_r - \tau_s = y_r + y_{r-1} + \cdots + y_{m+1} - x_m - x_{m-1} - \cdots - x_s$$

both x and y variables come into play yielding the following factors in the respective integrands:

$$\begin{cases} e^{-i\epsilon_p y_i}e^{-\eta y_i}, \text{ for } i = m+1, \ldots, r \\ e^{i\epsilon_p x_j}e^{\eta x_j}, \text{ for } j = s, s+1, \ldots, m \end{cases}$$

(*iii*) $m \geq r > s$: Here,

$$\tau_r - \tau_s = -x_s - x_{s+1} - \cdots - x_{r-1}$$

which leads to the factors

$$e^{i\epsilon_p x_j} e^{\eta x_j}, \text{ for } j = s, s+1, \ldots, r-1$$

in the integrands of the x-integrals.

To summarize: A particle line crossing two successive vertices contributes to the corresponding x- or y-integration. The orbital-energy factors in the respective integrals are

$$\begin{cases} e^{-i\epsilon_p y_i} \text{ for } y\text{-integrals} \\ e^{+i\epsilon_p x_i} \text{ for } x\text{-integrals} \end{cases} \tag{A.4.10}$$

The convergence factors contribute as follows

$$\begin{cases} e^{-\eta y_i} \text{ for } y\text{-integrals} \\ e^{\eta x_i} \text{ for } x\text{-integrals} \end{cases} \tag{A.4.11}$$

In the same way, we can analyze a *hole line*, $G_h^0(\tau_s, \tau_r)$, running from τ_r to τ_s. As above, we may distinguish the following three cases

(*i*) $r > s \geq m+1$;
(*ii*) $r > m \geq s$;
(*iii*) $m \geq r > s$, (or $\tau_r = 0$);

and analyze them separately. The finding is that any hole line crossing a pair of successive vertices contributes to the associated x- or y-integration, the orbital-energy factors being

$$\begin{cases} e^{i\epsilon_h y_i} \text{ for } y\text{-integrals} \\ e^{-i\epsilon_h x_i} \text{ for } x\text{-integrals} \end{cases} \tag{A.4.12}$$

Note that the orbital energies enter the exponential factors with signs being opposite to those in the particle case (A.4.10). By contrast, the convergence factors deriving from the hole lines agree with those from the particle lines, as given by (A.4.11).

Finally, we have to consider the exponential factor $e^{i\omega t}$ associated with the *Fourier transform*. As assumed in Fig. A.1, the vertex τ_v, $v \geq m+1$, is assigned to the original time argument t. This is consistent with the class (I) of time-orderings ($t > 0$) considered so far. Expressing τ_v in terms of new variables,

$$\tau_v = y_{m+1} + y_{m+2} + \cdots + y_v \tag{A.4.13}$$

shows that the factors $e^{i\omega y_i}$ enter the integrations over y_i, $i = m+1, \ldots, v$. Obviously, those y_i variables correspond to pairs of successive τ vertices crossed by the

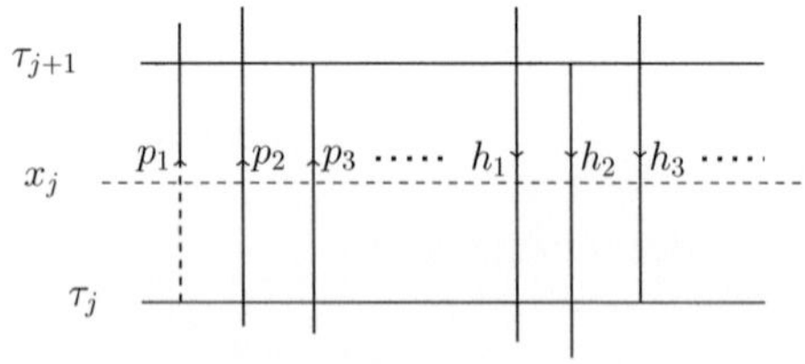

Fig. A.2 Particle and hole lines crossing the cut (dashed line) between vertex levels τ_j, τ_{j+1}

auxiliary ω line starting at the vertex $\tau_v = t$ and ending at the vertex $t' = 0$ (between τ_m and τ_{m+1}).

Now the x_i- and y_i-integrations can successively be performed. Let us consider a particular x-integration, say, over x_j $(1 \leq j \leq m)$. Let there be n_p particle lines with indices $p_1, p_2, \ldots$ running between the vertices τ_j and τ_{j+1} (or likewise crossing the "cut" between τ_j and τ_{j+1}), as well as n_h hole lines, $h_1, h_2, \ldots$. A schematic depiction of the situation is given in Fig. A.2. Each of the particle and hole lines contributes a factor to the integrand, yielding altogether

$$f(x_j) = e^{i(\epsilon_{p_1}+\epsilon_{p_2}+\cdots-\epsilon_{h_1}-\epsilon_{h_2}-\ldots)x_j} e^{(n_p+n_h)\eta x_j} \tag{A.4.14}$$

Note that the overall convergence factor $e^{(n_p+n_h)\eta x_j}$ guarantees that the integrand vanishes in the limit $x_j \to -\infty$, as $n_p + n_h > 0$ is a non-vanishing integer. The integral becomes

$$\int_{-\infty}^{0} \mathrm{d}x_j f(x_j) = i\left(\epsilon_{h_1} + \epsilon_{h_2} + \cdots - \epsilon_{p_1} - \epsilon_{p_2} + \cdots + i(n_p + n_h)\eta\right)^{-1} \tag{A.4.15}$$

Obviously, the infinitesimal $i(n_p + n_h)\eta$ in the denominator is no longer relevant and can be omitted.

A similar procedure applies to the y integrals. For the integration over y_j, $j = m+1, \ldots, n+1$, the integrand is of the form

$$f(y_j) = e^{i(\omega\sigma_j-\epsilon_{p'_1}-\epsilon_{p'_2}+\cdots+\epsilon_{h'_1}+\epsilon_{h'_2}+\ldots)y_j} e^{-(n'_p+n'_h)\eta y_j} \tag{A.4.16}$$

assuming here the presence of n'_p particle and n'_h hole lines. In addition to the contributions from the particle and hole lines crossing the cut between the τ_j and τ_{j-1} vertices, the factor $e^{i\omega y_j}$ comes into play if the auxiliary ω-line crosses the cut. The parameter $\sigma_j = 1, 0$ accounts for the two possibilities.

The integral can be evaluated to give

$$\int_{0}^{\infty} \mathrm{d}y_j f(y_j) = i\left(\omega\sigma_j + \epsilon_{h'_1} + \epsilon_{h'_2} + \cdots - \epsilon_{p'_1} - \epsilon_{p'_2} - \cdots + i(n'_p + n'_h)\eta\right)^{-1} \tag{A.4.17}$$

where the upper limit of the integral vanishes due to the cumulated convergence factor for the y_j-integration, $e^{-(n'_p+n'_h)\eta y_j}$. As above, the infinitesimal $i(n'_p + n'_h)\eta$

in the denominator can be discarded if the denominator is a constant ($\sigma_j = 0$); for ω-dependent denominators ($\sigma_j = 1$), the infinitesimal $i(n'_p + n'_h)\eta$ may be replaced by $i\eta$.

For the case $t > 0$ considered so far, the expressions (A.4.15), (A.4.17) establish rule (G3); note, moreover, that each integral introduces a factor i, as accounted for by rule (G4). The time-orderings of class (II), where $t < 0$, can be treated in a completely analogous way. The ω-dependent denominators in the time-ordered diagrams of class (I) and (II) are of the form $\omega + \cdots + i\eta$ and $\omega + \cdots - i\eta$, respectively.

Revisiting the Adiabatic Limit

After having established the rules for performing the time integrations in the Feynman diagrams, we now may revisit the issue of the adiabatic limit $\epsilon \to 0$ underlying the Gell-Mann and Low theorem (see Sect. 4.2). Here, ϵ is the parameter of the adiabatic switching function $e^{-\epsilon|t|}$ augmenting the interaction part of the hamiltonian (cf. Eq. 4.36).

As stated in Sects. 5.3 and 7.2, the adiabatic limit exists (and is trivial) for the *linked* Feynman diagrams. How can this claim be justified? Just take that limit a priori by skipping the n switching functions $e^{-\epsilon|t_i|}$, $i = 1, \ldots, n$, for the internal vertices of an nth order diagram and verify that a well-defined analytical expression can be assigned to that diagram according the Goldstone rules of Sect. 7.2. However, the analysis given above can easily be extended to take into account the adiabatic switching functions as well, allowing us to take the adiabatic limit a posteriori. Let us consider again a specific time-ordering, such as (A.4.3) belonging to class (I). Using the τ arguments, the product of the n switching functions can be written as

$$e^{-(|t_1|+\cdots+|t_n|)} = e^{\epsilon(\tau_1+\tau_2+\cdots+\tau_m)}e^{-\epsilon(\tau_{m+1}+\ldots\tau_{m+2}+\cdots+\tau_{n+1})+\epsilon\tau_v} \qquad \text{(A.4.18)}$$

Note that $\tau_v = t$ is the time argument of an external vertex and has to be excluded from the product on the right-hand side, more specifically, from the second factor (supposing the time-ordering $t > 0$). This is the achieved by the final term, $+\epsilon\tau_v$, in the exponent.

Using the transformations (A.4.5), (A.4.7), the τ-variables are replaced by the x- and y-variables. According to

$$\tau_1 + \tau_2 + \cdots + \tau_m = mx_m + (m-1)x_{m-1} + \cdots + x_1 \qquad \text{(A.4.19)}$$

the first factor on the right-hand side of Eq. (A.4.18) can be expressed by a product of factors $e^{\epsilon j x_j}$, $j = 1, \ldots, m$. Thus, for each x_j integral, the original convergence factor, $e^{\eta(n_p^{(j)}+n_h^{(j)})x_j}$, due to the hole and particle lines crossing the cut j is augmented by the factor $e^{\epsilon j x_j}$, acting as a "supporting" convergence factor in the limit $\epsilon \to 0$. This means that the ϵ-factors are redundant and can safely be omitted: There is no difference between taking the limit $\epsilon \to 0$ a priori or a posteriori.

In a similar way, the ϵ-functions of the second factor in (A.4.18) can be introduced in the y-integrations. Here, the relation between the sums in the exponent is

$$\tau_{m+1} + \cdots + \tau_{n+1} = (n-m+1)y_{m+1} + (n-m)y_{m+2} + \cdots + 3y_{n-1} + 2y_n + y_{n+1} \tag{A.4.20}$$

As noted above, the external time argument τ_v, being

$$\tau_v = y_{m+1} + \cdots + y_{v-1} + y_v \tag{A.4.21}$$

in terms of the y arguments, has to be discarded. Altogether, for each y_i integration, there is an additional convergence factor, $e^{-\epsilon\lambda_j y_j}$, with λ_j being a positive integer for $j = m+1, \ldots, n$, and $\lambda_{n+1} \geq 0$. In the limit $\epsilon \to 0$, the ϵ-functions act as convergence factors together with the original η convergence functions in eliminating the upper y-integration limits. As above, there is no difference between taking the adiabatic limit a priori or a posteriori.

Remark:
From the analysis given above, it follows that the argument of the redundancy of the ϵ-functions might also be reversed: One could keep the ϵ-functions and discard the η convergence factors in the G^0-expressions (A.4.9) as redundant. Here it is required, though, to introduce a convergence factor in the Fourier transform (A.4.1):

$$D(\omega) = \int_{-\infty}^{\infty} \mathrm{d}t\, e^{i\omega t} e^{-\eta|t|} D(t, 0) \tag{A.4.22}$$

Then, the evaluation of the time integrations for a given Feynman diagram can be based on the expression

$$D(\omega) = \lim_{\epsilon \to 0} \int_{-\infty}^{\infty} \mathrm{d}t \int_{-\infty}^{\infty} \mathrm{d}t_n \int_{-\infty}^{\infty} \mathrm{d}t_{n-1} \ldots \int_{-\infty}^{\infty} \mathrm{d}t_1\, e^{i\omega t} e^{-\eta|t|} D(t, 0; t_1, \ldots, t_n) e^{-\epsilon(|t_1| + \cdots + |t_n|)} \tag{A.4.23}$$

where all η factors in $D(t, 0; t_1, \ldots, t_n)$ have been omitted. The general procedure used to establish the Goldstone rules (G1)–(G4) can be applied in a completely analogous way, leading to identical results. Here, the infinitesimal ϵ and the remaining infinitesimal η ensure that the respective upper and lower integration limits vanish. In the denominators resulting from the x- and y-integrals, the limit $\epsilon \to 0$ is trivial, while the η infinitesimal enters the ω dependent denominators in the form $(\omega \cdots \pm i\eta)$ for time-orderings of class (I) or (II), respectively.

A.5 Dyson Expansion Method for the Static Self-Energy Part

In this Appendix, we briefly describe an approximation method for the static part of the self-energy referred to as Dyson expansion method (DEM) [8, 9].

According to Eq. (8.18), the self-energy is the sum of a static and a dynamic part,

$$\boldsymbol{\Sigma}(\omega) = \boldsymbol{\Sigma}(\infty) + \boldsymbol{M}(\omega) \tag{A.5.1}$$

Here, the static part, $\boldsymbol{\Sigma}(\infty)$, can be expressed directly in terms of density matrix elements (Eq. 8.34), or likewise, using Eq. (3.34), in terms of electron propagator integrals,

$$\Sigma_{pq}(\infty) = \sum_{u,v} V_{pu[qv]} \left(\rho_{vu} - \rho_{vu}^{(0)}\right) = \sum_{u,v} V_{pu[qv]} \frac{1}{2\pi i} \oint \mathrm{d}\omega \left(G_{vu}(\omega) - G_{vu}^{0}(\omega)\right) \tag{A.5.2}$$

The static part is related to the dynamic part and can be determined consistently once the dynamic part $\boldsymbol{M}(\omega)$ or an approximation to it has been established. This is seen by writing the Dyson equation (8.10) more explicitly as

$$\boldsymbol{G}(\omega) = \boldsymbol{G}^0(\omega) + \boldsymbol{G}^0(\omega)(\boldsymbol{\Sigma}(\infty) + \boldsymbol{M}(\omega))\boldsymbol{G}(\omega) \tag{A.5.3}$$

and inserting this form in Eq. (A.5.2). The result is an implicit equation for $\boldsymbol{\Sigma}(\infty)$. For a given $\boldsymbol{M}(\omega)$, the associated result for $\boldsymbol{\Sigma}(\infty)$ can be obtained, in principle, via an obvious iteration scheme. However, such a self-consistent procedure may not be very practical, and it is more advisable to relinquish the quest for a fully self-consistent solution for $\boldsymbol{\Sigma}(\infty)$ and rather resort to an approximation such as the DEM considered below.

Dyson Expansion

The starting point is the truncation of the Dyson expansion (8.12) after the linear term in $\boldsymbol{\Sigma}(\omega)$:

$$\boldsymbol{G}(\omega) = \boldsymbol{G}^0(\omega) + \boldsymbol{G}^0(\omega)(\boldsymbol{\Sigma}(\infty) + \boldsymbol{M}(\omega))\boldsymbol{G}^0(\omega) + \ldots \tag{A.5.4}$$

Note that the truncation error here is of fourth order because the perturbation expansion of $\boldsymbol{\Sigma}(\omega)$ begins in second order. Inserting the truncated Dyson expansion in Eq. (A.5.2) yields the following linear set of equations for the matrix elements of $\boldsymbol{\Sigma}(\infty)$:

$$\Sigma_{pq}(\infty) = \sum_{r,s} V_{pr[qs]} \frac{1}{2\pi i} \oint G_s^0(\omega)\Sigma_{sr}(\infty)G_r^0(\omega)d\omega + b_{pq} + O(5) \tag{A.5.5}$$

where $O(5)$ indicates that the truncation error for $\Sigma_{pq}(\infty)$ is of fifth order. The inhomogenities b_{pq} are given by

$$b_{pq} = \sum_{r,s} V_{pr[qs]} Q_{sr} \tag{A.5.6}$$

where Q_{sr} denote the contour integrals

$$Q_{sr} = \frac{1}{2\pi i} \oint G_s^0(\omega) M_{sr}(\omega) G_r^0(\omega) d\omega \tag{A.5.7}$$

based on the matrix elements of the dynamic self-energy part $\boldsymbol{M}(\omega)$. Performing the contour integrations in Eq. (A.5.5) allows us to write the linear equations in the more explicit form

$$\Sigma_{pq}(\infty) = \sum_{r,s} V_{pr[qs]} \frac{n_s \bar{n}_r - \bar{n}_s n_r}{\epsilon_s - \epsilon_r} \Sigma_{sr}(\infty) + b_{pq} \tag{A.5.8}$$

Here the DE truncation error is no longer indicated.

First, we discuss the solution of the set of linear equations for given inhomogenities b_{pq}. Obviously, the linear equations for the p-h and h-p matrix elements of $\boldsymbol{\Sigma}(\infty)$ are decoupled from those for the h-h and p-p elements:

$$\Sigma_{ak}(\infty) = \sum_{b,l} \left(V_{al[kb]} \frac{1}{\epsilon_l - \epsilon_b} \Sigma_{bl}(\infty) + V_{ab[kl]} \frac{1}{\epsilon_l - \epsilon_b} \Sigma_{lb}(\infty) \right) + b_{ak} \tag{A.5.9a}$$

$$\Sigma_{ka}(\infty) = \sum_{b,l} \left(V_{kl[ab]} \frac{1}{\epsilon_l - \epsilon_b} \Sigma_{bl}(\infty) + V_{kb[al]} \frac{1}{\epsilon_l - \epsilon_b} \Sigma_{lb}(\infty) \right) + b_{ka} \tag{A.5.9b}$$

Supposing real orbitals, the linear equations can be restricted to the p-h components $\Sigma_{ak}(\infty)$:

$$\Sigma_{ak}(\infty) = \sum_{b,l} \frac{1}{\epsilon_l - \epsilon_b} (V_{al[kb]} + V_{ab[kl]})\, \Sigma_{bl}(\infty) + b_{ak} \tag{A.5.10}$$

Introducing the p-h coefficient matrix $\boldsymbol{A}^{ph}$,

$$A^{ph}_{ak,bl} = \frac{1}{\epsilon_l - \epsilon_b} (V_{al[kb]} + V_{ab[kl]}) \tag{A.5.11}$$

and corresponding (column) vectors $\boldsymbol{\Sigma}^{ph}(\infty)$ and $\boldsymbol{b}^{ph}$, Eq. (A.5.9) can be written in a more compact form as

$$\boldsymbol{\Sigma}^{ph}(\infty) = \boldsymbol{A}^{ph} \boldsymbol{\Sigma}^{ph}(\infty) + \boldsymbol{b}^{ph} \tag{A.5.12}$$

Thus, the formal solution of the linear set of equations takes on the form

$$\boldsymbol{\Sigma}^{ph}(\infty) = (\mathbf{1} - \boldsymbol{A}^{ph})^{-1}\boldsymbol{b}^{ph} \tag{A.5.13}$$

Once the p-h and h-p components of $\boldsymbol{\Sigma}(\infty)$ have been obtained by solving the Eqs. (A.5.9) or (A.5.10), the h-h and p-p matrix elements can be evaluated according to

$$\Sigma_{ij}(\infty) = \sum_{b,l} \frac{1}{\epsilon_l - \epsilon_b}(V_{il[jb]}\Sigma_{bl}(\infty) + V_{ib[jl]}\Sigma_{lb}(\infty)) + b_{ij} \tag{A.5.14}$$

$$\Sigma_{ac}(\infty) = \sum_{b,l} \frac{1}{\epsilon_l - \epsilon_b}(V_{al[cb]}\Sigma_{bl}(\infty) + V_{ab[cl]}\Sigma_{lb}(\infty)) + b_{ac} \tag{A.5.15}$$

Treatment of the Inhomogenities

The inhomogenities b_{pq}, as given by Eqs. (A.5.6), (A.5.7), establish the connection with the dynamic part $\boldsymbol{M}(\omega)$ of the self-energy via the contour integrals (A.5.7). If $\boldsymbol{M}(\omega)$ is available in the explicit form of spectral representation (8.19), the contour integrals Q_{rs} can readily be evaluated. Here, it is useful to distinguish between p-p, h-h, and p-h elements of $\boldsymbol{Q}$. In the p-p case, the calculation is as follows:

$$\begin{aligned} Q_{ab} &= \frac{1}{2\pi i}\oint G_a^0(\omega)M_{ab}(\omega)G_b^0(\omega)\,\mathrm{d}\omega \\ &= \sum_{\nu\in\{N-1\}} \frac{1}{2\pi i}\oint \frac{1}{\omega - \epsilon_a + i\eta}\,\frac{m_a^{(\nu)}m_b^{(\nu)*}}{\omega - \omega_\nu - i\eta}\,\frac{1}{\omega - \epsilon_b + i\eta}\,\mathrm{d}\omega && \text{(A.5.16)} \\ &= \sum_{\nu\in\{N-1\}} \frac{m_a^{(\nu)}m_b^{(\nu)*}}{(\epsilon_a - \omega_\nu)(\epsilon_b - \omega_\nu)} && \text{(A.5.17)} \end{aligned}$$

Note that here only the $\boldsymbol{M}^-(\omega)$ part contributes. In a similar way, one obtains the h-h and p-h integrals

$$Q_{kl} = -\sum_{\mu\in\{N+1\}} \frac{m_k^{(\mu)}m_l^{(\mu)*}}{(\epsilon_k - \omega_\mu)(\epsilon_l - \omega_\mu)} \tag{A.5.18}$$

$$Q_{ak} = -\sum_{\mu\in\{N+1\}} \frac{m_a^{(\mu)}m_k^{(\mu)*}}{(\epsilon_a - \epsilon_k)(\epsilon_k - \omega_\mu)} - \sum_{\nu\in\{N-1\}} \frac{m_a^{(\nu)}m_k^{(\nu)*}}{(\epsilon_a - \epsilon_k)(\epsilon_a - \omega_\nu)} \tag{A.5.19}$$

In the latter case, both parts $\boldsymbol{M}^{\pm}(\omega)$ of the self-energy contribute, whereas only $\boldsymbol{M}^{+}(\omega)$ is involved in the h-h results. Note that $Q_{ka} = Q_{ak}^*$.

The computation of the Q_{rs} integrals according to the explicit expressions above presupposes that the full spectral information of $\boldsymbol{M}(\omega)$ is available. Usually, however, the acquisition of that information within a given approximation is computationally expensive and hardly expedient. In the ADC approximation for $\boldsymbol{M}(\omega)$ presented in Chap. 9, the explicit computation of the self-energy pole positions and residue amplitudes can be circumvented as described in the following.

According to Eq. (9.2), each of the $\boldsymbol{M}^{\pm}(\omega)$ parts can be written in the ADC form

$$M^{\pm}_{pq}(\omega) = (\boldsymbol{U}^{\pm}_{p})^{\dagger}\left(\omega - \boldsymbol{K}^{\pm} - \boldsymbol{C}^{\pm}\right)^{-1}\boldsymbol{U}^{\pm}_{q} \tag{A.5.20}$$

where $\boldsymbol{K}^{\pm} + \boldsymbol{C}^{\pm}$ and $U^{\pm}_{p}$ are the secular matrices and coupling vectors as given at the respective ADC level. The associated ADC eigenvalue equations (9.27)

$$(\boldsymbol{K}^{\pm} + \boldsymbol{C}^{\pm})\boldsymbol{Y}^{\pm} = \boldsymbol{Y}^{\pm}\boldsymbol{\Omega}^{\pm},\ (\boldsymbol{Y}^{\pm})^{\dagger}\boldsymbol{Y}^{\pm} = \mathbf{1} \tag{A.5.21}$$

determine the pole positions as the eigenvalues ω_n, and the residue amplitudes (9.28) according to

$$m_{np} = (\boldsymbol{Y}^{\pm}_{n})^{\dagger}\boldsymbol{U}^{\pm}_{p} \tag{A.5.22}$$

In view of the size of the ADC secular matrices, their full diagonalization, needed to obtain all eigenvalues and residue amplitudes, is obviously not a viable computational strategy. For an alternative approach, we introduce the (column) vectors

$$\boldsymbol{V}^{+}_{k} = (\epsilon_k\mathbf{1} - \boldsymbol{K}^{+} - \boldsymbol{C}^{+})^{-1}\boldsymbol{U}^{+}_{k} \tag{A.5.23}$$

for occupied orbitals, k, and

$$\boldsymbol{V}^{-}_{a} = (\epsilon_a\mathbf{1} - \boldsymbol{K}^{-} - \boldsymbol{C}^{-})^{-1}\boldsymbol{U}^{-}_{a} \tag{A.5.24}$$

for unoccupied orbitals, a. With the help of these vectors, the Q_{rs} integrals can be written simply as

$$Q_{ab} = (\boldsymbol{V}^{-}_{a})^{\dagger}\boldsymbol{V}^{-}_{b} \tag{A.5.25}$$

$$Q_{kl} = (\boldsymbol{V}^{+}_{k})^{\dagger}\boldsymbol{V}^{+}_{l} \tag{A.5.26}$$

$$Q_{ak} = -\frac{1}{\epsilon_a - \epsilon_k}(\boldsymbol{U}^{+}_{a})^{\dagger}\boldsymbol{V}^{+}_{k} - \frac{1}{\epsilon_a - \epsilon_k}(\boldsymbol{U}^{-}_{a})^{\dagger}\boldsymbol{V}^{-}_{k} \tag{A.5.27}$$

To verify these expressions, consider the relations

$$\begin{aligned}\boldsymbol{V}_r &= (\epsilon_r\mathbf{1} - \boldsymbol{K} - \boldsymbol{C})^{-1}\boldsymbol{U}_r \\ &= \boldsymbol{Y}(\epsilon_r\mathbf{1} - \boldsymbol{\Omega})^{-1}\boldsymbol{Y}^{\dagger}\boldsymbol{U}_r = \boldsymbol{Y}(\epsilon_r\mathbf{1} - \boldsymbol{\Omega})^{-1}\boldsymbol{m}_r\end{aligned}$$

which apply to both types of vectors; for notational ease, the $\pm$ superscripts have been dropped. The term $\boldsymbol{m}_r$ in the last line denotes the (column) vector of residue amplitudes $m^{(n)}_{r}$ of $\boldsymbol{M}^{+}(\omega)$ or $\boldsymbol{M}^{-}(\omega)$.

Now the computational task consists in solving linear equations of the type

$$\boldsymbol{V}_r = (\epsilon_r \mathbf{1} - \boldsymbol{K} - \boldsymbol{C})^{-1} \boldsymbol{U}_r \tag{A.5.28}$$

for $\boldsymbol{K}^+ + \boldsymbol{C}^+$ and occupied orbitals ϵ_k, as well as for $\boldsymbol{K}^- + \boldsymbol{C}^-$ and unoccupied orbitals ϵ_a. In both cases, the elements of the diagonal matrix $\epsilon_r \mathbf{1} - \boldsymbol{K}$ are of the order of a double excitation energy, $\pm(\epsilon_a + \epsilon_b - \epsilon_k - \epsilon_l)$. This suggests an iterative solution of the matrix inversions as follows. Let

$$\boldsymbol{\Delta}_r = \epsilon_r \mathbf{1} - \boldsymbol{K} \tag{A.5.29}$$

denote the respective diagonal matrix and rewrite $\epsilon_r \mathbf{1} - \boldsymbol{K} - \boldsymbol{C}$ as

$$\epsilon_r \mathbf{1} - \boldsymbol{K} - \boldsymbol{C} = \boldsymbol{\Delta}_r - \boldsymbol{C} = (\mathbf{1} - \boldsymbol{C}\boldsymbol{\Delta}_r^{-1})\boldsymbol{\Delta}_r \tag{A.5.30}$$

Then, the inverse matrix can be expanded in a geometrical series,

$$\begin{aligned}(\epsilon_r \mathbf{1} - \boldsymbol{K} - \boldsymbol{C})^{-1} \boldsymbol{U}_r =& (\Delta_r - \boldsymbol{C})^{-1} \boldsymbol{U}_r \\ =& \boldsymbol{\Delta}_r^{-1} \left(\mathbf{1} + \boldsymbol{C}\boldsymbol{\Delta}_r^{-1} + (\boldsymbol{C}\boldsymbol{\Delta}_r^{-1})^2 + \ldots\right) \boldsymbol{U}_r\end{aligned}$$

which in turn defines an obvious iteration scheme

$$\begin{aligned}\boldsymbol{V}_r(0) =& \boldsymbol{\Delta}_r^{-1} \boldsymbol{U}_r \\ \boldsymbol{V}_r(n+1) =& \boldsymbol{V}_r(0) + \Delta_r^{-1} \boldsymbol{C} \boldsymbol{V}_r(n)\end{aligned} \tag{A.5.31}$$

for the $\boldsymbol{V}_r$ vector itself.

An alternative computational approach is the Lanczos diagonalization [3, 4] of the secular matrices $\boldsymbol{K}^\pm + \boldsymbol{C}^\pm$. Here, the p-p integrals (A.5.16), for example, are approximated by the expressions

$$Q_{ab} \approx \sum_{s=1}^{L} \frac{\tilde{m}_{sa}^* \tilde{m}_{sb}}{(\epsilon_a - \tilde{\omega}_s)(\epsilon_b - \tilde{\omega}_s)} \tag{A.5.32}$$

based on the Lanczos pseudo-spectrum for $\boldsymbol{K}^- + \boldsymbol{C}^-$ upon L Lanczos steps. Here, $\tilde{\omega}_s$ are the corresponding eigenvalues, and the amplitudes are obtained according to $\tilde{m}_{sa} = \boldsymbol{Z}_s^\dagger \boldsymbol{U}_a^-$ from the Lanczos vectors $\boldsymbol{Z}_s$. A more detailed report on the use of the Lanczos method in the present context has been given in Ref. [9].

A.6 Proofs of Order Relations

A.6.1 *Diagrammatic Perturbation Theory for Ground-State CI and CC Amplitudes*

To explain the order relations (17.8),

$$t_I \sim O([I]-1),\ [I]>1 \tag{A.6.1}$$

for the CC amplitudes we have to take a closer look at the diagrammatic PT for the ground state. In Sect. 4.4, we have considered the Gell-Mann and Low expression (4.42) for the ground state,

$$|\Psi_0\rangle = \lim_{\epsilon\to 0} \frac{\hat{U}_\epsilon(0,-\infty)|\Phi_0\rangle}{\langle\Phi_0|\hat{U}_\epsilon(0,-\infty)|\Phi_0\rangle} \tag{A.6.2}$$

and discussed how the time integrations in the time-evolution operator ,

$$\hat{U}_\epsilon(0,-\infty) = \sum_{n=0}^{\infty}\frac{(-i)^n}{n!}\int_{-\infty}^{0} dt_1 e^{-\epsilon|t_1|}\ldots\int_{-\infty}^{0} dt_n\, e^{-\epsilon|t_n|}\hat{\mathcal{T}}\left[\hat{H}_I(t_1)\ldots\hat{H}_I(t_n)\right] \tag{A.6.3}$$

could explicitly be performed resulting in the closed-form expression (4.70). Then, the connection to the familiar RSPT series (see Sect. A.1) could be established by inserting these results both in the numerator and the denominator in Eq. (A.6.2) and expanding their ratio. Alternatively, one could have proceeded along the lines of Chaps. 5–7 to devise a diagrammatic PT formulation for the ground-state or the ground-state energy. Let us consider in particular the diagrammatic PT expansions for individual ground-state amplitudes (CI coefficients),

$$x_I = \langle\Phi_0|\hat{C}_I^\dagger|\Psi_0\rangle = \lim_{\epsilon\to 0}\frac{\langle\Phi_0|\hat{C}_I^\dagger\hat{U}_\epsilon(0,-\infty)|\Phi_0\rangle}{\langle\Phi_0|\hat{U}_\epsilon(0,-\infty)|\Phi_0\rangle} \tag{A.6.4}$$

in the ground-state CI expansion (17.10),

$$|\Psi_0\rangle = |\Phi_0\rangle + \sum_I x_I\hat{C}_I\,|\Phi_0\rangle \tag{A.6.5}$$

The respective excitation operator, $\hat{C}_I^\dagger$, $I \equiv ab\ldots kl$, can be supplied with the time argument $t=0$, that is, $\hat{C}_I^\dagger(0) = c_l^\dagger(0)c_k^\dagger(0)\ldots c_b(0)c_a(0)$ (see Eq. 3.50). Now $\hat{C}_I^\dagger(0)$ can be incorporated in the time-ordered products of the time-evolution operator (4.37) according to

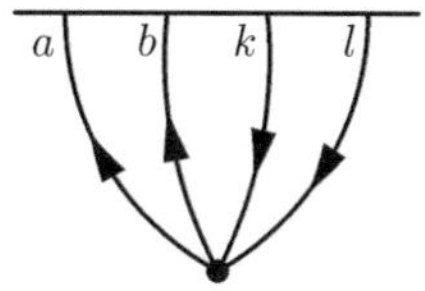

Fig. A.3 First-order Feynman diagram (Abrikosov form) for the ground-state amplitude x_{abkl}

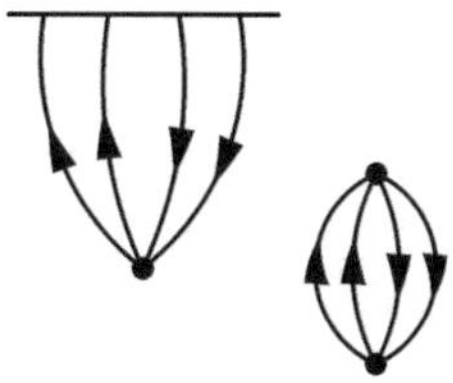

Fig. A.4 Disconnected third-order Feynman diagram (Abrikosov form) for the ground-state amplitude x_{abkl}

$$\hat{C}_I^\dagger(0)\hat{\mathcal{T}}\left[\hat{H}_I(t_1)\dots\hat{H}_I(t_n)\right] = \hat{\mathcal{T}}\left[\hat{C}_I^\dagger(0)\hat{H}_I(t_1)\dots\hat{H}_I(t_n)\right] \tag{A.6.6}$$

since the time argument $t = 0$ is always larger than the other time arguments. Finally, the expectation values $\langle\Phi_0|\dots|\Phi_0\rangle$ can be evaluated using Wick's theorem as described in Sect. 5.1. The operator $C_I^\dagger$ represents a fixed external vertex in the resulting Feynman diagrams for the numerator in Eq. (A.6.2).

As a simple example, we consider the first-order diagram for the double excitation $\hat{C}_{abkl} = c_a^\dagger c_b^\dagger c_k c_l$ shown in Fig. A.3. The corresponding analytical expression reads (as the reader should check)

$$x_{abkl}^{(1)} = (-i)\int_{-\infty}^{0} \mathrm{d}t_1 e^{\epsilon t_1} V_{ab[lk]} G_a^0(0,t_1) G_b^0(0,t_1) G_k^0(t_1,0) G_l^0(t_1,0)$$

which after performing the time integration (and the limit $\epsilon \to 0$) yields the familiar first-order CI coefficient (A.1.16),

$$x_{abkl}^{(1)} = \frac{V_{ab[kl]}}{\epsilon_a + \epsilon_b - \epsilon_k - \epsilon_l} \tag{A.6.7}$$

A linked-cluster theorem of the kind discussed in Sect. 5.3 applies to the PT expansion (A.6.4). Here the external vertex $C_I^\dagger$ allows one to distinguish contraction schemes or diagrams as being entirely connected (to the vertex $C_I^\dagger$) or having parts not connected to $C_I^\dagger$. Figure A.4 depicts a simple disconnected diagram of third order. As a result of the linked-cluster theorem, the PT expansion (A.6.4) of x_I simplifies to

$$x_I = \lim_{\epsilon\to 0} \langle\Phi_0|\hat{C}_I^\dagger \hat{U}_\epsilon(0,-\infty)|\Phi_0\rangle_C \tag{A.6.8}$$

where the subscript C indicates that only connected diagrams are taken into account.

An interesting new feature can be seen in Fig. A.5, which shows a second-order diagram for the $4p$-$4h$ (quadrupel) amplitude $x_{abcd;ijkl}$. While this is a connected

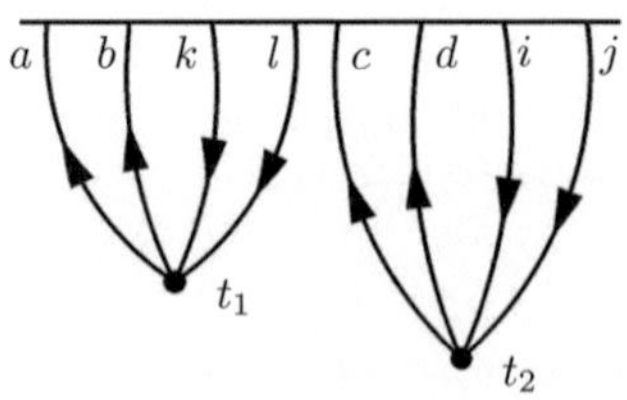

Fig. A.5 Factorizing Feynman diagram (second order) for the quadruple excitation amplitude $x_{abcd;ijkl}$

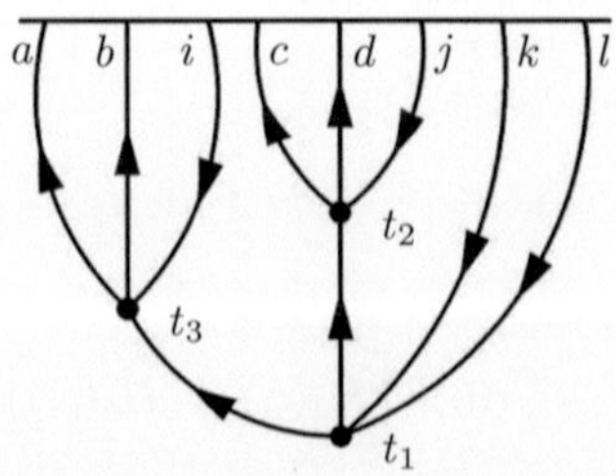

Fig. A.6 Strictly connected (non-factorizing) Feynman diagram (third order) for the quadruple excitation amplitude $x_{abcd;ijkl}$

diagram in the original sense, that is, containing no sub-units not linked to the external vertex, $C^\dagger_{abcd;ijkl}$, it nevertheless factorizes into the product of two first-order double excitation diagrams, which is enabled by the multiplicative structure of the external vertex $C^\dagger_{abcd;ijkl}$. The diagram in Fig. A.5, denoted $Q_1(abcd;ijkl)$), entails two Goldstone (time-ordered) diagrams which can be combined as follows:

$$\begin{aligned} Q_1(abcd;ijkl) =& \frac{1}{\epsilon_a+\epsilon_b+\epsilon_c \cdots -\epsilon_k-\epsilon_l}\left(V_{ab[ij]}\frac{V_{cd[kl]}}{\epsilon_c+\epsilon_d-\epsilon_k-\epsilon_l} + V_{cd[kl]}\frac{V_{ab[ij]}}{\epsilon_a+\epsilon_b-\epsilon_i-\epsilon_j}\right) \\ =& x^{(1)}_{abij}\,x^{(1)}_{cdkl} \end{aligned} \quad \text{(A.6.9)}$$

which makes the product form of $Q_1(abcd;ijkl)$ explicit. Note that there are altogether 18 distinct factorizing diagrams $Q_\nu(abcd;ijkl)$, $\nu = 1, \ldots, 18$, contributing to the second-order quadruple amplitude $x^{(2)}_{abcd;ijkl}$. The four unoccupied orbitals, a, b, c, d, and the four occupied ones, i, j, k, l, allow for 36 different double excitations, forming 18 pairs of complementary double excitations, such as $(acjl)$ and $(bdik)$, entailing $\hat{C}_{abcd;ijkl} = \hat{C}_{acjl}\hat{C}_{bdik}$. The diagrammatic results can directly be verified by evaluating $x^{(2)}_{abcd;ijkl}$ with the second-order RSPT expression (A.1.20) for the ground-state (see Exercise 17.2).

As the factorizing diagrams for the quadrupole amplitudes show, the concept of connectivity can be extended: A diagram for the ground state amplitudes x_I is termed **strictly connected** if it has no unlinked sub-units (original meaning) and, moreover, cannot be disassembled into disjoint parts by cutting the external vertex line. A strictly connected diagram, pertaining to the quadrupel amplitude $x_{abcd;ijkl}$, is shown Fig. A.6.

A generalized linked-cluster theorem based on the concept of strictly connected diagrams was established by J. Hubbard [10]. However, the diagrams considered by Hubbard relate directly to the time-evolution operator (A.6.3) rather than the

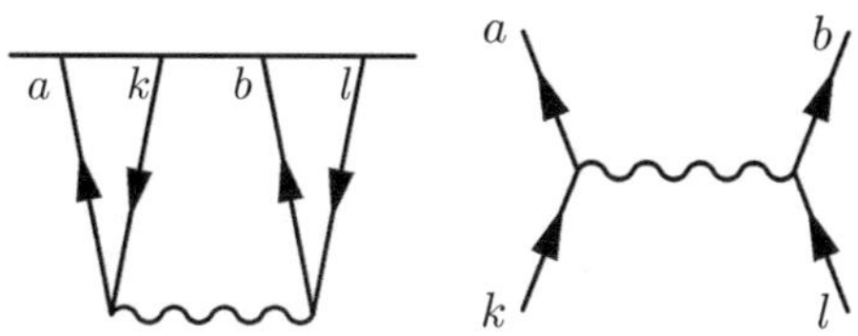

Fig. A.7 First-order Feynman diagram for the ground-state amplitude x_{abkl} (left) and the associated Hubbard diagram (right)

amplitudes (A.6.4). In devising these diagrams, Hubbard makes use of Wick's theorem (5.16) in a more general way than in Chap. 5. Unlike in Eq. (5.17), the operator identities are analyzed with regard to their application to the HF ground state: $\hat{\mathcal{T}}\left[\hat{a}_i\hat{a}_j\hat{a}_k\hat{a}_l\ldots\hat{a}_r\hat{a}_s\hat{a}_t\right]|\Phi_0\rangle$. Thereby, all contributions retaining only unphysical operators drop out, whereas $\hat{\mathcal{N}}$ products in Eq. (5.16) that feature solely physical operators persist and give rise to non-vanishing diagrammatic contributions. Those surviving operators are referred to by Hubbard as *edges* of the respective diagram. As a consequence, any Hubbard diagram is either a c-number (corresponding to a fully contracted term) or a product

$$D_I^h = d_I\hat{C}_I \tag{A.6.10}$$

of a c-number d_I and an excitation operator $\hat{C}_I$ from the set (14.10). For the Hubbard diagrams of the latter form, an obvious one-to-one mapping can be established onto the diagrams for the ground-state amplitudes (A.6.4). Consider an arbitrary diagram D_I for the amplitude $x_I = \langle\Phi_0|\hat{C}_I^\dagger|\Psi_0\rangle$, and let δ_I denote the corresponding analytical expression. Then, there is a corresponding Hubbard diagram D_I^h (obtained by identifying the external lines of D_I with edges in the Hubbard diagram) so that $D_I^h = \delta_I\hat{C}_I$. This is illustrated in Fig. A.7, showing the first-order Feynman diagram for the $2p$-$2h$ amplitude x_{abkl} and the associated Hubbard diagram.

Summing up all Hubbard diagrams of the form (A.6.10) yields the CI operator expansion,

$$\lim_{\epsilon\to 0}\left[\hat{U}_\epsilon(0,-\infty)\right]_C^h = \sum_I x_I\hat{C}_I \tag{A.6.11}$$

and the corresponding representation of the exact ground state according to Eq. (A.6.5). As indicated by the subscript C, the (somewhat symbolical) expression on the left side comprises only connected diagrams (not having disconnected c-number sub-units) as a result of the conventional linked-cluster theorem. Note that any individual coefficient x_I can likewise be obtained according to Eq. (A.6.8) from the diagrams for the corresponding ground-state amplitude (A.6.4).

Like the diagram shown in Fig. A.5, a general Hubbard diagram D_I^h contributing to the term $x_I\hat{C}_I$ may consist of two (or more) disconnected sub-units, say D_K^h and D_L^h, resulting in a corresponding factorization of the analytical expression,

$$D_I^h = d_I\hat{C}_I = d_K\hat{C}_K \times d_L\hat{C}_L \tag{A.6.12}$$

As opposed to these unlinked diagrams, there are **strictly connected** (SC) Hubbard diagrams, $D_I^{h|SC}$, forming a subset of the Hubbard diagrams of the form (A.6.10). Summing up all SC diagrams for a given configuration (or edge structure) I,

$$\sum D_I^{h|SC} = t_I \hat{C}_I$$

defines a new amplitude, t_I, which differs from x_I. As should be noted, the t_I can likewise be obtained from the diagrams for the ground-state amplitude (A.6.8) by discarding all diagrams that are not strictly connected:

$$t_I = \lim_{\epsilon \to 0} \langle \Phi_0 | \hat{C}_I^\dagger \hat{U}_\epsilon(0, -\infty) | \Phi_0 \rangle_{SC} \tag{A.6.13}$$

In analogy to Eq. (A.6.11), the sum of all strictly connected Hubbard diagrams establishes a particular operator expansion,

$$\lim_{\epsilon \to 0} \left[\hat{U}_\epsilon(0, -\infty) \right]_{SC}^h = \sum_I t_I \hat{C}_I \tag{A.6.14}$$

and this operator expansion allows for a representation of the ground state in the form

$$|\Psi_0\rangle = e^{\sum_I t_I \hat{C}_I} |\Phi_0\rangle \tag{A.6.15}$$

This is the central result of Hubbard's extended linked-cluster theorem. The latter two equations can be seen as a foundation of the CC ansatz (17.1), and, moreover, establish a diagrammatic PT approach to the CC amplitudes.

The validity of the order relations (A.6.1) for the t-amplitudes can be seen by inspecting the construction principle underlying the strictly connected diagrams as shown in Fig. A.6. Beginning with a double excitation, $[I] = 2$, $O(t_I) = 1$, it takes each one interaction point (order 1) to increase the excitation class by 1, that is, by a $1p$-$1h$ excitation. This means it takes (at least) $\nu - 2$ interaction points to arrive at an excitation of class $\nu > 2$. The $\nu - 2$ interaction points have a cumulated order of $\nu - 2$, to which the order 1 of the initial double excitation vertex has to be added.

A.6.2 *Proof of Canonical Order Relations*

Originally, a proof of the canonical order relations obeyed by the ISR-ADC secular matrix elements was given in Ref. [11]. However, a more elegant proof [12] is enabled by recourse to the closed-form expressions of the biorthogonal coupled cluster (BCC) representation discussed in Chap. 17. Following the latter concept, we shall first derive the BCC order relations and then use these results to establish the canonical ISR-ADC order relations.

BCC Order Relations

In the derivation of the canonical order relations (17.45) for the LL part of $\boldsymbol{M}^{cc}$, we shall essentially follow the presentation in App. 1 of Ref. [13], which itself is based on the general proof given in Ref. [12]. An alternative approach was pursued by Christiansen et al. [14] and Hald et al. [15].

Consider an LL matrix element M_{IJ}^{cc} with $[I] > [J]$. What is to be shown is

$$M_{IJ}^{cc} \sim O([I] - [J]) \tag{A.6.16}$$

which means that in the PT expansion of M_{IJ}^{cc} the non-vanishing contributions do not begin before order $n = [I] - [J]$. Let us write the matrix element somewhat more conveniently as

$$\begin{aligned} M_{IJ}^{cc} =& \langle \Phi_0 | \hat{C}_I^\dagger e^{-\hat{T}} [\hat{H}, \hat{C}_J] e^{\hat{T}} | \Phi_0 \rangle \\ =& \langle \Phi_0 | \hat{C}_I^\dagger e^{-\hat{T}} \hat{K}_J \, e^{\hat{T}} | \Phi_0 \rangle \end{aligned} \tag{A.6.17}$$

where

$$\hat{K}_J = [\hat{H}, \hat{C}_J] \tag{A.6.18}$$

denotes the commutator of the hamiltonian and the excitation operator $\hat{C}_J$. Obviously, $\hat{K}_J$ can be partitioned according to

$$\hat{K}_J = \hat{K}_J^{(0)} + \hat{K}_J^{(1)} \tag{A.6.19}$$

into a zeroth-order contribution, $\hat{K}_J^{(0)} = [\hat{H}_0, \hat{C}_J]$, and a first-order contribution, $\hat{K}_J^{(1)} = [\hat{H}_I, \hat{C}_J]$. The considered matrix element is of the form

$$M_{IJ}^{cc} \sim \langle \Phi_0 | \hat{C}_I^\dagger \hat{O}_J | \Phi_0 \rangle \tag{A.6.20}$$

where the operator

$$\hat{O}_J = e^{-\hat{T}} \hat{K}_J \, e^{\hat{T}} \tag{A.6.21}$$

is subject to a PT expansion, $\hat{O}_J = \hat{O}_J^{(0)} + \hat{O}_J^{(1)} + \ldots$. What we have to show is that the matrix element vanishes for the orders $\nu = 0, \ldots, [I] - [J] - 1$, of this series. To this end, the concept of the **rank** of an operator is essential. For a (charge-conserving) operator given by a product of fermion operators, the rank is the number of creation operators ($c^\dagger$) in the product. If an operator is a sum of such fermion operator products, the rank is defined as the maximal rank of its constituents. For example, the rank of $\hat{H}_0$ and $\hat{H}_I$ is 1 and 2, respectively. The rank of $\hat{C}_I$ is its excitation class number, $r = [I]$, and the class-specific $\hat{T}_\mu$ operators are of rank μ. Why is the operator rank of interest? Because the nth order contribution to the matrix element (A.6.20) necessarily vanishes,

$$\langle\Phi_0|\hat{C}_I^\dagger \hat{O}_J^{(n)}|\Phi_0\rangle = 0 \tag{A.6.22}$$

if the rank r of the operator $\hat{O}_J^{(n)}$ is lower than the rank of $C_I^\dagger$, that is, if $r < [I]$. Thus, the intended proof can be carried out by inspecting the rank and order of successive contributions to the operator $\hat{O}_J$.

The two parts $\hat{K}_J^{(0)}$ and $\hat{K}_J^{(1)}$ of the commutator (A.6.18) have the ranks $[J]$ and $[J]+1$, respectively. More general, the commutator $[\hat{A}, \hat{B}]$ of two operators $\hat{A}$ and $\hat{B}$ with definite ranks, r and r', respectively, is of rank $r + r' - 1$. Another commutator feature relates to the number of unphysical fermion operators in an operator product: A commutator $[\hat{A}, \hat{B}]$ of a general product operator $\hat{A}$ and a physical operator $\hat{B}$, as in $\hat{K}_J^{(1)}$, reduces the number of unphysical fermion operators in $\hat{A}$ by 1. Accordingly, the terms of $\hat{K}_J^{(1)}$ exhibit at most three unphysical c-operators compared to the four unphysical operators in $\hat{H}_I$. As a particular case, the zeroth-order commutator $\hat{K}_J^{(0)}$ consists only of physical c-operators, reflecting the fact that $\hat{H}_0$ is a diagonal one-particle operator. As a consequence, $\hat{K}_J^{(0)}$ commutes with the CC operator $\hat{T}$ so that

$$e^{-\hat{T}}\hat{K}_J^{(0)}e^{\hat{T}} = \hat{K}_J^{(0)} \tag{A.6.23}$$

Accordingly, the matrix element (A.6.17) can be written as the sum

$$M_{IJ}^{cc} = \langle\Phi_0|\hat{C}_I^\dagger \hat{K}_J^{(0)}|\Phi_0\rangle + \langle\Phi_0|\hat{C}_I^\dagger e^{-\hat{T}}\hat{K}_J^{(1)}e^{\hat{T}}|\Phi_0\rangle \tag{A.6.24}$$

of a zeroth-order contribution and a contribution involving $\hat{K}_J^{(1)}$, being at least of first order. Recalling that the rank of $\hat{K}_J^{(0)}$ is $[J]$, the zeroth-order term, $M_{IJ}^{cc(0)}$, necessarily vanishes unless $[I] = [J]$; that is, both I and J belong to the same excitation class; to put it differently, zeroth-order contributions arise only in the diagonal blocks of $\boldsymbol{M}^{cc}$.

To analyze the higher-order contribution, we make use of the Baker–Hausdorff (BH) expansion according to

$$e^{-\hat{T}}\hat{K}_J^{(1)}e^{\hat{T}} = \hat{K}_J^{(1)} + [\hat{K}_J^{(1)}, \hat{T}] + \frac{1}{2}[[\hat{K}_J^{(1)}, \hat{T}], \hat{T}] + \frac{1}{6}[[[\hat{K}_J^{(1)}, \hat{T}], \hat{T}], \hat{T}] \tag{A.6.25}$$

which here terminates after the threefold commutator since each successive commutator eliminates one of the original three unphysical c-operators in $\hat{K}_J^{(1)}$. The BH expansion (A.6.25) begins with $\hat{K}_J^{(1)}$, being of the rank $[J]+1$ and representing the only first-order contribution in the operator expansion (A.6.25). This means that there is a non-vanishing first-order contribution if (and only if) $[I] = [J]+1$. Stated differently, the sub-diagonal blocks of $\boldsymbol{M}^{cc}$ obey the order relation $\boldsymbol{M}_{\mu+1\mu}^{cc} \sim O(1)$; all other LL blocks are at least of second order.

Now consider matrix elements M_{IJ}^{cc} where I and J differ by more than one excitation class: $[I] = [J] + \mu,\ \mu \geq 2$. Thus, the minimal rank in the operator (A.6.25) required to yield a non-vanishing matrix element is $r = [J] + \mu$. In the first commutator, the required rank comes with the CC operator for class μ: $[\hat{K}_J^{(1)}, \hat{T}_\mu]$ is of rank $r = [J] + \mu$, while its PT order is given by μ (being the sum of the order $\mu - 1$ of $\hat{T}_\mu$ according to Eq. (A.6.1) and the order 1 of $\hat{K}_J^{(1)}$). Likewise, the double commutator $[[\hat{K}_J^{(1)}, \hat{T}_2], \hat{T}_{\mu-1}]$, involving a lower class CC operator, yields terms of rank $[J] + \mu$, but again the resulting PT order is μ. One may inspect all other possibilities of generating operator terms of rank $r = [J] + \mu$ and one will find that the concomitant PT order is always equal (or larger) than μ. This proves the order relations (A.6.16) for the LL blocks $\boldsymbol{M}_{\mu\nu}^{cc}$ with $\mu \geq \nu + 2$.

Some complementary notes are of interest:

1. Obviously, the order relations (A.6.16) apply also to the BCC representation of the hamiltonian itself,

$$H_{IJ}^{cc} = \langle \overline{\Phi}_I | \hat{H} | \Psi_J^0 \rangle \tag{A.6.26}$$

as $\boldsymbol{H}^{cc}$ differs from $\boldsymbol{M}^{cc}$ only in the diagonal elements,

$$\boldsymbol{M}^{cc} = \boldsymbol{H}^{cc} - E_0 \mathbf{1} \tag{A.6.27}$$

As an instructive exercise, one may explicitly establish the non-trivial order relation for the $3p$-$3h$/$1p$-$1h$ matrix elements of $\boldsymbol{H}^{cc}$,

$$H_{abcjkl,aj}^{cc} \sim O(2) \tag{A.6.28}$$

by verifying that various first-order contributions cancel each other here. (See also Exercise 17.3.)

2. In a similar (and even simpler) way, the BCC representation of a general one-particle operator $\hat{D}$

$$D_{IJ}^{cc} = \langle \overline{\Phi}_I | \hat{D} | \Psi_J^0 \rangle \tag{A.6.29}$$

can be analyzed. The matrix elements in the LL part of $\boldsymbol{D}^{cc}$ obey the relations

$$D_{IJ}^{cc} \sim O([I] - [J] - 1), \quad [I] > [J] \tag{A.6.30}$$

The full order structure is depicted in Fig. A.8.

3. The left CC transition moments

$$D_I^{(l)} = \langle \overline{\Phi}_I | \hat{D} | \Psi_0 \rangle \sim O([I] - 1) \tag{A.6.31}$$

can be obtained as a special case ($[J] = 0$) of Eqs. (A.6.29), (A.6.30).

Proof of the ISR-ADC Order Relations

After having established the order relations for the BCC representation, the proof of the canonical order relations (12.1),

Fig. A.8 Order structure of the BCC representation $\boldsymbol{D}^{cc}$ of a one-particle operator $\hat{D}$

	$1p$-$1h$	$2p$-$2h$	$3p$-$3h$	$4p$-$4h$	$5p$-$5h$	...
$1p$-$1h$	0	0	-	-	-	...
$2h$-$2p$	0	0	0	-	-	...
$3p$-$3h$	1	0	0	0	-	-
$4p$-$4h$	2	1	0	0	0	-
$5p$-$5h$	3	2	1	0	0	0
⋮	⋮	⋮	⋮	⋮	⋮	⋱

$$M_{IJ} \sim O(|[I] - [J]|) \quad \text{(A.6.32)}$$

for the ISR-ADC secular matrices (Eqs. 11.16, 14.3),

$$M_{IJ} = \langle \tilde{\Psi}_I | \hat{H} - E_0 | \tilde{\Psi}_J \rangle \quad \text{(A.6.33)}$$

is straightforward. Since $\boldsymbol{M}$ is hermitian, we can confine ourselves to states where $[I] \geq [J]$. The case $[I] = [J]$ is trivial, as $\boldsymbol{M}_{\mu\mu} \sim O(0)$ is seen by the construction of the intermediate states. Accordingly, we may suppose $[I] > [J]$ in the following and simplify Eq. (A.6.33) accordingly:

$$M_{IJ} = \langle \tilde{\Psi}_I | \hat{H} | \tilde{\Psi}_J \rangle \quad \text{(A.6.34)}$$

To relate the ISR-ADC matrix elements (A.6.34) to the BCC representation, we make use of the biorthogonal resolution of the identity (RI),

$$\hat{\mathbb{1}} = \sum_K |\Psi_K^0\rangle\langle\overline{\Phi}_K| \quad \text{(A.6.35)}$$

in terms of the CC states (17.12) and their biorthogonal counterparts (17.13). Note that in the case of N-electron excitations, there is a ground-state term $K = 0$, that is, $|\Psi_0\rangle\langle\Phi_0|$. Inserting the biorthogonal RI twice, the matrix element (A.6.34) can be written as

$$M_{IJ} = \sum_{K,L} \langle \tilde{\Psi}_I | \Psi_K^0 \rangle \langle \overline{\Phi}_K | \hat{H} | \Psi_L^0 \rangle \langle \overline{\Phi}_L | \tilde{\Psi}_J \rangle \quad \text{(A.6.36)}$$

By construction, the intermediate state $|\tilde{\Psi}_I\rangle$ is orthogonal to the CE states $|\Psi_K^0\rangle$ of lower classes, $[K] < [I]$. Therefore, the summation over K can be restricted to $[K] \geq [I]$. An opposite restriction, namely $[L] \leq [J]$, applies to the summation over L. To see this, consider the matrix element

$$\langle\overline{\Phi}_L|\tilde{\Psi}_J\rangle = \langle\Phi_0|\hat{C}_L^\dagger e^{-\hat{T}}|\tilde{\Psi}_J\rangle \tag{A.6.37}$$

and recall that by construction $|\tilde{\Psi}_J\rangle$ is of the form of

$$|\tilde{\Psi}_J\rangle = \sum_{[L']\leq[J]} z_{L'}\hat{C}_{L'}|\Psi_0\rangle \tag{A.6.38}$$

where the excitation classes of the operators $\hat{C}_{L'}$ are restricted to $[L'] \leq [J]$. As a consequence,

$$\langle\overline{\Phi}_L|\tilde{\Psi}_J\rangle = \sum_{[L']\leq[J]} z_{L'}\langle\Phi_0|\hat{C}_L^\dagger\hat{C}_{L'}|\Phi_0\rangle = 0, \quad \text{for } [L] > [J] \tag{A.6.39}$$

Here, $e^{-\hat{T}}$ has been commuted to the right, yielding $e^{-\hat{T}}|\Psi_0\rangle = |\Phi_0\rangle$. Altogether, the summations in Eq. (A.6.36) are restricted according to $[K] \geq [I] > [J] \geq [L]$. Since $\langle\overline{\Phi}_K|\hat{H}|\Psi_L^0\rangle \sim O([K]-[L])$ according to the BCC order relations (A.6.16), we may conclude that every summation term and thus the entire sum (A.6.36) is at least of the order $[I]-[J]$. The minimum value $[I]-[J]$ is associated with the case $[K]=[I]$ and $[L]=[J]$, as here the overlap integrals $\langle\tilde{\Psi}_I|\Psi_I^0\rangle$ and $\langle\overline{\Phi}_J|\tilde{\Psi}_J\rangle$ are of zeroth order (in fact, of the form $1+O(2)$).

ISR of Operators and Transition Moments

The foregoing proof can easily be transferred to the ISR of an arbitrary one-particle operator, $\hat{D}$,

$$\tilde{D}_{IJ} = \langle\tilde{\Psi}_I|\hat{D}|\tilde{\Psi}_J\rangle \tag{A.6.40}$$

as considered in Sects. 11.3 and 12.1 for $(N-1)$-electron excitations (Eq. 11.51) and Sect. 14.3 for N-electron excitations (Eq. 14.46). As above, one may insert the biorthogonal RI before and after the operator $\hat{D}$ and use the order structure (A.6.30) in the related BCC matrix $\hat{D}^{cc}$. The order relations thereby established read (see Eq. 12.17 and Fig. 12.2)

$$\tilde{D}_{IJ} \sim O(|[I]-[J]|-1), \quad [I] \neq [J] \tag{A.6.41}$$

The transition moments

$$F_I(D) = \langle\tilde{\Psi}_I|\hat{D}|\Psi_0\rangle \sim O([I]-1) \tag{A.6.42}$$

can be treated in a similar way, using here Eq. (A.6.31). This establishes the order relations (11.20) or (14.4) for the transition amplitudes, being

$$f_{I,rs} \sim O([I]-1) \tag{A.6.43}$$

in the N-electron case.

Order Structure of the ISR Eigenvalue Matrix

As an immediate consequence of the order structure (12.1) of the ISR secular matrix $\boldsymbol{M}$, the canonical order relations apply to the eigenvector matrix $\boldsymbol{X}$ as well (see Eq. 12.15),

$$X_{Jn} = \langle \tilde{\Psi}_J | \Psi_n \rangle \sim O(|[J]-[n]|) \tag{A.6.44}$$

Here, it is assumed that the final state $|\Psi_n\rangle$ has a PT origin in a specific HF configuration, say $|\Phi_I\rangle$, of class $[I]$:

$$|\Psi_n\rangle \leftarrow |\Phi_I\rangle \tag{A.6.45}$$

Accordingly, $|\Psi_n\rangle$ is said to belong to the excitation class $[n] := [I]$.

The emergence of the order relations (A.6.44) can be explained as follows (for a different, more stringent derivation see Ref. [13]). Let $\boldsymbol{X}_n$ be the eigenvector for a state $|\Psi_n\rangle$ of class $[n]$,

$$\boldsymbol{M}\boldsymbol{X}_n = \omega_n \boldsymbol{X}_n \tag{A.6.46}$$

and assume that $|\Psi_n\rangle$ derives from a HF configuration $|\Phi_I\rangle$, $[I]=[n]$. Now we consider an eigenvector component X_{Jn}, where J denotes an excitation from another excitation class, $[J] \neq [I]$. Applying a simple matrix perturbation theory (MPT), where the diagonal matrix elements, M_{KK}, furnish the "zeroth-order" level, and the non-diagonal elements, M_{KL}, $K \neq L$, define the perturbation, one obtains a "first-order" contribution to X_{Jn} of the form

$$X_{Jn} \leftarrow \frac{M_{JI}}{M_{JJ}-M_{II}} \tag{A.6.47}$$

This contribution results from the coupling of the configurations $|\Phi_I\rangle$ and $|\Phi_J\rangle$ via the secular matrix element M_{JI}. The actual PT order of this term is equal to the order of M_{JI},

$$\frac{M_{JI}}{M_{JJ}-M_{II}} \sim M_{JI} \sim O(|[J]-[I]|) = O(|[J]-[n]|) \tag{A.6.48}$$

since the denominator is of zeroth order, $M_{JJ}-M_{II} \sim O(0)$. In fact, there are no contributions of lower order than that. Consider an "interaction path" $J \leftarrow K \leftarrow I$ encountered in "second order" of MPT. The corresponding contribution is given by

$$X_{Jn} \leftarrow \frac{M_{JK}M_{KI}}{\Delta_{JI}\Delta_{KI}} \tag{A.6.49}$$

where $\Delta_{LI} = M_{LL} - M_{II}$. Again, the order is determined by the numerator so that

$$M_{JK}M_{KI} \sim O(|[J]-[K]|) + O(|[K]-[I]|) \geq O(|[J]-[I]|) \tag{A.6.50}$$

Note that the equals sign applies if the classes of the interacting states are ordered according to $[J] > [K] > [I]$ or $[J] < [K] < [I]$. In a similar way, one may analyze MPT interaction paths of "third" and "higher order." The resulting contribution to X_{Jn} is always at least of the PT order $O(|[J]-[n]|)$.

The Order of Truncation Errors

The order relations (A.6.44) can be directly used to analyze the errors in the excitation energies (or transition moments) arising in (systematic) truncations of the secular expansion manifold. To this end, we write the energy ω_n as an expectation value

$$\omega_n = \boldsymbol{X}_n^\dagger \boldsymbol{M} \boldsymbol{X}_n = \sum_{JK} X_{Jn}^* M_{JK} X_{Kn} \tag{A.6.51}$$

and inspect the orders of the individual contributions. The diagonal secular matrix elements, M_{JJ}, are of zeroth order. Hence, the full expression for ω_n can be replaced with the sum of dominant diagonal contributions,

$$\omega_n \sim \sum_J X_{Jn}^* X_{Jn} \tag{A.6.52}$$

This means that configurations of class $[J]$ give rise to contributions to ω_n that are, according to Eq. (A.6.44), of the order

$$X_{Jn}^* X_{Jn} \sim 2(O([J]-[n]), \quad [J] > [n] \tag{A.6.53}$$

Conversely, if the configuration manifold is truncated after class μ ($\geq [n]$), the truncation error is of the order

$$O_{TE}(\mu) = 2(\mu + 1 - [n]), \quad \mu + 1 > [n] \tag{A.6.54}$$

that is, the order of contributions related to class $\mu + 1$, being the lowest of the disregarded configuration classes.

A.7 Linear Response Theory and the Polarization Propagator

The polarization propagator discussed in Chap. 13 is closely related to polarizabilities or, more general, to the linear response of properties, as, e.g., dipole moments, to a time-dependent "external" perturbation [5]. In this Appendix, we briefly outline that relation and show how the ADC approach can be used to treat linear response properties. For a more comprehensive presentation, general references, and a discussion of the computational performance, the reader is referred to Ref. [16].

Let $\hat{H}_x(t)$ denote the hamiltonian for an external perturbation,

$$\hat{H}_x(t) = \hat{B} f(t) \tag{A.7.1}$$

where

$$\hat{B} = \sum B_{rs} c_r^\dagger c_s \tag{A.7.2}$$

is a (constant) one-particle operator and $f(t)$ a time-dependent function describing the time-dependence of the perturbation. The perturbation is assumed to be turned on at t_0; that is, $f(t) = 0$ for $t \le t_0$. The full hamiltonian is given by

$$\hat{H}'(t) = \hat{H} + \hat{H}_x(t) \tag{A.7.3}$$

where $\hat{H}$ is the hamiltonian of the unperturbed system. Initially, that is, at and prior to t_0, the system is assumed to be in the ground state $|\Psi_0\rangle$ of $\hat{H}$.

To construct the solution of the TDSE, we may resort to the procedure presented in Sect. 4.1, where now $\hat{H}$ (rather than $\hat{H}_0$) is the time-independent part of the full hamiltonian. As in Eq. (4.13), the time-dependent wave function can be written in the form

$$|\Psi(t)\rangle = e^{-i\hat{H}t}|\Psi_x(t)\rangle \tag{A.7.4}$$

where $|\Psi_x(t)\rangle$ denotes the wave function in the interaction picture with $\hat{H}$ acting as the unperturbed part of the hamiltonian. Analogous to Eq. (4.19), we may write

$$|\Psi_x(t)\rangle = \hat{U}(t, t_0)|\Psi_0\rangle \tag{A.7.5}$$

where the time-evolution operator in the interaction picture, $\hat{U}(t, t_0)$, is subject to an integral equation of the type featured in Eq. (4.26). Solving this integral equation in an iterative way generates a perturbation expansion (see Eq. 4.27). In the present case, where the time-dependent part of the hamiltonian is given by Eq. (A.7.1), the expansion through first order reads

$$\hat{U}(t, t_0) = \hat{\mathbb{1}} - i \int_{t_0}^{t} \hat{B}_I(t') f(t')\, \mathrm{d}t' + \dots \tag{A.7.6}$$

Here,

$$\hat{B}_I(t) = e^{i\hat{H}t}\hat{B}e^{-i\hat{H}t} \tag{A.7.7}$$

is the perturbation operator in the (time-dependent) interaction picture.

Now we consider a property associated with a one-particle operator, say $\hat{A} = \sum A_{rs}c_r^\dagger c_s$. For $t \geq t_0$, the expectation value of $\hat{A}$, being

$$\bar{A}_0 = \langle\Psi_0|\hat{A}|\Psi_0\rangle \tag{A.7.8}$$

at t_0, evolves in time according to

$$\bar{A}(t) = \langle\Psi(t)|\hat{A}|\Psi(t)\rangle = \langle\Psi_x(t)|\hat{A}_I(t)|\Psi_x(t)\rangle \tag{A.7.9}$$

where

$$\hat{A}_I(t) = e^{i\hat{H}t}\hat{A}e^{-i\hat{H}t} \tag{A.7.10}$$

is representation of $\hat{A}$ in the interaction picture. The difference

$$\Delta A(t) = \bar{A}(t) - \bar{A}_0 \tag{A.7.11}$$

is referred to as the response of the property A to the time-dependent perturbation $\hat{H}_x(t)$. Using Eqs. (A.7.5), (A.7.6), the function $\Delta A(t)$ can be expanded in a PT series in the perturbing potential, beginning in first order (**linear response**):

$$\begin{aligned}\Delta A^{(1)}(t) =& \langle\Psi_x^{(1)}(t)|\hat{A}(t)|\Psi_0\rangle\rangle + \langle\Psi_0|\hat{A}(t)|\Psi_x^{(1)}(t)\rangle \\ =& \int_{t_0}^{t} i\langle\Psi_0|[\hat{B}(t'), \hat{A}(t)]|\Psi_0\rangle f(t')\,\mathrm{d}t'\end{aligned} \tag{A.7.12}$$

Introducing the so-called *linear response function* $R_{AB}(t, t')$,

$$R_{AB}(t, t') = i\theta(t - t')\langle\Psi_0|[\hat{B}(t'), \hat{A}(t)]|\Psi_0\rangle \tag{A.7.13}$$

the response can be written in the compact form

$$\Delta A^{(1)}(t) = \int_{t_0}^{\infty} R_{AB}(t, t')f(t')\,\mathrm{d}t' \tag{A.7.14}$$

Here, $R_{AB}(t, t')$ represents the generic part of the response, while the particular time-dependence of the respective perturbation enters via $f(t')$.

The energy representation of the response function is obtained via the Fourier transformation

$$R_{AB}(\omega) = \int_{-\infty}^{\infty} e^{i(\omega+i\eta)t} R_{AB}(t, 0)\,\mathrm{d}t \tag{A.7.15}$$

Here, the convergence factor $e^{-\eta t}$ has been introduced to ensure the definiteness of the Fourier integral at the upper limit, $t = \infty$ (see Eq. 3.13 in Sect. 3.1); note that the lower bound of the integral actually is $t = 0$ due to the step function $\theta(t)$.

Like in the spectral representation of the electron propagator (see Sect. 3.1), the Fourier transformation (A.7.15) can explicitly be performed upon employing the resolution of the identity in terms of the exact eigenstates $|\Psi_n\rangle$, $n = 0, 1, \ldots$ in the commutator expression (A.7.13). The result is the *sum-over-states* (SOS) formula

$$R_{AB}(\omega) = \sum_{n \neq 0} \frac{\langle \Psi_0 | \hat{A} | \Psi_n \rangle \langle \Psi_n | \hat{B} | \Psi_0 \rangle}{\omega - E_n + E_0 + i\eta} - \sum_{n \neq 0} \frac{\langle \Psi_0 | \hat{B} | \Psi_n \rangle \langle \Psi_n | \hat{A} | \Psi_0 \rangle}{\omega + E_n - E_0 + i\eta} \tag{A.7.16}$$

Note that the restriction $n \neq 0$ in the two sums reflects the fact that the original $n = 0$ contributions to the first and second term cancel each other. The response function can be written in a compact resolvent-type form as follows:

$$\begin{aligned} R_{AB}(\omega) = {} & \langle \Psi_0 | \hat{A} (\omega - \hat{H} + E_0 + i\eta)^{-1} \hat{B} | \Psi_0 \rangle \\ & - \langle \Psi_0 | \hat{B} (\omega + \hat{H} - E_0 + i\eta)^{-1} \hat{A} | \Psi_0 \rangle \end{aligned} \tag{A.7.17}$$

The equivalence of the latter form with the SOS expression (A.7.16) can easily be verified by using again the resolution of the identity in an appropriate way.

According to Eq. (A.7.16), the linear response function consists of two contributions,

$$R_{AB}(\omega) = R^{I}_{AB}(\omega) + R^{II}_{AB}(\omega) \tag{A.7.18}$$

which, in fact, are redundant as the relation

$$R^{II}_{AB}(\omega) = R^{I}_{AB}(-\omega)^* \tag{A.7.19}$$

shows.

The connection to the polarization propagator can easily be established by expanding the operators $\hat{A}$ and $\hat{B}$ (as in Eq. A.7.2) in the SOS expression (A.7.16) and comparing the result with the spectral representation (13.1) of the polarization propagator. This shows that $R^{I}_{AB}(\omega)$ can be written as

$$R^{I}_{AB}(\omega) = \sum_{r,s,r',s'} A_{rs} \Pi^{+}_{rs,r's'}(\omega) B_{r's'} \tag{A.7.20}$$

in terms of matrix elements of $\boldsymbol{\Pi}^{+}(\omega)$.

Now the ADC representation of $R^{I}_{AB}(\omega)$ is simply obtained by using the ADC form of $\boldsymbol{\Pi}^{+}(\omega)$ according to Eqs. (14.2)–(14.4). The resulting expression reads

$$R^{I}_{AB}(\omega) = \boldsymbol{F}(A)^{\dagger} (\omega - \boldsymbol{M})^{-1} \boldsymbol{F}(B) \tag{A.7.21}$$

Here, the infinitesimal $i\eta$ has been skipped; $\boldsymbol{F}(A)$ denotes a vector of components $F_J(A)$ formed from the ADC transition amplitudes $f_{J,rs}$ and operator matrix elements according to

$$F_J(A) = \sum f_{J,rs} A_{rs} \tag{A.7.22}$$

Of course, the ADC form of the response function could also have been obtained by applying to Eq. (A.7.17) the resolution of the identity in terms of the ECO intermediate states (Eqs. 14.12, 14.13). The resulting expression for the full response function reads

$$R_{AB}(\omega) = \boldsymbol{F}(A)^\dagger(\omega - \boldsymbol{M})^{-1}\boldsymbol{F}(B) - \boldsymbol{F}(B)^\dagger(\omega + \boldsymbol{M})^{-1}\boldsymbol{F}(A) \tag{A.7.23}$$

Using the ADC(n) approximations for the secular matrix $\boldsymbol{M}$ and the transition amplitudes $f_{J,rs}$, Eq. (A.7.23) establishes computational schemes for frequency-dependent response properties being consistent through nth order of perturbation theory (in the residual electron-electron interaction). As noted in Sect. A.1, the computational method of choice in dealing with Eq. (A.7.23) is the Lanczos diagonalization algorithm.

A.8 Superoperator Approach to the Electron Propagator

The superoperator formulation [17] allows for a very compact and seemingly intuitive derivation of algebraic propagator equations. However, being hardly self-evident to the uninitiated, it should be helpful to explain this approach by relating its workings to more familiar concepts.

A.8.1 Superoperator Definitions

First, we consider the superoperator form of the electron propagator and show its equivalence with the original definition.

For a given operator $\hat{A}$, a *superoperator* (indicated by a double hat) can be defined according to

$$\hat{\hat{A}}\hat{X} = [\hat{X}, \hat{A}] \tag{A.8.1}$$

Of actual interest is the hamiltonian superoperator, $\hat{\hat{H}}$,

$$\hat{\hat{H}}\hat{X} = [\hat{X}, \hat{H}] \tag{A.8.2}$$

and the identity superoperator $\hat{\hat{I}}$ defined by

$$\hat{\hat{I}}\hat{X} = \hat{X} \tag{A.8.3}$$

Obviously, there is a notational benefit of the superoperator concept, allowing one, for example, to write an n-fold nested commutator compactly as

$$\hat{\hat{H}}^n\hat{X} = [\ldots[[\hat{X}, \hat{H}], \hat{H}], \ldots \hat{H}] \tag{A.8.4}$$

Another definition needed in the superoperator context is the "binary product" of two operators $\hat{A}$ and $\hat{B}$, being the ground-state expectation value of the anticommutator $\{\hat{A}^\dagger, \hat{B}\}$,

$$(\hat{A}|\hat{B}) = \langle 0|\{\hat{A}^\dagger, \hat{B}\}|0\rangle \tag{A.8.5}$$

Here and in the following, we use the abridged notation $|0\rangle \equiv |\Psi_0\rangle$. A corresponding "binary" matrix element is defined by

$$(\hat{A}|\hat{O}|\hat{B}) = \langle 0|\{\hat{A}^\dagger, \hat{O}\hat{B}\}|0\rangle = \langle 0|\hat{A}^\dagger\hat{O}\hat{B}|0\rangle + \langle 0|\hat{O}\hat{B}\hat{A}^\dagger|0\rangle \tag{A.8.6}$$

Note that this symmetrical notation supposes that $\hat{O}$ is hermitian and $|0\rangle$ an eigenstate of $\hat{O}$ (otherwise $(\hat{A}|\hat{O}\hat{B}) \neq (\hat{O}\hat{A}|\hat{B})$).

The superoperator formulation of the electron propagator reads

$$G^S_{pq}(\omega) = (c_q|(\omega\hat{\hat{I}} - \hat{\hat{H}})^{-1}|c_p) \tag{A.8.7}$$

where

$$\hat{\hat{R}}(\omega) = (\omega\hat{\hat{I}} - \hat{\hat{H}})^{-1} \tag{A.8.8}$$

is referred to as the superoperator resolvent. Expanding the binary matrix element yields

$$G^S_{pq}(\omega) = \langle 0|c^\dagger_q \hat{\hat{R}}(\omega)c_p|0\rangle + \langle 0|(\hat{\hat{R}}(\omega)c_p)c^\dagger_q|0\rangle \tag{A.8.9}$$

which makes explicit that

$$G^S_{pq}(\omega) = G^{S(-)}_{pq}(\omega) + G^{S(+)}_{pq}(\omega) \tag{A.8.10}$$

is composed of an $(N-1)$- and an $(N+1)$-electron part.

Equivalence with the Original Definition

It is not obvious that the superoperator form of the electron propagator is equivalent to the usual definition, as given by Eqs. (3.24), (3.25). To verify that equivalence, one may expand the superoperator resolvent

$$(\omega\hat{\hat{I}} - \hat{\hat{H}})^{-1} = \frac{1}{\omega}\sum_{\nu=0}^{\infty}\frac{\hat{\hat{H}}^\nu}{\omega^\nu} \tag{A.8.11}$$

and compare the resulting nested commutator expressions with the terms obtained by applying the Baker–Hausdorff expansion

$$c_p[t] = e^{i\hat{H}t}c_p e^{-i\hat{H}t} = c_p + it[\hat{H}, c_p] + \frac{(it)^2}{2!}[\hat{H}, [\hat{H}, c_p]] + \ldots \tag{A.8.12}$$

in the definition (3.3) of the electron propagator and then performing the time integration of the Fourier transform (3.10).

A simpler proof of the equivalence is as follows [7]. According to Eq. (A.8.8), the superoperator resolvent $\hat{\hat{R}}(\omega)$ is defined as the inverse of the superoperator $(\omega\hat{\hat{I}} - \hat{\hat{H}})$. Thus, the action of $\hat{\hat{R}}(\omega)$ on an arbitrary operator $\hat{A}$

$$\hat{\hat{R}}(\omega)\hat{A} = \hat{X} \tag{A.8.13}$$

generates an ordinary operator $\hat{X}$, such that

$$(\omega\hat{\hat{I}} - \hat{\hat{H}})\hat{X} = \hat{A} \tag{A.8.14}$$

To determine $\hat{X}$, we take matrix elements on both sides,

$$\langle m|(\omega\hat{\hat{I}} - \hat{\hat{H}})\hat{X}|n\rangle = \langle m|\hat{A}|n\rangle \tag{A.8.15}$$

where $|m\rangle$ and $|n\rangle$ denote exact energy eigenstates of the hamiltonian for systems of possibly distinct electron numbers $N, N \pm 1, \ldots$. For example, in the case $\hat{A} \equiv c_p$, non-vanishing matrix elements are obtained if $|m\rangle$ and $|n\rangle$ are $(N-1)$- and N-electron states, respectively. Equation (A.8.15) can be evaluated further:

$$\langle m|\hat{A}|n\rangle = \langle m|\omega\hat{X} - [\hat{X}, \hat{H}]|n\rangle = (\omega + E_m - E_n)\langle m|\hat{X}|n\rangle \tag{A.8.16}$$

which can be resolved for $\langle m|\hat{X}|n\rangle$, yielding

$$\langle m|\hat{X}|n\rangle = \langle m|\hat{\hat{R}}(\omega)\hat{A}|n\rangle = \frac{\langle m|\hat{A}|n\rangle}{\omega + E_m - E_n} \tag{A.8.17}$$

Now we consider the superoperator expression (A.8.9) and insert resolutions of the identity (RI),

$$G^S_{pq}(\omega) = \sum_m \langle 0|c_q^\dagger|m\rangle\langle m|\hat{\hat{R}}(\omega)c_p|0\rangle + \sum_n \langle 0|\hat{\hat{R}}(\omega)c_p|n\rangle\langle n|c_q^\dagger|0\rangle \tag{A.8.18}$$

where the states $|m\rangle$ and $|n\rangle$ are $(N-1)$- and $(N+1)$-electron states, respectively. With Eq. (A.8.17), we obtain

$$G^S_{pq}(\omega) = \sum_{m\in\{N-1\}} \frac{\langle 0|c_q^\dagger|m\rangle\langle m|c_p|0\rangle}{\omega + E_m - E_0} + \sum_{n\in\{N+1\}} \frac{\langle 0|c_p|n\rangle\langle n|c_q^\dagger|0\rangle}{\omega - E_n + E_0} \tag{A.8.19}$$

which is just the spectral representation (3.17) of the original electron propagator, $G_{pq}(\omega)$. This shows that $G^S_{pq}(\omega) \equiv G_{pq}(\omega)$ supposing that the infinitesimals $\pm i\eta$ in the denominators can be ignored (if needed they could be "smuggled" into Eq. (A.8.9)).

A.8.2 Superoperator Equations

Having established that Eq. (A.8.7) is a legitimate definition of the electron propagator, we now discuss the superoperator derivation of the basic algebraic equations for the propagator and show their equivalence with the secular equations of the EOM approach discussed in Sect. 16.1.

Operator-Based Resolution of the Identity for Binary Products

The use of operator-based resolution of the identity in binary products, being a basic ingredient in the superoperator approach to the electron propagator, is not self-explanatory. To better understand the mathematical structure of the ensuing operator algebra, one may first visit the Excursus at the end of this Appendix, reviewing analogous procedures in the simpler case of state representations.

Consider two operators, $\hat{A}$, $\hat{B}$, generating $(N-1)$-electron states, $\hat{A}|0\rangle$ and $\hat{B}|0\rangle$. (Likewise, one could proceed with operators for $(N+1)$-electron states.) What we want to show is that a RI in terms of basis operators can be applied to the binary product (A.8.5) in the following form:

$$(\hat{A}|\hat{B}) = (\hat{A}|\boldsymbol{h})(\boldsymbol{h}|\boldsymbol{h})^{-1}(\boldsymbol{h}|\hat{B}) \tag{A.8.20}$$

Here,

$$\boldsymbol{h} \equiv \{\hat{O}_I\} \tag{A.8.21}$$

is a short notation for the set of the basis operators (16.6),

$$\{\hat{O}_I\} = \{c_k;\, c_a^\dagger c_k c_l;\, \dots\} \cup \{c_a;\, c_k^\dagger c_b c_a;\, \dots\} \tag{A.8.22}$$

which were introduced in Sect. 16.1. The binary product can be rewritten as

$$\begin{aligned}(\hat{A}|\hat{B}) =& \langle 0|\{\hat{A}^\dagger, \hat{B}\}|0\rangle \\ =& \langle 0|\hat{A}^\dagger \hat{B}|0\rangle + \langle 0|\hat{B}\hat{A}^\dagger|0\rangle \\ =& \langle 0|\hat{A}^\dagger \hat{B}|0\rangle + \langle 0|\hat{A}\hat{B}^\dagger|0\rangle \\ =& \langle 0|(\hat{A}^\dagger + \hat{A})(\hat{B}^\dagger + \hat{B})|0\rangle \end{aligned} \tag{A.8.23}$$

Scalar products of states with different particle numbers vanish, so that

$$\langle 0|\hat{A}\hat{B}|0\rangle = \langle 0|\hat{A}^\dagger \hat{B}^\dagger|0\rangle = 0 \tag{A.8.24}$$

which has been used to arrive at the last equation. In the third equation, it is assumed that $\hat{A}$, $\hat{B}$ are real-valued operators and the ground-state wave function is real-valued too, so that

$$\langle 0|\hat{B}\hat{A}^\dagger|0\rangle = \langle 0|\hat{B}\hat{A}^\dagger|0\rangle^* = \langle 0|\hat{A}\hat{B}^\dagger|0\rangle \tag{A.8.25}$$

In the case of complex operators, one may instead resort to the identity $(\hat{A}|\hat{B}) = \langle 0|(\hat{A}^{*\dagger} + \hat{A})(\hat{B}^{*\dagger} + \hat{B})|0\rangle$ (see Sect. 16.2 and Exercises 16.2 and 16.3).

Now we may bring in the $(N \pm 1)$-electron (Fock space) states (16.27) discussed in Sect. 16.2,

$$|\Theta_I\rangle = (\hat{O}_I^\dagger + \hat{O}_I)|0\rangle \tag{A.8.26}$$

Obviously, the overlap matrix elements (16.29) are just the binary products, $(\hat{O}_I|\hat{O}_J)$:

$$S_{IJ} = \langle\Theta_I|\Theta_J\rangle = \langle 0|\{\hat{O}_I^\dagger, \hat{O}_J\}|0\rangle = (\hat{O}_I|\hat{O}_J) \tag{A.8.27}$$

Analogous to Eq. (A.8.46), the states $|\Theta_J\rangle$ establish a conventional RI, based on a non-orthonormal basis,

$$\hat{\mathbb{1}} = \sum_{IJ} |\Theta_I\rangle (S^{-1})_{IJ} \langle\Theta_J| \tag{A.8.28}$$

Applying this in the last line of Eq. (A.8.23) gives

$$(\hat{A}|\hat{B}) = \sum_{IJ} \langle 0|(\hat{A}^\dagger + \hat{A})|\Theta_I\rangle (S^{-1})_{IJ} \langle\Theta_J|(\hat{B}^\dagger + \hat{B})|0\rangle \tag{A.8.29}$$

Here, the matrix elements on both sides of S^{-1} can again be recast into the compact binary product form,

$$\langle 0|(\hat{A}^\dagger + \hat{A})|\Theta_I\rangle = \langle 0|(\hat{A}^\dagger + \hat{A})(\hat{O}_I^\dagger + \hat{O}_I)|0\rangle = (\hat{A}|\hat{O}_I) \tag{A.8.30}$$

so that Eq. (A.8.29) takes on the form

$$(\hat{A}|\hat{B}) = \sum_{IJ} (\hat{A}|\hat{O}_I)(\boldsymbol{O}|\boldsymbol{O})^{-1}|_{IJ}(\hat{O}_J|\hat{B}) \tag{A.8.31}$$

where $(\boldsymbol{O}|\boldsymbol{O})$ is used instead of S according to Eq. (A.8.27). Obviously, this is just a somewhat more explicit form of Eq. (A.8.20).

"Inner Projection" of the Superoperator Resolvent

Now we come back to the definition (A.8.7) of the electron propagator, which can be written in an obvious matrix notation as

$$\boldsymbol{G}(\omega) = (\boldsymbol{c}|\hat{\hat{R}}(\omega)|\boldsymbol{c}) \tag{A.8.32}$$

Analogous to Eq. (A.8.29), we may apply the RI in the form of Eq. (A.8.28) to obtain

$$(\boldsymbol{c}|\hat{\hat{R}}(\omega)\boldsymbol{c}) = (\boldsymbol{c}|\boldsymbol{h})(\boldsymbol{h}|\boldsymbol{h})^{-1}(\boldsymbol{h}|\hat{\hat{R}}\boldsymbol{c}) \tag{A.8.33}$$

The RI can be used once more in the operator product $\hat{\hat{R}}\,\boldsymbol{c}$,

$$\hat{\hat{R}}\,\boldsymbol{c} = \hat{\hat{R}}(\omega)|\boldsymbol{h})(\boldsymbol{h}|\boldsymbol{h})^{-1}(\boldsymbol{h}|\boldsymbol{c} \tag{A.8.34}$$

so that Eq. (A.8.33) takes on the form

$$(\boldsymbol{c}|\hat{\hat{R}}(\omega)|\boldsymbol{c}) = (\boldsymbol{c}|\boldsymbol{h})(\boldsymbol{h}|\boldsymbol{h})^{-1}(\boldsymbol{h}|\hat{\hat{R}}(\omega)|\boldsymbol{h})(\boldsymbol{h}|\boldsymbol{h})^{-1}(\boldsymbol{h}|\boldsymbol{c}) \tag{A.8.35}$$

Analogous to Eq. (A.8.51), one can derive the relation

$$(\boldsymbol{h}|(\omega\hat{\hat{1}} - \hat{\hat{H}})|\boldsymbol{h})^{-1} = (\boldsymbol{h}|\boldsymbol{h})^{-1}(\boldsymbol{h}|\hat{\hat{R}}(\omega)|\boldsymbol{h})(\boldsymbol{h}|\boldsymbol{h})^{-1} \tag{A.8.36}$$

for the generalized matrix representations of $(\omega\hat{\hat{1}} - \hat{\hat{H}})$ and its inverse, $\hat{\hat{R}}(\omega) = (\omega\hat{\hat{1}} - \hat{\hat{H}})^{-1}$. This allows us finally to write Eq. (A.8.35) as

$$\boldsymbol{G}(\omega) = (\boldsymbol{c}|\hat{\hat{R}}(\omega)|\boldsymbol{c}) = (\boldsymbol{c}|\boldsymbol{h})(\boldsymbol{h}|(\omega\hat{\hat{1}} - \hat{\hat{H}})|\boldsymbol{h})^{-1}(\boldsymbol{h}|\boldsymbol{c}) \tag{A.8.37}$$

The expression on the right-hand side is often referred to as the "inner projection" of the superoperator resolvent, implying a connection to Löwdin's inner projection concept [18].

To make contact with the EOM secular equations, we just expand the compact superoperator form into the underlying conventional expressions, so that Eq. (A.8.37) takes on the form

$$G_{pq}(\omega) = \sum_{I,J} \langle 0|\{c_q^\dagger, \hat{O}_I\}|0\rangle \left(\omega\langle 0|\{\hat{O}_K^\dagger, \hat{O}_L\}|0\rangle + \langle 0|\{\hat{O}_K^\dagger, [\hat{H}, \hat{O}_L]\}|0\rangle\right)^{-1}\Big|_{IJ} \langle 0|\{\hat{O}_J^\dagger, c_p\}|0\rangle \tag{A.8.38}$$

As the comparison with Eqs. (16.12)–(16.14) shows, the poles of $G_{pq}(\omega)$ are given - up to a sign change - by the roots (16.19) of the EOM secular equations, while the spectroscopic factors are obtained from the eigenvectors as in Eq. (16.25).

A.8.3 *Excursus: Matrix Representations of an Operator Inverse*

Consider an operator $\hat{A}$, and let $\boldsymbol{A}$ denote the matrix representation of $\hat{A}$,

$$A_{kl} = \langle k|\hat{A}|l\rangle \tag{A.8.39}$$

with respect to a complete set of orthonormal states, $|k\rangle$, $k = 1, 2, \ldots$. As is easily seen, the matrix representation of the inverse operator $\hat{A}^{-1}$

$$\hat{A}\hat{A}^{-1} = \hat{\mathbb{1}} \tag{A.8.40}$$

is just the inverse of the matrix $\boldsymbol{A}$. Let $\boldsymbol{A}'$ denote the matrix representation of $\hat{A}^{-1}$,

$$A'_{kl} = \langle k|\hat{A}^{-1}|l\rangle \tag{A.8.41}$$

Consider an arbitrary matrix element of Eq. (A.8.40) und use the resolution of the identity (RI),

$$\hat{\mathbb{1}} = \sum_k |k\rangle\langle k| \tag{A.8.42}$$

as follows:

$$\delta_{nm} = \langle n|\hat{A}\hat{A}^{-1}|m\rangle = \sum_k \langle n|\hat{A}|k\rangle\langle k|\hat{A}^{-1}|m\rangle = (\boldsymbol{A}\boldsymbol{A}')_{nm} \tag{A.8.43}$$

This shows that

$$\boldsymbol{A}' = \boldsymbol{A}^{-1} \tag{A.8.44}$$

The situation is less obvious if the representations is based on non-orthonormal states, $|\tilde{k}\rangle, k = 1, 2, \ldots$. Let $\boldsymbol{S}$ denote the corresponding overlap matrix,

$$S_{kl} = \langle\tilde{k}|\tilde{l}\rangle \tag{A.8.45}$$

so that the RI can be written as

$$\hat{\mathbb{1}} = \sum_{k,l} |\tilde{k}\rangle(S^{-1})_{kl}\langle\tilde{l}| \tag{A.8.46}$$

Again, we consider matrix representations $\tilde{\boldsymbol{A}}$ of $\hat{A}$ and $\tilde{\boldsymbol{A}}'$ of $\hat{A}^{-1}$:

$$\tilde{A}_{kl} = \langle\tilde{k}|\hat{A}|\tilde{l}\rangle, \; \tilde{A}'_{kl} = \langle\tilde{k}|\hat{A}^{-1}|\tilde{l}\rangle \tag{A.8.47}$$

To establish the relation between $\tilde{\boldsymbol{A}}'$ and $\tilde{\boldsymbol{A}}$, one can proceed as above, yielding

$$S_{nm} = \langle\tilde{n}|\hat{A}\hat{A}^{-1}|\tilde{m}\rangle = \sum_{k,l} \langle\tilde{n}|\hat{A}|\tilde{k}\rangle(S^{-1})_{kl}\langle\tilde{l}|\hat{A}^{-1}|\tilde{m}\rangle \tag{A.8.48}$$

or in matrix form,

$$\boldsymbol{S} = \tilde{\boldsymbol{A}}\boldsymbol{S}^{-1}\tilde{\boldsymbol{A}}' \tag{A.8.49}$$

This relation can be resolved for $\tilde{\boldsymbol{A}}'$

$$\tilde{\boldsymbol{A}}' = \boldsymbol{S}\tilde{\boldsymbol{A}}^{-1}\boldsymbol{S} \tag{A.8.50}$$

Note also the relation

$$\tilde{\boldsymbol{A}}^{-1} = \boldsymbol{S}^{-1}\tilde{\boldsymbol{A}}'\boldsymbol{S}^{-1} \tag{A.8.51}$$

for the inverse of $\tilde{\boldsymbol{A}}$.

A.9 Compilation of ADC Expressions

In the explicit ADC expressions given below, the indices $i, j, k, l, \ldots$ and $a, b, c, \ldots$ refer to occupied and unoccupied orbitals, respectively, with respect to the HF ground state. The subscripts $p, q, r \ldots$ label both occupied and unoccupied orbitals. The short notation

$$v_{rsuv} = \frac{V_{rs[uv]}}{\epsilon_r + \epsilon_s - \epsilon_u - \epsilon_v} \tag{A.9.1}$$

is used throughout. Note that $v_{uvrs} = -v^*_{rsuv}$.

A.9.1 ADC(3) Expressions for the Dynamical Self-energy

According to Eq. (9.2), the affinity and ionization part $\boldsymbol{M}^+(\omega)$ and $\boldsymbol{M}^-(\omega)$, respectively, of the dynamical self-energy can be written in the ADC form

$$M^{\pm}_{pq}(\omega) = \boldsymbol{U}^{\pm\dagger}_p(\omega\mathbf{1} - \boldsymbol{K}^{\pm} - \boldsymbol{C}^{\pm})^{-1}\boldsymbol{U}^{\pm}_q \tag{A.9.2}$$

At the ADC(3) level, the matrices $\boldsymbol{K}^{\pm}$, $\boldsymbol{C}^{\pm}$, and $\boldsymbol{U}^{\pm}_q$ are defined with respect to $2p$-$1h$ configurations, (jab), $a < b$, in the affinity part (+); and $2h$-$1p$ configurations, (akl), $k < l$, in the ionization part (−).

$$K^+_{jab,j'a'b'} = (-\epsilon_j + \epsilon_a + \epsilon_b)\,\delta_{jj'}\delta_{aa'}\delta_{bb'} \tag{A.9.3}$$

$$K^-_{akl,a'k'l'} = (-\epsilon_a + \epsilon_k + \epsilon_l)\,\delta_{aa'}\delta_{kk'}\delta_{ll'} \tag{A.9.4}$$

$$C^+_{jab,j'a'b'} = \delta_{jj'}V_{ab[a'b']} - (\,\delta_{aa'}V_{j'b[jb']} + \delta_{bb'}V_{j'a[ja']}\,) + (a' \leftrightarrow b') \tag{A.9.5}$$

$$C^-_{akl,a'k'l'} = -\delta_{aa'}V_{kl[k'l']} + (\,\delta_{kk'}V_{a'l[al']} + \delta_{ll'}V_{a'k[ak']}\,) - (k' \leftrightarrow l') \tag{A.9.6}$$

$$U^+_{jab,q} = V_{ab[qj]} - \tfrac{1}{2}\sum_{k,l} v_{abkl}V_{kl[qj]} + (\sum_{k,c} v_{ackj}V_{kb[qc]}\,) - (a \leftrightarrow b) \tag{A.9.7}$$

$$U^-_{akl,q} = V_{kl[qa]} + \tfrac{1}{2}\sum_{b,c} v_{klbc}V_{bc[qa]} - (\sum_{i,b} v_{kiba}V_{bl[qi]}\,) + (k \leftrightarrow l) \tag{A.9.8}$$

Here, $(r \leftrightarrow s)$ means the preceding term (in brackets) upon interchanging the indices r and s.

A.9.2 Direct ADC Expressions for the Ionization Part of the Electron Propagator

In the following, direct ADC expressions (Chap. 10) are compiled for the $\boldsymbol{G}^{-}(\omega)$ part of the electron propagator, covering specifically the secular matrix $\boldsymbol{K}+\boldsymbol{C}$ at the third-order (ADC(3)) level, and the effective transition moments $\boldsymbol{f}$ at the ADC(2) level. Here, the ADC configuration manifold comprises the $1h$ and $2h$-$1p$ configurations.

ADC(3) Secular Matrix

$1h$ diagonal block:

$$K_{kk'} = \epsilon_k \delta_{kk'} \tag{A.9.9}$$

$$C_{kk'} = \sum_{a<b,j} v_{abkj} v_{k'jab} \left(\epsilon_a + \epsilon_b - \epsilon_j - \tfrac{1}{2}\epsilon_k - \tfrac{1}{2}\epsilon_{k'}\right) \tag{A.9.10}$$

$$+ C^{(A)}_{kk'} + C^{(B)}_{kk'} + C^{(C)}_{kk'} + C^{(D)}_{kk'} + \Sigma^{(3)}_{kk'}(\infty) \tag{A.9.11}$$

where

$$C^{(A)}_{kk'} = \tfrac{1}{4} \sum_{\substack{a,b,c,d\\ l}} v_{abkl} v^*_{cdk'l} V_{cd[ab]} \tag{A.9.12}$$

$$C^{(B)}_{kk'} = \sum_{\substack{a,b,c\\ l,m}} v_{abkl} v^*_{ack'm} V_{lc[bm]} \tag{A.9.13}$$

$$C^{(C)}_{kk'} = \tfrac{1}{4} \sum_{\substack{a,b\\ l,m,j}} v_{ablm} v^*_{abjk'} V_{lm[jk]} + h.c. \tag{A.9.14}$$

$$C^{(D)}_{kk'} = \sum_{\substack{a,b,c\\ l,m}} v_{ablm} v^*_{bck'm} V_{lc[ka]} + h.c. \tag{A.9.15}$$

The last term in Eq. (A.9.11) is the third-order contribution to the constant self-energy part, $\Sigma_{kk'}(\infty)$ (see Sect. 8.2). In actual ADC(3) computations, it is recommendable to go beyond the strict third-order level here, which is afforded by the Dyson expansion method (DEM) presented in Sect. A.5. The actual third-order expressions can readily be derived from diagram $T3$ in Fig. 8.9 (see Exercise 8.2). Alternatively, one may resort to Eq. (8.34),

$$\Sigma^{(3)}_{pq}(\infty) = \sum_{r,s} V_{pr[qs]} \rho^{(2)}_{sr} \tag{A.9.16}$$

and relate the second-order density matrix elements via Eq. (10.37) to the effective transition amplitudes,

$$\rho^{(2)}_{kk'} = f^{(2)}_{kk'} + f^{(2)*}_{kk'} \tag{A.9.17}$$

$$\rho^{(2)}_{ka} = f^{(2)}_{ka} \tag{A.9.18}$$

$$\rho^{(2)}_{aa'} = \sum_{b,k,l} f^{(1)*}_{bkl,a} f^{(1)}_{bkl,a'} \tag{A.9.19}$$

The ADC(2) expressions needed here are below (Eqs. A.9.22–A.9.25).

$1h/2h$-$1p$ block:

$$C_{j,akl} = V_{kl[aj]} + \tfrac{1}{2}\sum_{b,c} v^*_{bckl} V_{bc[ja]} + (\sum_{b,i} v^*_{abli} V_{kb[ji]}) - (k \leftrightarrow l). \tag{A.9.20}$$

$2h$-$1p$ diagonal block:

$$\begin{aligned} K_{akl,a'k'l'} =& (-\epsilon_a + \epsilon_k + \epsilon_l)\delta_{aa'}\delta_{kk'}\delta_{ll'} \\ C_{akl,a'k'l'} =& -\delta_{aa'} V_{k'l'[kl]} + (\delta_{kk'} V_{al'[a'l]} + \delta_{ll'} V_{ak'[a'k]}) - (k' \leftrightarrow l'). \end{aligned} \tag{A.9.21}$$

ADC(2) Expressions for the Effective Transition Amplitudes

Usually, transition amplitudes and spectroscopic factors are not needed at utmost accuracy, and for most purposes, a practical and satisfactory approximation is obtained by combining ADC(3) eigenvectors with the ADC(2) expressions for the effective transition amplitudes. The complete ADC(3) expressions can be found in Ref. [19].

$1h$ part:

$$f_{kl} = \delta_{kl} - \tfrac{1}{2} \sum_{a<b,j} v_{abkj} v_{ljab} \tag{A.9.22}$$

$$f_{ka} = \frac{1}{\epsilon_a - \epsilon_k}\left(\tfrac{1}{2}\sum_{b,c,j} v_{bckj} V_{aj[bc]} - \tfrac{1}{2}\sum_{b,i,j} v_{abij} V_{ij[kb]}\right) \tag{A.9.23}$$

Note that the $f_{kk'}$ submatrix is hermitian.

$2h$-$1p$ part:

$$f_{akl,j} = 0 \tag{A.9.24}$$

$$f_{akl,b} = -v_{abkl} \tag{A.9.25}$$

A.9.3 ADC(2) Expressions for the Polarization Propagator

The direct ADC procedure for the $\mathbf{\Pi}^+(\omega)$ part of the polarisation propagator was discussed in Chap. 14. Here, the secular matrix elements and the effective transition amplitudes constituting the second-order ADC(2) scheme are listed. The configuration manifold is spanned by the $1p$-$1h$ and $2p$-$2h$ configurations.

Secular Matrix

$1p$-$1h$ diagonal block:

$$K_{ak,a'k'} = (\epsilon_a - \epsilon_k)\delta_{aa'}\delta_{kk'}, \tag{A.9.26}$$

$$C_{ak,a'k'} = -V_{ak'[a'k]} + C^{(A)}_{ak,a'k'} + C^{(B)}_{ak,a'k'} + C^{(C)}_{ak,a'k'} \tag{A.9.27}$$

where

$$C^{(A)}_{ak,a'k'} = \delta_{kk'}\frac{1}{2}\sum_{c,i,j} v_{acij}v_{ija'c}(\epsilon_i + \epsilon_j - \epsilon_c - \tfrac{1}{2}\epsilon_a - \tfrac{1}{2}\epsilon_{a'}), \tag{A.9.28}$$

$$C^{(B)}_{ak,a'k'} = \delta_{aa'}\frac{1}{2}\sum_{c,d,i} v_{cdki}v_{k'icd}(\tfrac{1}{2}\epsilon_k + \tfrac{1}{2}\epsilon_{k'} + \epsilon_i - \epsilon_c - \epsilon_d), \tag{A.9.29}$$

$$C^{(C)}_{ak,a'k'} = \sum_{c,i} v_{k'ia'c}v_{acik}(\tfrac{1}{2}\epsilon_k + \tfrac{1}{2}\epsilon_{k'} - \tfrac{1}{2}\epsilon_a - \tfrac{1}{2}\epsilon_{a'} + \epsilon_i - \epsilon_c). \tag{A.9.30}$$

$1p$-$1h/2p$-$2h$ block:

$$C^{(1)}_{ak,a'b'k'l'} = \delta_{aa'}V_{k'l'[kb']} - \delta_{ab'}V_{k'l'[ka']} - \delta_{kk'}V_{al'[a'b']} + \delta_{kl'}V_{ak'[a'b']}, \tag{A.9.31}$$

$2p$-$2h$ diagonal block:

$$K_{abkl,a'b'k'l'} = (\epsilon_a + \epsilon_b - \epsilon_k - \epsilon_l)\delta_{aa'}\delta_{bb'}\delta_{kk'}\delta_{ll'} \tag{A.9.32}$$

The $2p$-$2h$ diagonal block can be extended with the first-order secular matrix elements anticipated from the ADC(3) level:

$$\begin{aligned} C^{(1)}_{abkl,a'b'k'l'} =& \delta_{kk'}\delta_{ll'}V_{ab[a'b']} + \delta_{aa'}\delta_{bb'}V_{k'l'[kl]} \\ &- \left(\delta_{bb'}\delta_{ll'}V_{ak'[a'k]} + \delta_{bb'}\delta_{kk'}V_{al'[a'l]} + \delta_{aa'}\delta_{ll'}V_{bk'[b'k]} + \delta_{aa'}\delta_{kk'}V_{bl'[b'l]}\right) \\ &+ (k' \leftrightarrow l') + (a' \leftrightarrow b') - (k' \leftrightarrow l', a' \leftrightarrow b'). \end{aligned} \tag{A.9.33}$$

While this extension (ADC(2)-x) improves the treatment of doubly excited states, it does not afford consistently better results for the single excitations.

Effective Transition Amplitudes

$1p$-$1h$ part, p-h amplitudes:

$$f_{ak,a'k'} = \delta_{aa'}\delta_{kk'} + f^{(A)}_{ak,a'k'} + f^{(B)}_{ak,a'k'} + f^{(C)}_{ak,a'k'} \tag{A.9.34}$$

where

$$f^{(A)}_{ak,a'k'} = \frac{1}{4} \sum_{c,i,j} v_{acij} v_{ija'c} \delta_{kk'}, \tag{A.9.35}$$

$$f^{(B)}_{ak,a'k'} = \frac{1}{4} \sum_{c,d,i} v_{cdki} v_{k'icd} \delta_{aa'}, \tag{A.9.36}$$

$$f^{(C)}_{ak,a'k'} = -\frac{1}{2} \sum_{c,i} v_{acki} v_{k'ia'c}, \tag{A.9.37}$$

$1p$-$1h$ part, p-p amplitudes:

$$f_{ak,a'b'} = \frac{\delta_{aa'}}{\epsilon_{b'} - \epsilon_k} \Big(-\frac{1}{2} \sum_{c,i,j} v_{b'cij} V_{ij[kc]} + \frac{1}{2} \sum_{c,d,i} v_{cdki} V_{b'i[cd]} \Big) \tag{A.9.38}$$

$1p$-$1h$ part, h-h amplitudes:

$$f_{ak,k'l'} = \frac{\delta_{kl'}}{\epsilon_a - \epsilon_{k'}} \Big(-\frac{1}{2} \sum_{c,d,i} v_{cdik'} V_{ia[cd]} + \frac{1}{2} \sum_{c,i,j} v_{caij} V_{ij[ck']} \Big) \tag{A.9.39}$$

$1p$-$1h$ part, h-p amplitudes:

$$f_{ak,k'a'} = v_{aa'k'k} + \sum_{i=1}^{6} f^{(2,i)}_{ak,k'a'} (\epsilon_a + \epsilon_{a'} - \epsilon_k - \epsilon_{k'})^{-1} \tag{A.9.40}$$

where

$$f^{(2,1)}_{ak,k'a'} = \sum_{c,i} v_{ca'k'i} V_{ai[ck]}, \tag{A.9.41}$$

$$f^{(2,2)}_{ak,k'a'} = \sum_{c,i} v_{acik} V_{ia'[k'c]}, \tag{A.9.42}$$

$$f^{(2,3)}_{ak,k'a'} = -\sum_{c,i} v_{ca'ik} V_{ai[k'c]}, \tag{A.9.43}$$

$$f^{(2,4)}_{ak,k'a'} = -\sum_{c,i} v_{ack'i} V_{ia'[ck]}, \tag{A.9.44}$$

$$f^{(2,5)}_{ak,k'a'} = -\frac{1}{2} \sum_{c,d} v_{cdk'k} V_{aa'[cd]}, \tag{A.9.45}$$

$$f^{(2,6)}_{ak,k'a'} = -\frac{1}{2} \sum_{i,j} v_{aa'ij} V_{ij[k'k]}. \tag{A.9.46}$$

$2p$-$2h$ part:

$$\begin{aligned} f_{abkl,k'l'} &= \delta_{kl'}\, v_{ablk'} - \delta_{ll'}\, v_{abkk'} \\ f_{abkl,a'b'} &= \delta_{ba'}\, v_{ab'kl} - \delta_{aa'}\, v_{bb'kl} \end{aligned} \tag{A.9.47}$$

References

1. March NH, Young WH, Sampanthar S (1967) The many-body problem in quantum mechanics. Cambridge University Press, Cambridge
2. Davidson ER (1975) J Comput Phys 17:87
3. Parlett BN (1980) The symmetric eigenvalue problem. Prentice Hall, Englewood Cliffs
4. Cullum JK, Willoughby RA (1985) Lanczos algorithm for large symmetric eigenvalue computations. Birkhäuser, Boston
5. Fetter AL, Walecka JD (1971) Quantum theory of many-particle systems. McGraw-Hill, New York
6. Wick GC (1950) Phys Rev 80:268
7. Mertins F (1991) unpublished notes
8. von Niessen W, Schirmer J, Cederbaum L (1984) Comput Phys Rep 1:57
9. Schirmer J, Angonoa G (1989) J Chem Phys 91:1754
10. Hubbard J (1957) Proc R Soc A 240:539
11. Schirmer J (1991) Phys Rev A 43:4647
12. Mertins F, Schirmer J (1996) Phys Rev A 53:2140
13. Schirmer J, Mertins F (2009) Theor Chem Acc 125:145
14. Christiansen O, Koch H, Jørgensen P (1995) J Chem Phys 103:7429
15. Hald K, Jørgensen P, Olsen J, Jaszuński M (2001) J Chem Phys 115:671
16. Trofimov AB, Krivdina IL, Weller J, Schirmer J (2006) Chem Phys 329:1
17. Goscinski D, Lukman B (1970) Chem Phys Lett 7:573
18. Löwdin PO (1965) Phys Rev A 139:357
19. Schirmer J, Trofimov AB, Stelter G (1998) J Chem Phys 109:4734

Index

J. Schirmer, *Many-Body Methods for Atoms, Molecules and Clusters*, Lecture Notes in Chemistry 94, https://doi.org/10.1007/978-3-319-93602-4

MIX
Papier aus verantwortungsvollen Quellen
Paper from responsible sources
FSC® C105338

If you have any concerns about our products,
you can contact us on
ProductSafety@springernature.com

In case Publisher is established outside the EU,
the EU authorized representative is:
Springer Nature Customer Service Center GmbH
Europaplatz 3, 69115 Heidelberg, Germany

Printed by Libri Plureos GmbH
in Hamburg, Germany